Organic Chemistry I

FOR DUMMIES®

A Wiley Brand

2nd Edition

Organic Chemistry I

FOR DUMMIES®

A Wiley Brand

2nd Edition

by Arthur Winter, PhD

FOR DUMMIES®
A Wiley Brand

Organic Chemistry I For Dummies®, 2nd Edition

Published by: **John Wiley & Sons, Inc.,** 111 River Street, Hoboken, NJ 07030-5774, www.wiley.com

Copyright © 2014 by John Wiley & Sons, Inc., Hoboken, New Jersey

Published simultaneously in Canada

For general information on our other products and services, please contact our Customer Care Department within the U.S. at 877-762-2974, outside the U.S. at 317-572-3993, or fax 317-572-4002. For technical support, please visit www.wiley.com/techsupport.

Wiley publishes in a variety of print and electronic formats and by print-on-demand. Some material included with standard print versions of this book may not be included in e-books or in print-on-demand. If this book refers to media such as a CD or DVD that is not included in the version you purchased, you may download this material at http://booksupport.wiley.com. For more information about Wiley products, visit www.wiley.com.

Library of Congress Control Number: 2013954080

ISBN 978-1-118-82807-6 (pbk); ISBN 978-1-118-82796-3 (ebk);
ISBN 978-1-118-82813-7 (ebk)

Manufactured in the United States of America

10 9 8 7 6 5 4

Contents at a Glance

Table of Contents

Introduction

· ·

*R*egrettably, when many people think of chemicals, the first things that usually pop into their minds are substances of a disagreeable nature — harmful pesticides and chemical pollutants, nerve agents and chemical weapons, or carcinogens and toxins.

But most chemicals play roles of a more positive nature. For example, both water and sugar are chemicals. Why are these chemicals important? Well, for one thing, both are components of beer. The enzymes in yeasts are also important chemicals used in fermentation, a process that involves the break-down of sugars into beer. Ethyl alcohol is the all-important chemical responsible for beer's effect on the body. In my view, these three representative examples of chemicals thoroughly rebut the notion that all chemicals are bad.

In fact, those who have a bad opinion of all chemicals must suffer from the psychological condition of self-loathing, because human bodies are essentially large vats of chemicals. Your skin is made up of chemicals — along with your heart, lungs, kidneys, and all your other favorite organs and appendages. And most of the chemicals in your body — in addition to the chemicals in all other living things — are not just any kinds of chemicals, but organic chemicals. So, anyone who has any interest at all in the machinery of living things (or in the chemistry of beer and wine) will have to deal at some point and at some level with organic chemistry.

Of course, the natures of these dealings have historically not always been so pleasant. Pre-med students and bio majors (and even chemistry majors) have butted heads with organic chemistry for decades, and, regrettably, the winner of this duel has not always been the human.

Part of the problem, I think, comes from students' preconceptions of organic chemistry. I admit that, like many students, I had the worst preconceptions going into organic chemistry. When I thought of organic class, I thought of wearying trivia about the chemical elements, coma-inducing lectures delivered in a monotone, complex mathematical equations sprawling across mile-long chalkboards, and a cannon fire of structures and chemical reactions vomited one after the other in succession. The only successful students, I thought, would be those wearing thick spectacles, periodic-table ties, and imitation leather shoes with Velcro straps.

But if my preconceptions of organic lecture were bad, my preconceptions of organic labs were worse. I feared the organic laboratory course, certain the instant I would step into the lab, all the chemicals would instantly vaporize, condense on my unclothed extremities, and permeate my hair, pores, follicles, and nails. As a result, my skin would erupt in a rash. I would bald. My nails would yellow. The love of my life would take one look at my scarred physiognomy, sicken of men, and leave me sitting alone, Job-like, amongst the ashes of my existence, scratching my weeping sores with a broken potsherd.

Turns out I was wrong on that one. I was surprised to find that I actually *liked* organic chemistry. I really liked doing it — it was fun! And working in the laboratory making new substances was less toxic than I thought it would be and was instead interesting and even entertaining. I was wrong about the math, too: If you can count to 11 without taking off your shoes, you can do the math in organic chemistry. The turning point, really, was when I stopped fighting organic chemistry, stopped feeding my preconceptions, and changed my attitude. That was when I really started enjoying the subject.

I hope you choose not to fight organic chemistry from the beginning (as I did) and instead decide to just get along and become friends with organic chemistry. In that case, this book will help you get to know organic chemistry as quickly as possible (and as well as possible), so that when your professor decides to test you on how well you know your newfound comrade, you'll do just fine.

About This Book

With *Organic Chemistry I For Dummies,* I've written the book that I would've wanted when I was taking the first semester of organic chemistry. That means that this book is very practical. It doesn't try to mimic a textbook, or try to replace it. Instead, it's designed as a complement to a text, highlighting the most important concepts in your textbook. Whereas a textbook gives you mostly a "just-the-facts-ma'am" style of coverage of the material — and provides you with lots of problems at the ends of chapters to see if you can apply those facts — this book acts as an interpreter, translator, and guide to the fundamental concepts in the subject. This book also gets to the nuts and bolts of how to actually go about tackling certain problems in organic chemistry.

Tackling the problems is where the majority of students have the most trouble, in part because so many aspects of a problem must be considered. Where's the best place to start a problem? What should you be on the lookout for? What interesting features (that is, sneaky tricks) do professors like to slip into problems, and what's the best strategy for tackling a particular problem type? I answer some of these questions in this book. Although this book cannot possibly show you how to solve every kind of problem that you encounter in organic chemistry, it does provide guides for areas that, in my experience, students typically have trouble with.

Beyond the problem types covered, these guides should give you insight into how to logically go about solving problems in organic chemistry. They show you how to rationally organize your thoughts and illustrate the kind of thinking you need to perform when approaching new problems in organic chemistry. In this way, you see how to swim instead of just panicking after being shoved abruptly into the deep end of the pool.

Additionally, I make clear the most important underlying principles in organic chemistry. I use familiar and easy-to-understand language, along with a great many clarifying analogies, to make palatable the hard concepts and technical jargon that comes with the territory. While this book is designed for students taking a first-semester course in introductory organic chemistry, it should also be a solid primer for those who want to understand the subject independently of a course.

Foolish Assumptions

In this book, I assume that you've had at least some chemistry in the past, and that you're familiar with the basic principles of chemistry. For example, I assume that you're familiar with the periodic table, that you understand what atoms are and what they're made up of (neutrons, protons, and electrons), and that you have some knowledge of bonding and chemical reactions. You should also know about kinetics (like rate equations and rate constants) and chemical equilibria. If you've had a two-semester course in general chemistry, that's perfect. (If you feel that your general chemistry is a bit rusty, turn to Chapter 2 — there, I review the most important concepts that you need to know for organic chemistry.)

Icons Used in This Book

Icons are the helpful little pictures in the margins. I use them to give you a heads-up about the material. I use the following icons in this book:

I use this icon when giving timesaving pointers.

I double dip with this icon. I use it not only to jog your memory about something that you should have learned previously, but also for really important concepts that you should remember.

I use this icon to warn you of common traps that students can fall into when tackling certain problems.

I try to avoid getting too technical, so you won't see this icon very much. When I use it, I do so to mark a discussion of a concept that's a little more in depth (which you can skip if you want to).

Beyond the Book

In addition to the material in the print or e-book you're reading right now, this product also comes with some access-anywhere material on the web. For the common functional groups in organic chemistry and the periodic table of elements, check out the free Cheat Sheet at www.dummies.com/cheatsheet/organicchemistry.

You can find some articles online that tie together and offer new insights to the material you find in this book. Go to www.dummies.com/extras/organicchemistry for these informative articles.

Where to Go from Here

In short, from here you can go anywhere you want. All of the chapters in this book are designed to be modular, so you can hop-scotch around, reading the chapters in any order you find most suitable. Perhaps you're having trouble with a particular concept, like drawing resonance structures or solving for structures using NMR spectroscopy. In that case, skip straight to the chapter that deals with that particular topic. Or, if you want, you can read the book straight through, using it as a kind of interpreter and guide to the textbook.

If you get the gist of what organic is all about, and have a solid background in the critical concepts in general chemistry — like electron configuration, orbitals, and bonding — you may want to skip the first two chapters and dive right into Chapter 3, which explains how organic structures are drawn. Or you may want to just skim the first couple chapters as a quick introduction and memory refresher (summer vacations have a strange way of wiping your memory slate sparkly clean, particularly in the area of chemistry).

This book is yours, so use it in any way you think will help you the most.

Part I

Getting Started with Organic Chemistry

In this part . . .

✔ Get an introduction to organic chemistry.

✔ Speak organic chemistry using Lewis structures.

✔ See acids and bases and functional groups.

✔ Look at organic molecules in three dimensions.

Chapter 1

The Wonderful World of Organic Chemistry

*O*rganic chemistry is a tyrant you've heard about a lot. You've heard your acquaintances whisper about it in secret. It's mean, they say; it's brutish and impossibly difficult; it's unpleasant to be around (and smells sort of funny). This is the chapter where I introduce you to organic chemistry, and where, I hope, you decide to forget about the negative comments you've heard about the subject.

In this chapter, I show you that the nasty rumors about organic chemistry are (mostly) untrue. I also talk about what organic chemistry is, and why you should spend precious hours of your life studying it. I show you that discovering organic chemistry really is a worthwhile and enjoyable expedition. And the journey is not all uphill, either.

Shaking Hands with Organic Chemistry

Although organic is a very important and valuable subject, and for some it's even a highly enjoyable subject, I realize that organic chemistry is intimidating, especially when you first approach it. Perhaps you've already had what many old-timers refer to simply as The Experience, the one where you picked up the textbook for the first time. This is the time when you heaved the book off the shelf in the bookstore. When you strained your back trying to hold it aloft. When you felt The Dread creep down your spine as you scanned through the book's seemingly infinite number of pages and feared that, not only would you have to read all of it, but that reading it wouldn't be exactly like breezing through a Hardy Boys adventure or a Nancy Drew mystery.

No doubt, the material appeared strange. Opening to a page halfway through the book you saw bizarre chemical structures littering the page, curved arrows swooshing here and there like flocks of starlings, and data tables bulging with an inordinate number of values — values that you suspect you might be required to memorize. I admit that organic chemistry is a little frightening.

The soap opera of organic molecules

Organic molecules govern our life processes like metabolism, genetic coding, and energy storage. In nature, organic molecules also play out a crazy soap opera, acting as the medium for many twists and turns, deceptions, betrayals, strategic alliances, romances, and even warfare.

Take plants, for example. They seem so defenseless. When a predator comes to lunch on a plant's leaves, the plant can't just pack up its bags and take off. It's stuck where it is, so there's nothing it can do, right? But although plants may seem defenseless, they're not. Many plants produce *antifeedants*, nasty organic compounds that are unpleasant tasting or even toxic to those who would dare eat them. (As a kid, I always *knew* Brussels sprouts contained something like this.) Predators that have feasted on a plant rich in these unpleasant compounds make sure to refrain from eating them in the future.

To produce antifeedants to discourage being eaten is bad enough, but sometimes plants have defenses that seem evil. Certain species of plants, for example, can detect when a caterpillar has decided to munch on its leaves

(by detecting organic molecules present in the caterpillar's spit!). When the plant detects that a caterpillar has decided to make supper on its leaves, the plant emits volatile organic molecules into the air, chemicals designed to attract wasps. When the wasps buzz by to check out what's up, they see the caterpillars eating the plant and killing it. The wasps couldn't care less about the misfortunes of the plant, but the female wasps do need a comfortable spot to hatch their eggs. And what's a snugger nursery than the innards of a fat, juicy caterpillar?

When a wasp spots a caterpillar, it swoops down, makes a crash landing on the caterpillar's back, stings the caterpillar into paralysis, and lays its eggs *inside* of it! When the wasp larvae hatch shortly thereafter, they make the caterpillar their first meal, munching on it contentedly from the inside out. The wasp has now reproduced and has had its little offspring fed, and the plant is rid of its pest — a strange alliance between wasp and plant, all thanks to communication by organic molecules. And that's just one episode in this never-ending soap opera, produced, funded, and sponsored by organic molecules.

I think most students feel this way before they take this class, and probably even your professor did, as did her professor before her. *So you're not alone.* But you can take comfort in knowing that organic chemistry is not as hard as it looks. Those who put in the required amount of work — which, admittedly, is a lot — and don't fall behind, almost always do well. More than almost any other subject, organic chemistry rewards the hard workers (like you), and relentlessly punishes the slothful (the others in your class). I think understanding organic chemistry is not so much hard as it is hard work.

I hope all this talk about how intimidating the course is hasn't put a damper on your enthusiasm, because the subject of organic chemistry really is a doozy. To learn about organic chemistry is to learn about life itself, because living organisms are composed of organic molecules and use organic molecules to function. Swarms of organic molecules are at work in your body — fueling your brain, helping your neurons fire, and getting the muscles in your mouth to clench open and shut — and that's just a small sampling of the organic molecules needed in order for you to complain about your school's chemistry requirements.

Humans, in fact, are composed almost entirely of organic molecules (all the soft parts anyway), from our muscles, hair, and organs, to the fats that cushion our bellies and keep us toasty warm during sweltering summer nights (some people are more richly blessed in this regard than others). Organic molecules can also range in size from the very tiny, like the carbon dioxide you exhale that consists of only three atoms, to the staggeringly large, like DNA, which acts as your molecular instruction manual and is made up of millions of atoms.

What Are Organic Molecules, Exactly?

But what ties all of these molecules together? *What exactly makes a molecule organic?* The answer lies in a single, precious atom: carbon. All organic molecules contain carbon, and to study organic chemistry is to study molecules made of carbon and to see what kinds of reactions they undergo and how they're put together. When these principles are known, that knowledge can be put to good use, to make better drugs, stronger plastics, better materials to make smaller and faster computer chips, better paints, dyes, coatings, explosives, and polymers, and a million other things that help to improve our quality of life.

That said, I should also point out that the field of organic chemistry is essentially an arbitrary one, that the same fundamental laws of chemistry and physics that apply to inorganic molecules apply just as well to organic ones. This connectivity of the branches of chemistry is actually a relatively new idea, as organic molecules were once falsely thought to have a "vital life force" that other molecules didn't possess. Despite the destruction of this theory of *vitalism,* chemists still keep the old divisions of chemistry, divisions that define the branches of physical chemistry, inorganic chemistry, and biochemistry. But these barriers are slowly beginning to dissolve, and they're kept mainly to help students focus on the material taught in a given course.

Given the many elements present in the universe, it is fascinating that living things selected carbon as their building block. So, what makes carbon so special? What makes it better as the foundation for life than any of the other elements? What makes this atom so important that an entire subject focuses around this single atom, while the chemistries of all the other elements are tossed into a big mushy pile known as *inorganic chemistry?* Is carbon really, in fact, all that special compared to the many other elements that could have been selected?

In short, yes. Carbon is very special, and its usefulness lies in its versatility. Carbon has the capability of forming four bonds, so molecules that contain carbon can be of varied and intricate designs. Also, carbon bonds represent the perfect trade-off between stability and reactivity — carbon bonds are neither too strong nor too weak. Instead, they epitomize what chemists refer to affectionately as the *Goldilocks principle* — carbon bonds are neither "too hot" nor "too cold," but are "just right." If these bonds were too strong, carbon would be unreactive and useless to organisms; if they were too weak, they would be unstable and would be just as worthless. Instead, carbon bonds straddle the two extremes, being neither too strong nor too weak, making them fit for being the backbone of life.

Also, carbon is one of the very few elements that can form strong bonds to itself, in addition to being able to form bonds to a wide variety of other elements. Carbon bonds can even double back to form rings. Because of this ability to bond with itself and other elements, carbon can form an incredibly vast array of molecules. Millions of organic compounds have already been made and characterized, and undoubtedly many millions more will be discovered (perhaps, dear reader, by you!).

An Organic Chemist by Any Other Name . . .

Just as the field of chemistry can be broken down into different branches, so, too, can the field of organic chemistry be broken down into specialized areas of research. Those who work in these different areas — these specialized "organic chemists" — illustrate the diversity of the field of organic chemistry and its connection to other branches in chemistry, branches like physical chemistry, biochemistry, and inorganic chemistry.

Synthetic organic chemists

Synthetic organic chemists concern themselves with making organic molecules. In particular, synthetic chemists are interested in taking cheap and available starting materials and converting them into valuable products. Some synthetic chemists devote themselves to developing procedures that can be used by others in constructing complex molecules. These chemists want to develop general procedures that are flexible and can be used in synthesizing as many different kinds of molecules as possible. Others devote themselves to developing reactions that make certain kinds of bonds, such as carbon-carbon bonds.

Others use known procedures to tackle multistep syntheses — the making of complex compounds using many individual, known reactions. Performing these multistep syntheses tests the limits of known procedures. These multi-step syntheses force innovation and creativity on the part of the chemist, in addition to encouraging endurance and flexibility when a step in the synthe-sis goes wrong (things inevitably go wrong during the synthesis of complex molecules). Such innovation contributes to the body of knowledge of organic chemistry.

Synthetic organic chemists often flock to the pharmaceutical industry, mapping out efficient reaction pathways to make drugs and optimizing reactions to make complicated organic molecules as cheaply and efficiently as possible for use as pharmaceuticals. (Sometimes improving the yield of the reaction of a big-name drug by a few percentage points can save millions of dollars for a pharmaceutical company each year.) If you take a laboratory course in organic chemistry, you'll be doing a lot of organic synthesis.

Bioorganic chemists

Bioorganic chemists are particularly interested in the enzymes of living organisms. Enzymes are very large organic molecules, and are the worker bees of cells, catalyzing (speeding up) all the reactions in the cell. These enzymes range from the moderately important ones, such as the ones that keep us alive by breaking down food and storing energy, to the really important ones, like the ones in yeasts that are responsible for fermentation, or the breaking down of sugars into alcohol.

These catalysts work with an efficiency and selectivity that synthetic organic chemists can only envy (see the previous section). Bioorganic chemists are particularly interested in looking at these marvels of nature, these enzymes, and determining how they operate. When chemists understand the mechanisms of how these enzymes catalyze particular reactions in the cell, this knowledge can be used to design enzyme *inhibitors*, molecules that block the action of these enzymes.

Such inhibitors make up a great deal of the drugs on the market today. Aspirin, for example, is an inhibitor of the *cyclooxegenase* (COX) enzymes. These COX enzymes are responsible for making the pain transmitters in the body (called the *prostaglandins*). These transmitters are the messengers that tell your brain to feel pain in the thumb that you just smashed with a slip of your hammer. When the aspirin drug inhibits these COX enzymes from operating, the enzymes in your body can no longer make these pain-signaling molecules. In this way, the feeling of pain in the body is reduced. Many other examples of these kinds of inhibitor drugs exist today, and the process of designing these drugs is aided by bioorganic chemists.

Natural products chemists

Natural products chemists isolate compounds from living things. Organic compounds isolated from living organisms are called *natural products.* Throughout history, drugs have come from natural products. In fact, only recently have drugs been made synthetically in the lab. Penicillin, for example, is a natural product produced by a fungus, and this famous drug has saved millions of lives by killing harmful bacteria. The healing properties of herbs and teas and other "witches' brews" are usually the result of the natural products contained in the plants. Some Native American groups chewed willow bark to relieve pain, as the bark contained the active form of aspirin; other Native American groups engaged in the smoking of peyote, which contains a natural product with hallucinogenic properties. Smokers get a buzz from the natural product in tobacco called nicotine; coffee drinkers get their buzz from the natural product found in coffee beans called caffeine.

Even today, a great many of the drugs found on the shelves of pharmacies are derived from natural products. Once extracted from the living organism, natural products are often tested by chemists for biological activity. For example, a natural product might be tested to see if it can kill bacteria or cancer cells, or if it can act as an anti-inflammatory drug. Often when chemists find a "hit" — a compound that shows useful biological activity — the structures of these natural products are then modified by synthetic organic chemists to try to increase the potency of the compound or to reduce the number of harmful side effects produced by the natural product.

To take another example, after a few decades of use, the natural penicillin isolated from mold ceased to be as effective as an antibiotic, as bacteria developed mechanisms for resistance to this drug, including evolving enzymes that snipped the penicillin molecule into pieces within bacterial cells that rendered the drug ineffective. As a result, synthetic chemists had to synthesize new derivatives of penicillin that still killed bacteria, but bypassed their mechanisms of resistance. Because bacteria eventually evolve resistance to new molecules, we currently have what amounts to an escalating battle of chemical warfare between humans and bacteria. In this fight, bacteria develop resistance to known drugs and we develop new molecules for the next round of attack. Both synthetic chemists and natural products chemists play a collaborative role in developing more effective antibiotics.

Physical organic chemists

Physical organic chemists are interested in understanding the underlying principles that determine why atoms behave as they do. Physical organic chemists, in particular, study the underlying principles and behaviors of organic molecules. Some physical organic chemists are interested in modeling the behavior of chemical systems and understanding the properties and reactivities of molecules. Others study and predict how fast certain reactions will occur; this specialized area is called *kinetics*. Still others study the energies of molecules, and use equations to predict how much product a reaction will make at equilibrium; this area is called *thermodynamics*. Physical organic chemists are also interested in *spectroscopy* and *photochemistry*, both of which study the interactions of light with molecules. (Photosynthesis by plants is probably the most well-known example of light interacting with molecules in nature.)

Organometallic chemists

Organometallic chemists are interested in molecules that contain both metals and carbon. Such molecules are often used as catalysts for chemical reactions. (Catalysts speed up reactions.) Carbon-carbon bonds are strong compared to carbon-metal bonds, so these carbon-metal bonds are much more easily made and more easily broken than carbon-carbon bonds. As such, they're useful for catalyzing chemical transformations of organic molecules. Many organometallic chemists concern themselves with making and optimizing organometallic catalysts for specific kinds of reactions.

Computational chemists

With the recent advances in the speed of computers, chemists have rushed to use computers to aid their own studies of atoms and molecules. *Computational chemists* model compounds (both inorganic and organic compounds) to predict many different properties of these compounds. For example, computational chemists are often interested in the three-dimensional structure of molecules and in the energies of molecules.

The models generated by computational chemists are getting more and more sophisticated as computers increase in speed and as physical chemists create better models. Many drugs are now modeled on computers by computational chemists; this process is called *in silico* drug design, meaning that the drug is designed in the silicon-based computer. Typically, drugs work by blocking a receptor on an enzyme (see the earlier section on bioorganic chemists). *In silico* drug design focuses on modeling to see which compounds would best fit into the drug's target receptor. This allows for *rational drug design*, or the use of the brain and a molecular model to come up with the structure of a drug instead of simply using the "brute-force methods" that involve testing thousands of randomly selected compounds and looking for biological activity. Computational methods are not sophisticated enough that we can fire all the experimentalists yet (and, perhaps, they may never reach that level of sophistication), but they are useful as a partner to understand, explain, and predict the results from lab experiments.

Materials chemists

Materials chemists are interested in, well, materials. Plastics, polymers, coatings, paints, dyes, explosives — all these are of interest to the materials chemist. Materials chemists often work with both organic and inorganic materials, but many of the compounds of interest to materials chemists are organic. Teflon is an organic polymeric material that keeps things from sticking to surfaces, polyvinyl chloride (PVC) is a polymer used to make pipes, and polyethylene is a plastic found in milk jugs and carpeting.

Materials chemists also design environmentally safe detergents that retain their cleaning power. Organic materials are also required for photolithography to make smaller, faster, and more reliable computer chips. All these applications and millions of others are of interest to the materials chemist.

Chapter 2

Dissecting Atoms: Atomic Structure and Bonding

*I*n this chapter, you take apart an atom, study the most important pieces (being careful not to lose any), and then put it back together again, as if you were an atom mechanic. After you see all the pieces, including where they fit in an atom and how they work, you begin to see how atoms come together and bond, and you discover the different kinds of bonds. Here, you find out that atoms were not created equally: Some atoms are greedy, and they selfishly plunder the electrons in a bond for themselves, while others are more generous. I show you how to distinguish the altruistic atoms from the swine, and show you how this predictor can be used to see the separation of charge in a bond or molecule (this separation is called a *dipole*), which can be useful in understanding the reactivity of a molecule. I also dissect *orbitals* — the apartments that electrons reside in — and show how their overlap leads to bonding with other atoms.

So, prepare to get your hands greasy and have carbon grit etched under your fingernails. And don't worry about the mess.

Electron House Arrest: Shells and Orbitals

The soul of an atom is the number of protons it has in its nucleus; this number cannot be changed without changing the identity of the atom itself. You can determine how many protons an atom has by looking at its atomic number on the periodic table. Your friend carbon, for example, has an atomic number of 6, so it has six protons tucked away in its nucleus. Because protons are positively charged, an atom needs the same number of electrons (which are negatively charged) as it has protons to remain electrically neutral.

If an atom has more or fewer electrons than it has protons — in other words, when the number of positively charged parts doesn't balance the number of negatively charged parts — the atom itself becomes electrically charged and is called an *ion*. If an atom has more electrons than the number of protons in its nucleus, it becomes a negatively charged ion, called an *anion* (pronounced ANN-eye-on). If it has fewer electrons than protons, it becomes a positively charged ion called a *cation* (pronounced CAT-eye-on).

Unlike protons, electrons are not held tightly in the nucleus of an atom; instead, they're held in shells that surround the nucleus. In a qualitative way, you can think of the electron shells as being concentric spheres that surround the nucleus of the atom. The first shell is the closest to the nucleus of the atom, is of the lowest energy, and can hold up to two electrons. (You often see electrons abbreviated as e$^-$, so using this notation, the first shell can hold 2e$^-$). The second shell is higher in energy, is farther away from the nucleus, and can hold up to eight electrons. The third shell is higher yet in energy, and can hold up to 18 electrons. See Figure 2-1 for a diagram of these shells. I don't talk about the shells higher than the third (because you don't deal with them in organic chemistry), except to say that the higher the number of the shell, the farther it is from the nucleus, the more electrons it can hold, and the higher it is in energy.

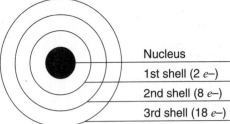

Nucleus

1st shell (2 $e-$)

2nd shell (8 $e-$)

3rd shell (18 $e-$)

Electron apartments: Orbitals

Electron shells are further subdivided into *orbitals*, or the actual location in which an electron is found within the shell. Quantum mechanics — that scary subject dealing with mathematical equations too complicated to cover in organic chemistry (yay!) — says that you can never know exactly where an electron is at a given moment, but you can know the region of space in which an electron will be found, and that is the electron's orbital.

So, what's the difference between a shell and an orbital? A shell indicates the energy level of a particular electron, and the orbital is the actual location in space where the electron resides. A shell that is full of electrons is spherical in shape (refer to Figure 2-1). The shell can be thought of as the floor in the apartment complex where an electron lives (the energy level), whereas the orbital is the actual apartment in which the electron resides.

You can take this analogy a step further to clarify what you know about the electron. All electrons in atoms are under house arrest. They can't be just anywhere in an atom — they're restricted to their particular orbital apartments. But quantum mechanics closes the doors and the windows to the apartment, so you can never peek in and know for sure exactly where the electron is at a given moment. (This uncertainty in knowing the locations of electrons is called the *Heisenberg uncertainty principle*. And now all you fans of *Breaking Bad* know where Walter White got his pseudonym.)

Although you can't know the exact location of an electron at any given moment, you can know the region in space in which an electron must be found, which is its orbital. And the shape of these apartments — these electron orbitals — becomes important in bonding. The atomic orbitals that you deal with in organic chemistry come in two kinds, the *s* orbitals and the *p* orbitals, and each kind has a distinctive shape. Drawings of these orbitals show where an electron in a particular orbital will be found more than 90 percent of the time. An *s* orbital is spherical in shape, whereas a *p* orbital is shaped like a dumbbell (sort of; see Figure 2-2). Each orbital can hold up to two electrons, but if there are two electrons in an orbital, they must have opposite spins. (You may have been taught that the *p* orbitals hold six electrons, but that's because there are three individual *p* orbitals in a *p* level, each of which holds two electrons.) The spin of an electron in an orbital is a somewhat abstract property that doesn't really have a counterpart in our big world, but you can think of these spins qualitatively as electrons spinning around the orbital like tops — one electron spins one way about the axis in the orbital, and the other spins the opposite way.

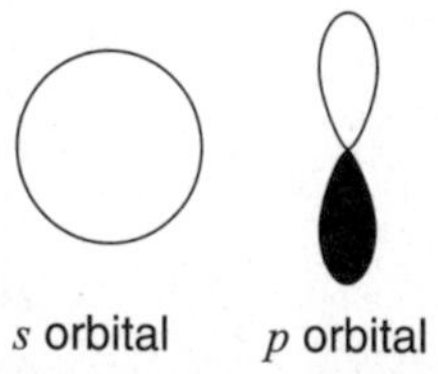

Figure 2-2:
The shapes
of the *s* and
p orbitals.

s orbital   *p* orbital

Chemists also use a specific syntax when referring to orbitals. A number is placed in front of an orbital type to designate which shell that orbital resides in. For example, the 2s orbital refers to the s orbital in the second shell. If the electron occupancy of that orbital is important, the number of electrons in that orbital is placed in a superscript following the number, as shown in Figure 2-3.

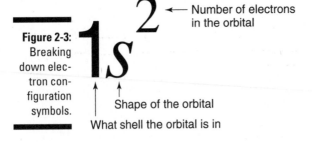

Figure 2-3:
Breaking
down elec-
tron con-
figuration
symbols.

Now that you know what orbitals are, you can see how the different kinds of orbitals fit into the electron shells. The 1s orbital is spherically symmetric, holds two electrons, and is the only orbital in the first shell. The second shell contains both s and p orbitals and holds up to eight electrons. The 2s orbital has the same spherical shape as the 1s orbital, but it's larger and higher in energy. The 2p level consists of three individual p orbitals — one orbital that points in the x direction (p_x), one that points in the y direction (p_y) and one that points in the z direction (p_z). Because each of these p orbitals is of equal energy, they're said to be *degenerate orbitals*, using organic-speak. See Figure 2-4 for pictures of the p orbitals. In general, the p levels can hold up to six electrons (because they have three individual p orbitals, each of which can hold two electrons), and the s levels can hold up to two electrons (because they have just one orbital that can hold up to two electrons).

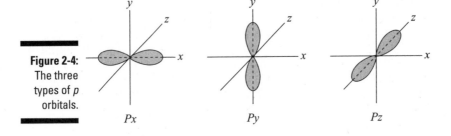

Figure 2-4:
The three
types of p
orbitals.

Electron instruction manual: Electron configuration

Chemists like to know which orbitals are occupied by electrons in an atom, because where the electrons are located in an atom often predicts that atom's reactivity. To build the *ground-state electron configuration*, or the list of orbitals occupied by electrons in a particular atom, you start by placing electrons into the lower energy orbitals and then build up from there. Nature, like human beings, is lazy and prefers to be in the lowest energy state possible. The Aufbau chart in Figure 2-5 (Aufbau is the German word for *building*) is helpful for remembering which orbitals fill first. Simply follow the arrows. The lowest-energy orbital is 1*s*, followed by 2*s*, 2*p*, 3*s*, 3*p*, 4*s*, and so on.

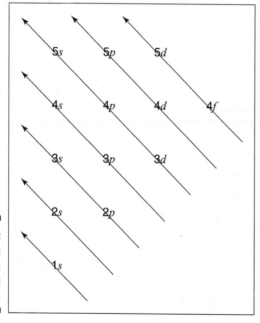

Figure 2-5:
The Aufbau electron-ordering scheme.

Filling orbitals with electrons is a fairly straightforward task — you just fill two electrons per orbital, starting with the lowest-energy orbitals and working up until you run out of electrons. But the last electrons you place into orbitals must sometimes be placed a little differently. *Hund's rule* tells you what to do when you come to the last of the electrons that you need to place into orbitals, and you're at an orbital level that will not be entirely filled. In such a case, Hund's rule states that the electrons should go into different orbitals with the same spin instead of pairing up into a single orbital with opposite spin. This rule applies, in part, because electrons repel each other and want to get as far away from each other as possible, and putting them into separate orbitals gets the two electrons farther away from each other.

Writing the electron configuration for an atom using Hund's rule will make more sense if you do one. Try determining the electron configuration of carbon, for example. Carbon has six electrons to put into orbitals. Because you always start by putting the electrons in the lowest-energy orbitals first and building up, you put the first two electrons in the $1s$ orbital, the next two into the $2s$, and the remaining two into the $2p$ orbitals. But because the $2p$ level can hold up to six electrons, you have to follow Hund's rule for these last two electrons and put the two electrons into different p orbitals with the same spin. As a result, these two electrons go into separate p orbitals, not the same one.

Therefore, the electron configuration of carbon is written $1s^2\, 2s^2\, 2p_x^1\, 2p_y^1\, 2p_z^0$ (not $1s^2\, 2s^2\, 2p_x^2\, 2p_y^0\, 2p_z^0$, which violates Hund's rule).

Atom Marriage: Bonding

Now that you know how electrons fit into atoms, you can see how atoms can come together and bond. But first, why do atoms make bonds? Aren't atoms happy by themselves? Aren't carbons happy with their carboniness, fluorines with their fluorininess, sodiums with their sodiuminess? Aren't they happy with the number of electrons allotted to them?

No, of course not! Atoms are like people; most of them aren't happy the way they are and would like to be like something else. Just as many people want to be like the rich, popular person down the street who throws big parties every night (rather than being like the poor chemistry nerd pecking away at his keyboard), atoms strive to be like the *noble gases*, the elements found in the eighth (and last) column of the periodic table. These noble gases (such as helium [He], neon [Ne], xenon [Xe], and argon [Ar]) are the Cary Grants and Marilyn Monroes of atoms — the atoms that all others wish they were and try to imitate. This desire of atoms to imitate the noble gases provides the driving force for many reactions.

So, why do atoms want to imitate the noble gases? What makes these particular atoms so attractive? The answer is their electronic structure. The noble gases are the only atoms that have their outermost shells filled with electrons, while all other atoms have shells that are only partially filled. And because a filled shell of electrons is the most stable possible electron configuration, it's always in style to have a full shell.

Among the atoms you encounter in organic chemistry, each shell in an atom can hold up to eight electrons, except for the first shell, which can hold two. (The third shell can actually hold 18 electrons, but often behaves as if it were full when it has eight.) The desire of atoms to have filled electron shells is often called the *octet rule*, referring to the desire of atoms in the second row of the periodic table to fill their outer shells with eight electrons, or to imitate those noble gases.

This desire of atoms to imitate the noble gases by filling up their shells is a major driving force of chemical reactions. In fact, the noble gases are so happy by themselves that they're almost completely unreactive. (They're so unreactive that they were called the "inert gases" until some smart-aleck chemists managed to get them to react under unusual conditions.)

The electrons in the outermost shell of an atom are referred to as the *valence electrons.* For bonding, the valence electrons are the most important, so you most often ignore the *core electrons* (the ones in the inner shells), because they don't participate in bonding. Instead, you focus entirely on the electrons in the valence shell.

To Share or Not to Share: Ionic and Covalent Bonding

Understanding the different kinds of bonding in molecules is important because the nature of the different bonds in a molecule often determines how the molecule will react. The two big categories of bonding are *ionic bonding*, in which the two electrons in a bond are not shared between the bonding atoms, and *covalent bonding*, in which the two electrons in a bond are shared between the two bonding atoms — and these classifications represent the extremes in bonding.

Mine! They're all mine! Ionic bonding

The following is an example of a reaction driven by this desire of atoms to imitate the noble gases. Sodium (Na) combines with chlorine (Cl) to make sodium chloride (NaCl), or table salt, as shown in Figure 2-6. Sodium is an atom found in the first column in the periodic table and has one electron in its outermost shell (one valence electron). Chlorine is in the second-to-last column of the periodic table (the column that contains the group VIIA elements) and has seven electrons in its outermost shell (or seven valence electrons). Often, to have an easier time understanding how a reaction is happening, the number of valence electrons an atom owns is represented by the number of dots around the atom. So, you give one dot to sodium because it has one valence electron, and seven dots to chlorine because it has seven.

Figure 2-6:
Making
NaCl.

Na· + ·Cl: ⟶ Na⁺ :Cl:⁻

To achieve its valence octet, sodium could either gain seven electrons or lose one. Likewise, to achieve its octet, chlorine could either lose seven electrons or gain one. Atoms generally don't like gaining or giving up more than three electrons, so sodium gives up its one valence electron to chlorine to leave sodium with only filled shells of electrons (by giving up its one electron in its outermost shell, it's left with only filled core shells), and chlorine accepts the electron from sodium and uses it to fill its octet. Because sodium has lost its one electron, it becomes a positively charged ion (a cation), and because the chlorine has accepted an extra electron, it becomes a negatively charged ion (an anion).

Sodium is happy to give up its electron, because when it has done so, it has imitated the electron configuration of the noble gas neon (Ne), which has a full valence shell. Similarly, chlorine, by gaining an electron, has imitated the valence shell of the noble gas argon (Ar). Having filled shells makes the atoms happy. When the sodium cation and the chlorine anion combine, you have stable sodium chloride (NaCl, or table salt), and (as far as these atoms are concerned), all is right with the world.

The attraction between the sodium cation and the chloride anion in sodium chloride is called an *ionic bond*. In an ionic bond, the electrons in the bond are shared like toys between siblings — which is to say not at all. The anionic species (chloride) has snatched the electron away from the cationic species (sodium). Because the electrons in the bond are not shared, the attraction is one of opposite (positive and negative) electrical charges. You've seen a similar kind of attraction if you've ever watched two magnets scooch together on a tabletop. The magnetic "bond" between the two magnets is similar to the ionic bond between sodium and chloride, albeit on a much larger scale.

The name's Bond, Covalent Bond

A different kind of bonding occurs when two hydrogen atoms come together to make hydrogen gas (H_2) as shown in Figure 2-7, although this reaction is driven by the same desire to imitate the noble gases, as in the reaction of sodium and chlorine.

Figure 2-7:
Making H_2.

$$H\cdot + H\cdot \longrightarrow H:H$$

A hydrogen atom has one electron, and so needs one electron to fill its shell. (Remember that the first shell can hold only two electrons, and the remaining outer shells can hold eight electrons.) Because both hydrogen atoms need one electron to fill the first shell, instead of one grabbing the electron from the other, they share their electrons equally. This molecular communism is called a *covalent bond*, a bond in which the electrons are shared between two atoms. Both hydrogen atoms are now happy, because each has achieved the electronic configuration of the noble gas helium (He).

Electron piggishness and electronegativity

TIP

So, how do you know whether a bond is going to be ionic or covalent? A good general tool is to look at the difference in electronegativities between the two atoms. The *electronegativity* of an atom is organic-speak for an atom's electron piggishness. An atom with a high electronegativity will hog the bonding electrons from an atom of low electronegativity. If the electronegativity difference is very large, the bond will be ionic because the atom with the larger electronegativity will essentially hog all the electrons. If the electronegativity difference is smaller, the bond can be thought of as being *polar covalent:* The electrons are shared, but not equally between the two atoms. And if the electronegativity difference is zero (as it is when two of the same atoms are joined together), the bond can be thought of as *purely covalent:* The electrons are shared equally between the two atoms. The general trend for electronegativities is that, as you go up and to the right in the periodic table, the electronegativity increases. Fluorine (F), therefore, is the biggest electron swine, because it's the most electronegative element (see Figure 2-8).

Figure 2-8:
The electro-
negativities
of some
elements. A
larger elec-
tronegativ-
ity value
indicates a
bigger elec-
tron pig.

H 2.1						
Li 1.0	Be 1.5		B 2.0	C 2.5	N 3.0	O 3.5	F 4.0
Na 0.9	Mg 1.2		Al 1.5	Si 1.8	P 2.1	S 2.5	Cl 3.0
K 0.8	Ca 1.0						Br 2.8
							I 2.5

Here are the general rules for determining whether a bond will be covalent or ionic:

- ✔ If there is no difference in the electronegativities of the two atoms, the bond will be purely covalent.
- ✔ If the electronegativity difference between the two atoms is between 0 and 2, the bond will be polar covalent.
- ✔ If the electronegativity difference between the two atoms is greater than 2, the bond will be ionic.

Table 2-1 shows some examples in which this rule is applied to different bonds.

Table 2-1	Classifying Bonds	
Bond	*Electronegativity Difference*	*Classification*
H-H	0	Purely covalent
Cl-Cl	0	Purely covalent
H-Cl	0.9	Polar covalent
C-N	0.5	Polar covalent
Li-F	3.0	Ionic
K-Cl	2.2	Ionic

Although ionic bonds are found most often in inorganic compounds (non-carbon-containing compounds), organic compounds are usually held together by covalent bonds. This trend makes sense from looking at the table of electronegativities (see Figure 2-8). Inorganic compounds are often formed when atoms from the left side of the periodic table bond with atoms from the right side of the periodic table. For example, you often see compounds like LiF, NaCl, KBr, and $MgBr_2$, where atoms from the first or second column bond with atoms found on the far-right side of the periodic table. Atoms on the left side of the periodic table have low electronegativities and atoms on the right side of the periodic table have high electronegativities, so many of these inorganic compounds are ionic, because the differences in their electronegativities are large.

Organic compounds, on the other hand, generally have bonds only between a few different kinds of atoms, and these atoms are generally found between them on the right side of the periodic table. Carbon (C), hydrogen (H), oxygen (O), and nitrogen (N) are the most prominent elements you see in organic compounds, but you also see phosphorous (P), sulfur (S), and most of the *halogens* (elements such as fluorine [F], chlorine [Cl], bromine [Br], and iodine [I] that are found in the second-to-last column). Because the electronegativity differences between these atoms are small, bonding between these atoms results in purely covalent or polar covalent bonds, where the electrons in the bond are shared between the two atoms involved in the bond. So, for the most part, the organic compounds I cover in this book are held together by covalent bonds.

Separating Charge: Dipole Moments

In a polar covalent bond, the electrons in the bond are not equally shared between the two atoms. Instead, the more electronegative atom bullies most of the bonding electrons away from the less electronegative atom, creating a separation of charge in the bond. This separation is called a *dipole moment*. Dipole moments are often used to explain how molecules react, so learning how to predict the dipole moment of any bond, or of a molecule, is a very important skill to add to your toolbox.

Consider, for example, hydrochloric acid (HCl), shown in Figure 2-9. A quick comparison of the hydrogen and chlorine electronegativities shows that chlorine is the more electronegative atom of the two (refer to Figure 2-8). That means that the electrons in the bond between hydrogen and chlorine are going to be hogged mostly by the chlorine. Because the electrons in the bond are going to be spending most of their time around chlorine and away from hydrogen, this puts a partially negative charge on the chlorine. (The symbol for the lowercase Greek letter delta, δ is used to mean "partial.") δ Because the electrons are going to be away from hydrogen, the hydrogen end of the molecule becomes partially positive. Using organic-speak, this separation of charge is called a dipole moment.

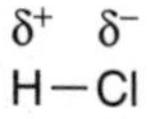

Figure 2-9:
Seeing bond
dipoles.

δ^+ δ^-
H—Cl

A strange-looking arrow, called the *dipole vector*, shown in Figure 2-10, is used to show the direction of the dipole moment, or the separation of charge. By convention, the head of the arrow points in the direction of the partial negative charge, while the tail that looks like a plus sign points in the direction of the partial positive charge.

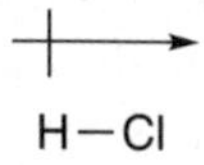

Figure 2-10:
Using dipole
vectors.

H—Cl

Problem solving: Predicting bond dipole moments

If you want to draw the dipole vector for a bond, you need to look at the electronegativities of the two atoms. The atom with the higher electronegativity becomes partially negative, because this atom is the greater electron hog, and the atom with the lower electronegativity becomes partially positive. You then draw the dipole vector with the head pointing toward the more electronegative atom and the tail pointing toward the partially positive atom. The size of the vector depends on the difference in electronegativity; draw a long vector for large differences in electronegativity, and a short vector for smaller differences.

Being able to predict the direction of the dipole moment is absolutely essential, because the dipole moment can be used to understand how molecules react (see the following section).

Problem solving: Predicting molecule dipole moments

Predicting the dipole moment for a molecule is slightly more complicated than predicting a dipole moment for a bond (see the preceding section). Look at the example of chloroform ($CHCl_3$) in Figure 2-11. To determine the dipole moment of chloroform, the first step you take is to find the dipole vectors of each of the individual bonds. (***Note:*** I've neglected the C-H bond because the electronegativity difference between hydrogen and carbon is so small that its contribution can usually be ignored.) I've drawn each of the vectors for the C-Cl bonds the same size, because each bond dipole is identical.

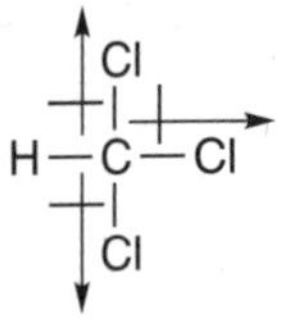

Figure 2-11:
The dipole
vectors for
chloroform
($CHCl_3$).

To determine the dipole moment for the molecule, then, you have to add up each of the individual bond vectors. To do this, you simply line up the vectors from head to tail (it doesn't matter which order you line them up in), as I've done in Figure 2-12. The dipole moment for the molecule points from where you started to where you ended up. In this case, it points to the right.

Figure 2-12:
Using the dipole vectors to predict the dipole moment for chloroform ($CHCl_3$).

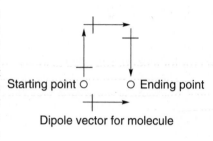

Starting point ○   ○ Ending point

Dipole vector for molecule

WARNING!

Just because individual bonds have dipole moments does not mean that the molecule has a dipole moment. Consider carbon dioxide (CO_2), shown in Figure 2-13. In this molecule, oxygen is more electronegative than carbon, so you draw the two dipole vectors pointing out. (Double bonds, like the ones between carbon and oxygen in carbon dioxide, are discussed more in Chapter 3.)

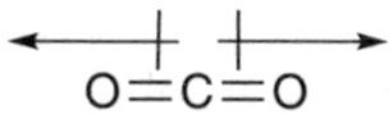

Figure 2-13:
Bond dipoles in CO_2.

O=C=O

To predict the dipole for the entire molecule, you have to add up all the vectors. In the case of carbon dioxide, however, even though there is a dipole for each of the individual bonds, the net dipole moment for the molecule is zero, because the oxygens are pulling in equal and opposite directions, and they cancel each other out, as shown in Figure 2-14. Imagine two equally strong men playing tug-of-war. Neither would win — it would simply be a stalemate (unless one of them tripped). Because the net dipole vector is zero, carbon dioxide has no dipole moment.

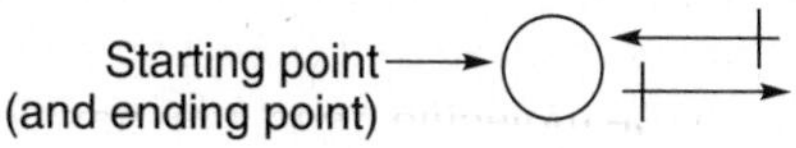

Figure 2-14:
CO_2 has no dipole moment.

Starting point ⟶ ○ ⇄
(and ending point)

Seeing Molecular Geometries

VSEPR theory — VSEPR is pronounced "vesper," and stands for *valence shell electron pair repulsion* — predicts the approximate geometry of bonds around an atom. This theory says that because electrons repel each other, bonds and *lone pairs* (non-bonding electron pairs) around an atom want to get as far away from each other as possible. As far as atoms are concerned, the electrons of other atoms are like clouds of unpleasant atom odor (I suppose this would be abbreviated a.o. rather than b.o.), so it's imperative for atoms to get as far away from the ungodly a.o. of other atoms as possible.

Extending this theory to molecules, an atom that has two bonds would want the bonds to be 180 degrees apart from each other in a linear geometry, giving the electrons in the bonds the largest separation possible (see Figure 2-15). For the same reason, an atom with three bonds would situate the bonds 120 degrees from each other in a *trigonal planar* geometry, and an atom with four bonds would situate the bonds 109.5 degrees away from each other, forming a pyramid-like *tetrahedron*. All these geometries put the bonds at the maximum distance apart that is possible. These three geometries (linear, trigonal planar, and tetrahedral) are the main geometries you need to think about in organic chemistry, because the atoms that form organic molecules generally form only four or fewer bonds.

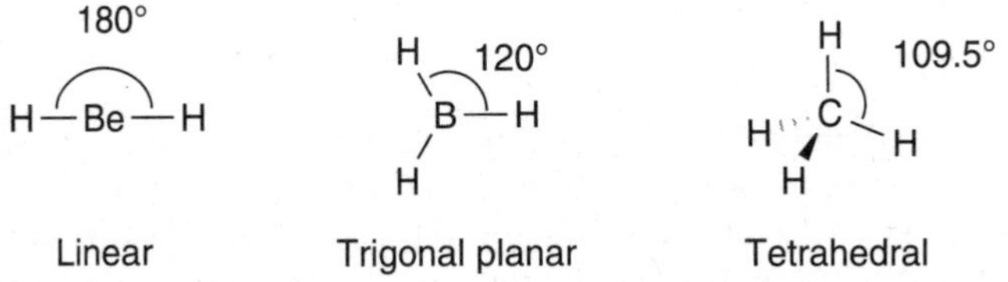

calc

Figure 2-15:
The three main geometries of organic molecules.

180°	120°	109.5°
H—Be—H	H, B—H, H	H, C, H, H, H
Linear	Trigonal planar	Tetrahedral

Mixing things up: Hybrid orbitals

When you know what bond angles are preferred around an atom (see the preceding section), you can see how orbitals' overlapping between atoms leads to bonding. Carbon is a handy model for how bonding works. Carbon has four valence electrons, so it wants to make four bonds so it can fill its octet and mimic the noble gas neon (Ne). But carbon's electron configuration ($1s^2 2s^2 2p_x^1 2p_y^1 2p_z^0$) shows that the 1s and 2s orbitals are completely filled, so only the two electrons in the p orbitals are available to be shared in a covalent bond. Likewise, carbon wants to have its four bonds oriented in a tetrahedral geometry around the carbon atom, with bond angles of 109.5 degrees. But p orbitals are oriented at 90-degree angles, perpendicular to each other, not at 109.5-degree angles. So, what's a carbon atom to do?

The first thing the carbon atom does is promote an electron from the filled $2s$ orbital into the last empty p orbital (see Figure 2-16). This leaves the atom with four orbitals, each of which contains one electron, perfect for making four covalent bonds. But why would carbon promote the electron? Doesn't putting an electron into a higher-energy orbital cost energy? It sure does, but this electron promotion also allows the formation of two additional bonds, which more than pays the cost. It's like investing a dollar and getting back a fiver. Still, carbon has that pesky problem of how to make the orbitals point in the right direction.

Figure 2-16:
Promoting an electron from the $2s$ orbital into the higher-energy $2p$ orbital allows carbon to form four bonds.

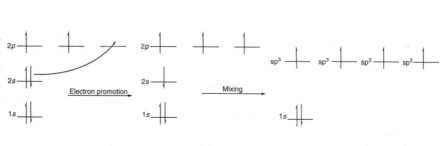

So, carbon does a most sneaky thing: It mixes the four orbitals together — the three $2p$ orbitals and the $2s$ orbital — and makes four new orbitals, called sp^3 *hybridized orbitals*, each identical, that point at 109.5-degree bond angles away from each other (how convenient!). These new sp^3 orbitals are called hybridized orbitals because they're hybrids of the original orbitals.

In the naming of hybridized orbitals, the superscript indicates the number of orbitals of each type that mix to form the hybrid. But if only one s or p orbital is involved in the hybridized orbital, the superscript is omitted. Thus, a hybridized orbital made from one s and three p orbitals is written as sp^3.

The mixed sp^3 orbitals are a weighted average of the orbitals that were tossed into the mixing pot. Mixing three p orbitals and one s orbital makes the four output sp^3 hybridized orbitals three-quarters p in character and one-quarter s in character (see Figure 2-17). It's like mixing jars of food coloring. Mix one jar of red food coloring and one jar of yellow food coloring, and you get two jars of orange food coloring, which is the "average" of the two colors.

Figure 2-17:
Four sp^3 hybridized orbitals are formed when the one s and three p orbitals are mixed.

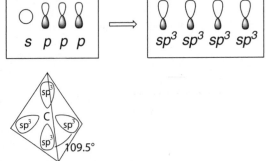

REMEMBER

With orbital mixing, the number of orbitals that are mixed must equal the number of hybridized orbitals that come out at the end. If you mix four atomic orbitals, you must get out four hybridized orbitals. Note that, for clarity, the small lobes of hybrid orbitals are often omitted in drawings.

What about an atom that has only three bonds to other atoms? In such a case, the four sp^3 hybrid orbitals would be no good because their bond angles are at 109.5 degrees, and you want 120-degree bond angles for an atom with three bonds so that the bonds can get maximum separation from each other. In that case, the orbitals mix a little differently. Instead of all four orbitals mixing, only three of them mix — the $2s$ orbital, and two of the p orbitals — while one of the p orbitals remains in its original unhybridized form (see Figure 2-18). Because you're mixing one s orbital and two p orbitals, the three hybrid orbitals that come out are said to be sp^2 hybridized orbitals, and these bonds are situated at 120-degree angles to each other.

Figure 2-18:
Three sp^2 hybridized orbitals are formed when the one s and two p orbitals are mixed.

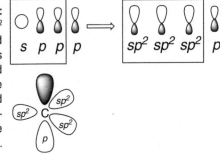

For an atom with two bonds, the ideal bond angle is 180 degrees, so only two of the orbitals mix — the s orbital and one of the p orbitals — and the two remaining p orbitals remain unchanged. These two orbitals are called sp hybridized orbitals (see Figure 2-19).

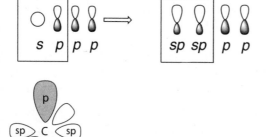

Figure 2-19:
Two sp hybridized orbitals are formed when one s and one p orbital are mixed.

Predicting hybridization for atoms

Predicting the hybridization of an atom is often as simple as counting the number of *substituents* (or number of atoms bonded to that particular atom) and lone pairs of electrons around that atom. For BeH_2 (refer to Figure 2-15), the beryllium (Be) has two substituents (two identical H atoms), so it's sp hybridized. For BH_3 (also refer to Figure 2-15), the boron (B) has three substituents (three H atoms), so it's sp^2 hybridized. And in methane, CH_4, which has four substituents, the carbon is sp^3 hybridized (refer to Figure 2-15). Knowing the hybridization of an atom can tell you the approximate bond angle and the geometry of the bonds around a given atom (see Table 2-2).

Table 2-2	Rule for Determining Hybridization		
Number of Substituents (Including Lone Pairs of Electrons)	**Hybridization**	**Approximate Bond Angle**	**Geometry**
2	sp	180 degrees	Linear
3	sp^2	120 degrees	Trigonal planar
4	sp^3	109.5 degrees	Tetrahedral

It's All Greek to Me: Sigma and Pi Bonding

Covalent bonds occur when the orbitals of bonding atoms overlap each other. Two kinds of covalent bonds can be formed in organic molecules — sigma (σ) and pi (π).

> ✔ *Sigma bonds* are bonds in which orbital overlap occurs between the two bonding nuclei.
>
> ✔ *Pi bonds* are bonds where orbital overlap occurs above and below the nuclei, and not directly between them.

Several different kinds of orbital overlaps can result in sigma bonds. For example, two *s* orbitals could overlap to make a sigma bond (such as in the bond between the two hydrogens in H_2), a hybridized orbital and an *s* orbital could overlap (such as in a C-H bond), or two hybridized orbitals could overlap (such as in a C-C bond). All these are sigma bonds, because orbital overlap takes place between the two bonding nuclei (see Figure 2-20).

Figure 2-20: Examples of orbital overlap for sigma and pi bonding.

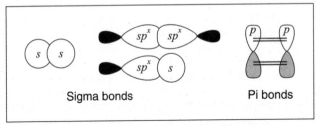

Sigma bonds

Pi bonds

Unlike sigma bonds, only one kind of orbital overlap makes pi bonds (well, only one kind you need to know about in organic chemistry, anyway), and this is the side-by-side overlap between *p* orbitals. With the side-by-side overlap of *p* orbitals, there is no overlap directly between the bonding nuclei, because the *p* orbitals have nodes in this region (a *node* is a region of zero electron density). There is, however, overlap above and below the nuclei. Pi bonds are less common than sigma bonds because they're found only in double and triple bonds, not in single bonds. While *single bonds* are bonds in which two electrons are used to make a bond between atoms, *double bonds* are bonds that contain four electrons; *triple bonds* are made up of six electrons shared between the two bonding atoms.

Now you can apply what you know about sigma and pi bonding and hybridization to draw the orbital diagram of a molecule. Being able to draw the orbital diagram of a molecule is important because this diagram shows you which kinds of orbitals are responsible for the different bonds in a molecule. (This orbital diagram can sometimes be helpful in explaining how certain kinds of molecules react, for example.) Use the three following steps to draw the orbital diagram of a molecule.

1. **Determine the hybridization of each of the atoms.**

   Note that hydrogen is the only atom whose orbitals remain unhybridized in organic compounds. (Recall that hydrogen contains only the 1s orbital in its valence shell.)

2. **Draw all the valence orbitals for each atom.**

3. **Determine which orbitals will overlap to make the bonds.**

   Double bonds consist of one sigma and one pi bond, and triple bonds consist of one sigma bond and two pi bonds. All single bonds are sigma bonds.

Consider ethylene (C_2H_4) as an example, shown in Figure 2-21.

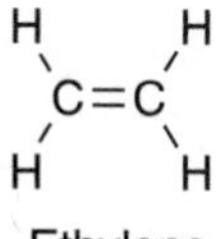

Figure 2-21:
Ethylene.

$$\begin{array}{ccc} H & & H \\ \diagdown & & \diagup \\ & C=C & \\ \diagup & & \diagdown \\ H & & H \end{array}$$

Ethylene

To determine the orbital picture for ethylene, first you want to determine the hybridization of each of the atoms. Because both carbons have three substituents (that is, each is attached to three other atoms), both of these atoms are sp^2 hybridized (refer to Table 2-2). This means that each carbon has three sp^2 orbitals plus one p orbital available for bonding. Hydrogen is not hybridized and, thus, has just the 1s orbital available for bonding. Hydrogen is the only atom that does not hybridize its orbitals for bonding, because it only has one valence orbital, the 1s orbital. Next, you draw each of the atoms with all its valence orbitals (ignoring all core orbitals, because they aren't important in bonding), as shown in Figure 2-22.

Figure 2-22:
The valence orbitals of each of the atoms in ethylene.

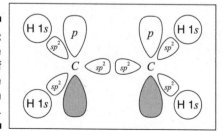

Next, you want to determine which orbitals overlap to make the bonds. For each of the C-H bonds, the bond will result from the overlap between an sp^2 orbital on the carbon and the $1s$ orbital on the hydrogen, making a sigma bond. For the double bond, one of the bonds comes from the two sp^2 hybridized orbitals overlapping between the carbon nuclei to make a sigma bond, while the other bond comes from the two p orbitals overlapping sideways to make a pi bond above and below the carbon nuclei.

This last step generates the orbital representation for ethylene (see Figure 2-23), because the orbitals that overlap to make each bond in the molecule are accounted for. How is knowing which orbitals make up a bond important? Often, the types of bonds in a molecule explain the reactions a molecule undergoes. For example, one of the bonds in ethylene's double bond is a sigma bond and one is a pi bond. Pi bonds are more reactive than sigma bonds (explained in Chapter 11), so one might suspect based on this orbital representation of ethylene that the pi bond is the most reactive bond in this molecule. This observation turns out, in fact, to be the case (as you also see in Chapter 11).

Figure 2-23:
The orbital representation of ethylene, showing which orbitals overlap to form the different bonds.

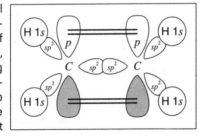

Chapter 3

Speaking with Pictures: Drawing Structures

I hope you've had the chance to listen in on a conversation between two chemists in the halls of a cavernous chemistry building as they discussed their experiments. Perhaps you overheard snatches of their conversation: "I don't get it," one might have said, "My proton NMR showed two multiplets one part per million upfield from the aromatic protons." Or maybe you heard, "I made the alkene via the Wittig, but for the life of me I simply could not get the reduced nitro group to react with the t-butyl chloride using an S_N1 in ethanol!" Or perhaps, "We return to planet Beldar with the earthlings in eight lunar cycles!" And you thought, "Now *what* language were they speaking?"

They were speaking organic, of course, which hasn't been classified yet as a foreign language, but perhaps could be, because learning organic chemistry is a lot like learning a foreign language. Organic chemistry has its own jargon, as the snatches of conversation you overheard show (though after you've read this book and are fluent in organic, you'll be able to understand what they were talking about). But I would argue that the similarity goes deeper than that. The words are different, too.

Before you can speak German or Japanese or Swahili — or any foreign language, for that matter — you need to learn what the different words mean. Similarly, before you can "speak organic," you need to learn how to draw structures. Structures are the "words" of the organic chemist, and becoming fluent in drawing structures is absolutely essential in order to do well in organic chemistry. Drawing structures may seem strange at first, and probably even intimidating, but with practice you'll soon get the hang of it. Like speaking a foreign language, with enough practice, drawing organic structures will simply become second nature (and kind of fun, too).

In this chapter, I give you the vocabulary for speaking organic chemistry by showing you how to draw organic structures using the standard Lewis convention. I also walk you through the ways that organic chemists abbreviate Lewis structures, both for clarity and for when they're feeling lazy (which, admittedly, is most of the time). Finally, I show you how to draw resonance structures, which are used to correct for a flaw in the Lewis structure model.

Models and molecules

People have a hard time comprehending objects that are either huge or tiny. That's because our minds think in terms of the sizes of the things we encounter in our daily lives. A kilometer is about as far as our walk across campus; 2 centimeters is about the length of our pinky tips. To us, these sizes make sense; we can relate to them. But comprehending gargantuan lengths, like the distance between the earth and a distant star millions of light-years away, or itty-bitty lengths, like the distance across an atom (a distance of only a few hundred picometers), is mind-boggling.

This difficulty creates a problem for chemists. How can they understand atoms and molecules if they're so small that it boggles chemists' minds just to think about them? How can chemists describe, organize, and classify things that they cannot see? In short, how can they be scientific about the study of chemistry?

In addition to our inability to comprehend such small things is the added complexity that atoms don't behave in the same way as objects of a larger scale; they don't behave like the things we see in our everyday lives — things that we can see or touch or throw, things like stink bombs or chemistry texts. Molecules behave in very bizarre ways as a result of their smallness, and human intuition based on the big world of everyday objects is no help in understanding the tiny world of molecules. For atoms, the classical physics that you've been taught since grade school fails disastrously. This is because something that is very, very small (like an electron) has been found to behave both like a particle (which you might expect) and as a wave (which you wouldn't). Electrons are small enough that they have the ability to "tunnel" through a barrier (the equivalent of a person walking through a wall), and they can exist in two places simultaneously, in addition to other very bizarre behaviors.

So, chemists must use models to describe molecules and their weird behavior. Chemists use models to describe the way a molecule is put together — where the electrons are and which atoms are attached together and in what way — and to represent how reactions might occur. The primary model used in organic chemistry is the Lewis structure. Although Lewis structures are only approximate models of how molecules look in actuality, they really do an excellent job of showing the connectivity of atoms. These models, however, are not always perfect for describing the exact locations of certain electrons, as I discuss in the section about resonance structures later in the chapter.

Picture-Talk: Lewis Structures

A Lewis structure is the chemist's way of depicting an infinitesimally small molecule on a macroscopic piece of paper. A Lewis structure shows what atoms are connected to each other, and it shows where the electrons in the molecule reside. Single bonds between two atoms are represented with a single line, signifying two shared electrons; double bonds are represented with a double line, signifying four shared electrons; and triple bonds are represented with a triple line, signifying six shared electrons (refer to Chapter 2 for more on bonding). Nonbonding electrons are indicated with dots on the atoms on which they reside.

Taking charge: Assigning formal charges

The first thing you want to be able to identify on a Lewis structure is which atoms have formal charges. Electrons are negatively charged, so an atom that is missing one or more electrons will have a positive charge. An atom that has one too many electrons will have a negative charge. Being able to quickly determine the charge on any atom in a given molecule is extremely important. So, here's a quick-and-dirty equation for determining the formal charge on an atom:

formal charge on an atom = valence electrons – dots – sticks

Here *dots* is simply the number of lone-pair electrons around the atom, while *sticks* is the number of bonds off the atom. (A single bond is one stick; a double is two; a triple, three.) The best way to see this is to try one. So, apply this equation to NH_2, shown in Figure 3-1, to calculate the formal charge on the nitrogen atom.

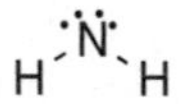

Figure 3-1:
The
amide ion.
$$H \overset{\cdot \cdot}{\underset{}{N}} H$$

Just plug the values into the formula. Nitrogen is in the fifth column of the periodic table and so has five valence electrons; it has four dots (four nonbonding electrons) and two sticks (two bonds).

formal charge = 5 valence electrons – 4 dots – 2 sticks = –1

The formal charge on the nitrogen, then, is –1, so in this molecule nitrogen is anionic. Using this same approach, you should be able to apply this formula quickly to all the atoms in a molecule. Still, even though this equation is fairly simple to use, applying it to every atom in a molecule is tedious. You want to be able to simply look at a structure and, within a couple seconds, determine which atoms have charges.

Fortunately, with enough practice, you'll quickly recognize certain patterns for different atoms that will indicate whether the atoms are neutral, positively charged, or negatively charged. For example, if carbon has four bonds, it will be neutral, if it has three bonds, it will be positively charged, and if it has three bonds and a lone pair, it will be negatively charged. You can see this generality to be true for yourself by plugging these different configurations for carbon into the equation for formal charge just shown.

Knowing that a neutral carbon atom always has four bonds lets you scan a molecule quickly to identify carbons that don't have four bonds, and, when you find one, say to yourself, "Aha! This carbon must have a charge." For nitrogen, you look to see which nitrogen atoms don't have three bonds and a lone pair; for oxygen, you look to see which ones don't have two bonds and two lone pairs, and so forth. This way, you won't need to apply the formula to every atom (although you can if you need to); you just look to see which atoms differ from the typical number of bonds found around an atom in its neutral form. I show these patterns for the most common elements in organic compounds in Figure 3-2.

Atom	Neutral	Cationic	Anionic
C	$-\overset{\displaystyle |}{\underset{\displaystyle |}{C}}-$	$-\overset{\displaystyle |}{\underset{\displaystyle |}{C}}{}^{+}$	$-\overset{\displaystyle |}{\underset{\displaystyle |}{C}}{:}^{-}$
N	$-\overset{\displaystyle |}{\underset{\displaystyle}{N}}{:}$	$-\overset{\displaystyle |}{\underset{\displaystyle |}{N}}{}^{+}-$	$\overset{\displaystyle}{N}{}^{-}$
O	$\overset{..}{\underset{..}{O}}$	$-\overset{\displaystyle |}{\underset{\displaystyle}{O}}{:}^{+}$	$-\overset{..}{\underset{..}{O}}{:}^{-}$
X	$-\overset{..}{\underset{..}{X}}{:}$	$-\overset{..}{\underset{..}{X}}{-}^{+}$	$:\overset{..}{\underset{..}{X}}{:}^{-}$

Figure 3-2: Some common charges on atoms.

X = F, Cl, Br, I

Chemists, in fact, use specific jargon to describe typical bonding patterns of atoms. The number of bonds around a neutral atom is referred to as that atom's *valency*. Carbon is tetravalent (*tetra* means four), because carbon has four bonds to other atoms when it's neutral; nitrogen is trivalent, because it has three bonds to other atoms (plus a lone pair) when it's neutral; oxygen is divalent, because it has two bonds to other atoms (plus two lone pairs) in its neutral form; and halogens (elements such as fluorine [F], chlorine [Cl], bromine [Br], and iodine [I]) are monovalent, because they have one bond to another atom (plus three lone pairs) when they're neutral.

Drawing structures

In addition to calculating charges, you also need to be able to quickly draw organic structures and interpret what the drawings mean, because these are the words you'll use to speak organic chemistry.

Organic chemists draw structures in several different ways, and because the different styles have different uses, you should be familiar with all of them. The most complete way of drawing organic structures is with the full Lewis structure (sometimes called a Kekulé structure). A Lewis structure explicitly draws out all the bonds in a molecule (an example is given in Figure 3-3). The lone pairs on atoms may or may not be shown in a Lewis structure (they often aren't), but formal charges are always shown. Later in this chapter, I show you how to determine the number of lone pairs on a given atom when these lone pairs aren't explicitly shown (don't worry, it's not difficult).

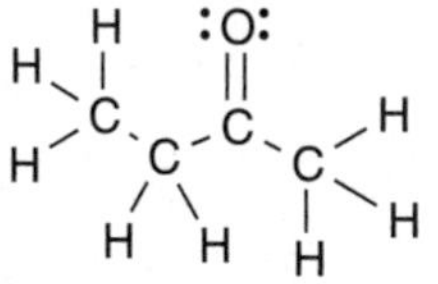

Figure 3-3:
The full
Lewis
structure for
butanone.

Full Lewis structures most faithfully show the structure of a molecule and how the molecule is put together, because every bond and every atom is shown. But this method of structure drawing can become tedious and overwhelming when you draw larger molecules. To make structure drawing quicker and the interpretation of change in a molecule easier to spot, chemists use shorthand notation, just as you may use abbreviations when you text your friends.

Atom packing: Condensed structures

One such structure abbreviation of the full Lewis structure is called a *condensed structure*. In a condensed structure, the bonds between carbon and hydrogens are not explicitly shown; instead, each carbon and the attached hydrogens are grouped together into a cluster (such as CH_2 or CH_3), and these clusters are written in a chain that shows the connectivity between the carbons. The carbon-carbon bonds may be explicitly shown, or they may simply be assumed. Condensed structures with shown and unshown bonds between the clusters are given in Figure 3-4 for the molecule butanone.

Figure 3-4:
Condensed structures of butanone.

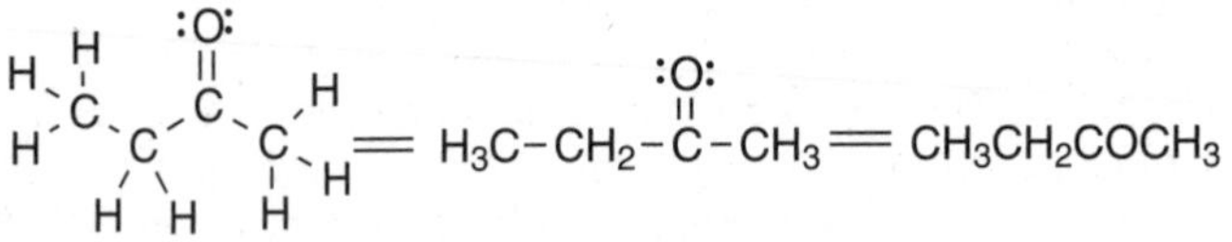

$$ = H_3C-CH_2-\overset{\displaystyle :O:}{\overset{\displaystyle \|}{C}}-CH_3 = CH_3CH_2COCH_3 $$

Condensed structures are most useful when the molecule consists of a straight chain of atoms; elaborate structures that contain rings are more difficult to portray with condensed structures. When two or more identical groups are attached to an atom, parentheses with a subscript can be used to abbreviate the structure even further. In this case, the subscript indicates the number of identical groups that are attached to the atom. An example of this grouping for condensed structures is given in Figure 3-5 for diethyl ether.

Figure 3-5:
Condensed structures for diethyl ether.

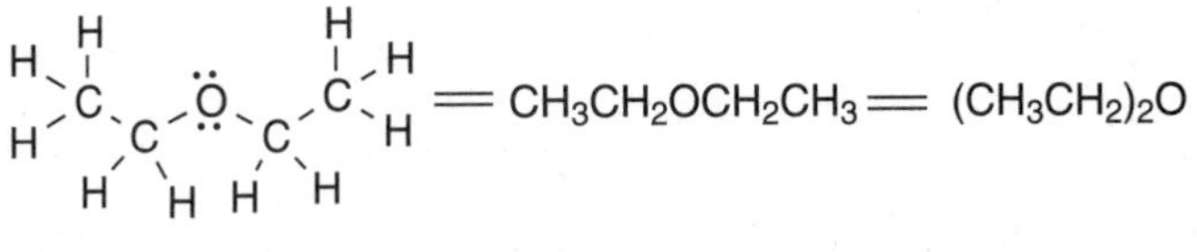

$$ = CH_3CH_2OCH_2CH_3 = (CH_3CH_2)_2O $$

Parentheses are also used when a chain becomes very long and a unit is repeated many times (for instance, when a CH_2 unit is repeated). Here, the subscript after the parentheses indicates how many times the unit is repeated, as shown for heptane in Figure 3-6.

Figure 3-6:
Heptane
condensed
structure.

$CH_3CH_2CH_2CH_2CH_2CH_2CH_3 = CH_3(CH_2)_5CH_3$

Structural shorthand: Line-bond structures

The most common method of structure drawing, however, is called the line-bond structure. In these structures, each point (or *node*) on a jagged line is assumed to be a carbon atom. Hydrogens attached to carbons are not explicitly shown in a line-bond structure. But because each carbon is presumed to be neutral unless a charge is explicitly written on the structure, this method of drawing structures assumes that you can mentally supply the number of hydrogens attached to each carbon by making the total number of bonds equal to four. So, if a neutral carbon is shown having two bonds to other carbon atoms, then it must also be bonded to two additional hydrogens that are not shown. Figure 3-7 shows the line-bond structure for isoheptane.

Figure 3-7:
Structures
for
isooctane.

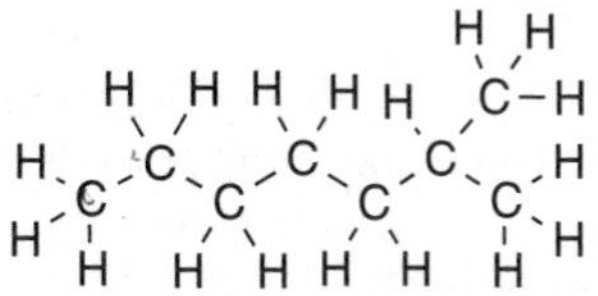

Full Lewis structure   Condensed structure   Line-bond structure

$CH_3(CH_2)_4CH(CH_3)_2$

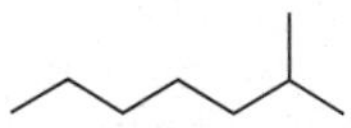

REMEMBER

The general rules for line-bond structures are as follows:

- ✔ Each point on a jagged line is assumed to be a carbon atom.

- ✔ The ends (tips) of lines are assumed to be carbon atoms.

- ✔ All hydrogens attached to non-carbon elements (like N, O, S, and so on) must be explicitly shown.

- ✔ All atoms are presumed to be neutral unless a charge is specifically shown.

Converting Lewis structures to line-bond structures

As shown in Figure 3-8, to convert a full Lewis structure to a line-bond structure, a jagged line is used to represent straight chain molecules, with each point (including the ends of the structure) representing a carbon atom.

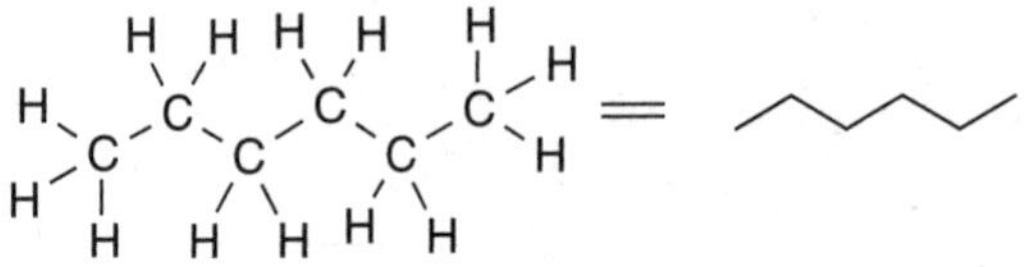

Figure 3-8:
Line-bond
structure for
hexane.

In line-bond structures, rings are represented by polygons, with each point representing a carbon atom, as shown for the five-membered ring in Figure 3-9. A triangle is a three-carbon ring (which, of course, is the smallest possible ring size); a square is a four-carbon ring; a pentagon is a five-carbon ring; and so forth.

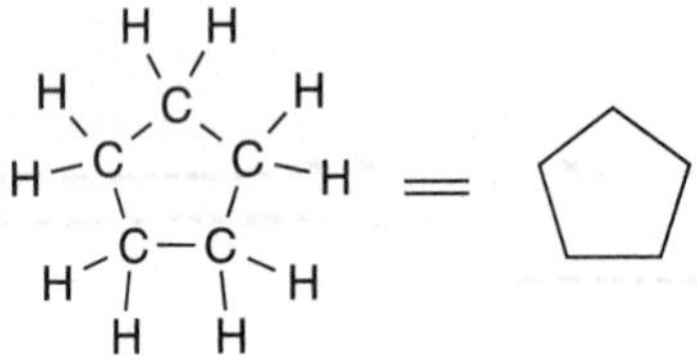

Figure 3-9:
Cyclo-
pentane
line-bond
structure.

Multiple bonds are explicitly drawn out. Triple bonds are drawn in a straight line, as shown in Figure 3-10, and are not jagged like single bonds (I talk about why in Chapter 11).

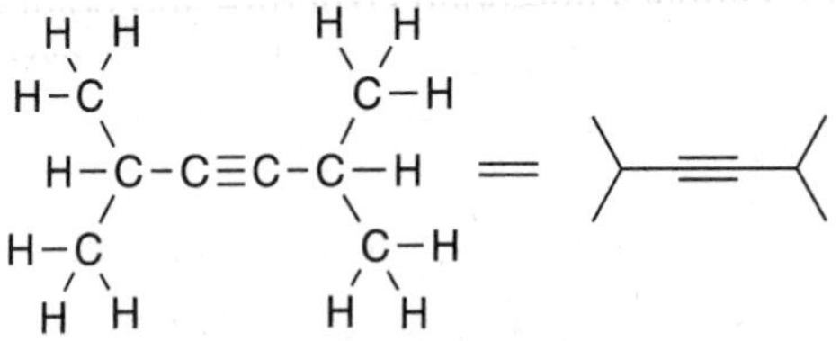

Figure 3-10:
Diisopropyl
acetylene.

Hydrogens attached to any atom other than carbon (atoms such as oxygen [O], sulfur [S], and nitrogen [N]) must be explicitly shown; only hydrogens attached to carbons are not shown. Figure 3-11 shows an example of how to draw non-carbon atoms in line-bond structures.

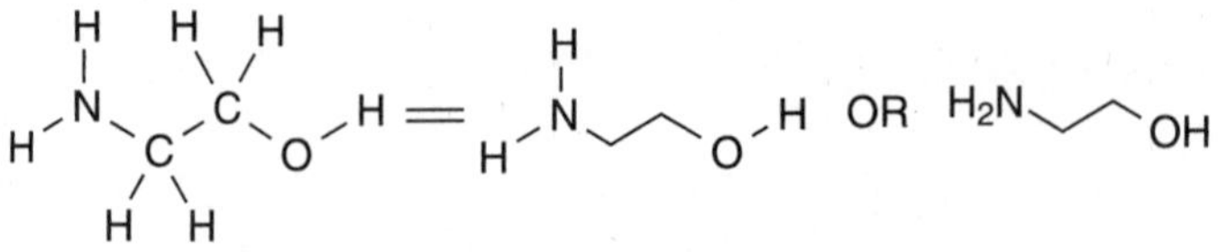

Figure 3-11:
Etha-
nolamine.

This line-bond method of structural representation is probably the method most commonly used by organic chemists — it's easy to draw, it's convenient, and it's uncluttered in comparison to the full Lewis structure. My guess is that you'll prefer, at least initially, to use full Lewis structures to represent molecules rather than this line-bond method, because Lewis structures don't require you to mentally supply the missing hydrogens.

At first, using full Lewis structures in your drawings is probably an excellent idea, because this structure has no structural abbreviations (just as you would want to learn how to speak with proper English before learning LOL and OMG). But you should familiarize yourself with line-bond drawings right away so you can quickly become comfortable using them. Your organic chemistry textbook — and your teacher — will likely use line-bond structures almost exclusively later in the course, so knowing how to draw and interpret line-bond structures will be to your advantage.

Determining the number of hydrogens on line-bond structures

Being able to quickly "see" how many hydrogens are attached to each carbon in a line-bond structure takes practice. But the math, at least, is simple. A neutral carbon has four bonds. So, if a line-bond structure shows a carbon that has three bonds to other atoms, it must have one implicit hydrogen; if it shows a carbon that has two bonds, it must have two implicit hydrogens; if it shows a carbon that has only one bond, it must have three implicit hydrogens. In Figure 3-12, I show a line-bond structure and point out the number of hydrogens assumed to be attached to each of the carbons.

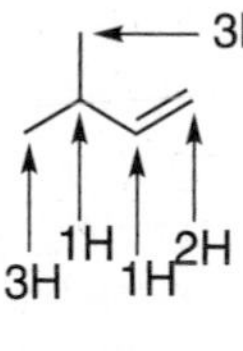

Figure 3-12:
Counting the assumed hydrogens.

Getting the hang of drawing line-bond structures takes a bit of work, but there's a nice payoff once you do because you'll be able to draw structures much faster (which is a significant issue when you're dealing with a large number of structures), and you'll be able to see chemical changes much more easily than if you were using the full Lewis structures (important when you begin chemical reactions of organic molecules). You'll also better understand the textbook and your instructor when they use these line-bond structures.

Quite often, you see structures drawn that combine these three different methods of structure drawing. In other words, it's not a requirement that an entire molecule be drawn using only one of the drawing methods. It's not unusual to see parts of a molecule drawn out with the full Lewis representation, parts drawn in condensed structure form, and parts drawn using the line-bond format. Chemists often have good reasons for doing this — to emphasize certain important parts of a molecule, for example — but at the beginning, when you're just trying to get everything straight in your head, you can easily get confused if a structure is drawn that combines the different structure drawing styles. Figure 3-13 shows an organic molecule that was drawn using a combination of structure drawing methods.

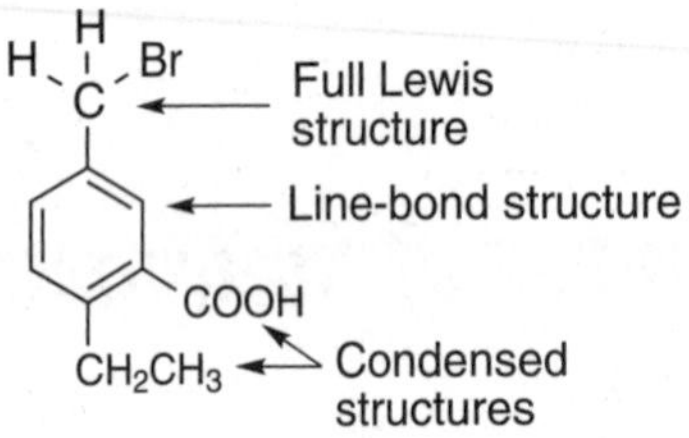

Figure 3-13:
A structure
drawing that
combines
the three
different
styles.

So lonely: Determining lone pairs on atoms

Most often, lone pairs on atoms are not shown explicitly because chemists assume that you can determine for yourself the number of lone pairs on the atom. One way to determine the number of lone pairs (if they aren't shown) is to simply plug the information into the rearranged equation for the formal charge, where *dots* is simply the number of nonbonding electrons.

dots = valence electrons – sticks – formal charge

The best way to master this is by practicing, so go ahead and try one. To determine the number of lone pairs on the nitrogen atom in NH_2^-, you plug into the equation the number of valence electrons for nitrogen (5), the number of sticks (2, because nitrogen has two bonds, one with each of the two hydrogens), and the formal charge on the atom (–1 in this case).

dots = 5 – 2 – (–1) = 4

So, NH_2^- has four nonbonding electrons, or two lone pairs of electrons (because there are two "dots" in every lone pair).

As is the case with distinguishing formal charges, after a while you'll become practiced enough at figuring out the number of lone pairs that you won't need to do this calculation, because you'll recognize the patterns of electrons and bonds. At that point, you'll know that a negatively charged nitrogen will have two lone pairs, that a negatively charged carbon will have one lone pair, that a neutral oxygen will have two lone pairs, and so on (refer to Figure 3-2 for the other patterns). But it's going to take some work before it becomes second nature, so practice, practice, practice!

Problem Solving: Arrow Pushing

Now for the squishy part. If structures are the words of organic chemistry, then arrow pushing is the grammar and syntax of organic chemistry. And I know how much everyone loves grammar. Knowing what the arrows represent and how to use them properly will help you understand organic chemistry. (Like English professors, chemistry professors are very picky about using this "grammar" correctly.)

In organic chemistry, you see five major types of arrows (see Figure 3-14):

✔ **Resonance arrow:** Used to indicate movement between resonance structures (which I talk about in the next section).

✔ **Equilibrium arrow:** Used to show reactions governed by equilibria (see Chapter 10). Making one of the equilibrium arrows longer shows the direction of the equilibrium in a reaction (in other words, which side of the reaction — reactants or products — is favored).

✔ **Reaction arrow:** Used to show the change of molecules by a reaction.

✔ **Full-headed arrow:** Used to show the movement of two electrons.

✔ **Half-headed arrow:** Used to show the movement of one electron.

Figure 3-14:
The types of arrows used in organic chemistry.

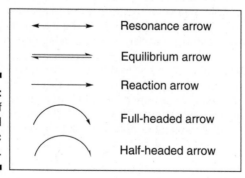

Resonance arrow

Equilibrium arrow

Reaction arrow

Full-headed arrow

Half-headed arrow

Electrons, electrons, electrons: These little guys are the keys to chemistry. They're what organic chemists care most dearly about. The neutrons and protons stay fixed in their little round shack at the nucleus, but the electrons are adventurous, bustling about making new bonds and breaking old ones, so organic chemists concern themselves mostly with what happens to the electrons. Because electrons form and break bonds (and protons and neutrons do not), the focus throughout organic chemistry is on where the electrons are and where they're going in a chemical reaction. So, for the most part, you ignore the protons and neutrons in the nucleus, because these remain constant, and focus your attention on what's happening with the electrons.

Organic chemists use half- and full-headed arrows to show the movement of electrons (sometimes half-headed arrows are called *fishhook arrows* because they look like fishhooks). Full-headed arrows are much more common than half-headed arrows, simply because most reactions involve the movement of lone pairs and bonds — each of which contains two electrons. Half-headed arrows are used for describing free radical reactions (described in Chapter 8), because these reactions involve the movement of single electrons. You'll need to become as good as Robin Hood at using these types of arrows.

Mastery of so-called "arrow pushing" is essential to mastering organic chemistry — and, importantly, to doing well in the class! Arrow pushing is something that can't be learned in one sitting. This kind of grammar takes lots and lots of practice to master. Arrow pushing is organic-speak for showing *how* a chemical change takes place. To do this, organic chemists use these single- and full-headed arrows, which show how you get from one structure to the next in a reaction (or in a resonance structure, which I describe in the next section) by showing the movement of electrons.

Full-headed arrows, by convention, are used to show the movement of electrons. Therefore, you always *draw arrows from electrons toward where they're going,* never the other way around. I think most students would prefer to have arrows show the movement of atoms rather than show the movement of electrons. But using an arrow to show the movement of atoms is wrong, and if you make your arrows do that on an exam, it will almost certainly cost you points. If, for example, you want to show water being protonated (receiving an H^+ ion) by acid, you would show one of the lone pairs on water's oxygen attacking the proton (the H^+), not the proton moving onto the water.

Because H^+ has no electrons (hydrogen has lost its one electron to become a positively charged cation), you may ask when would it be correct to show an arrow coming from an H^+. As King Lear may have said had he been an organic chemist: "Never! never! never! never! never!" Figure 3-15 shows the correct and incorrect ways to show the movement of electrons using a double-headed arrow.

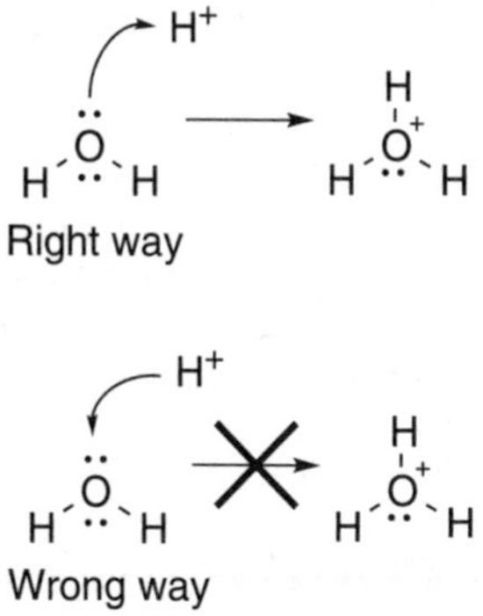

Figure 3-15:
The right
and wrong
ways to use
full-headed
arrows
to show
electron
movement.

Right way

Wrong way

Drawing Resonance Structures

Lewis structures show the locations of electrons correctly most of the time — but not always. To correct for a flaw in showing the locations of certain electrons in Lewis structures, chemists use *resonance structures*. Organic chemists use these resonance structures to account for a flaw in showing the locations of certain lone-pair and pi electrons in Lewis structures (pi electrons are electrons found in pi bonds; refer to Chapter 2). You can't get through organic chemistry without being able to draw resonance structures and without understanding what these structures mean.

The Lewis structure for a carboxylate anion (RCO_2^-, where R stands for the rest of the molecule) provides a good example of how Lewis structures are not always the best way to describe certain molecules. This Lewis structure (see Figure 3-16) predicts one double bond between carbon and one of the oxygens, and one single bond between carbon and the oxygen that contains the negative charge. According to this Lewis structure, the two C-O bonds are different. Instead, the actual structure is one in which both of the carbon-oxygen bonds are identical — with each C-O bond having some single-bond qualities and some double-bond qualities — and each of the oxygens shares the negative charge equally.

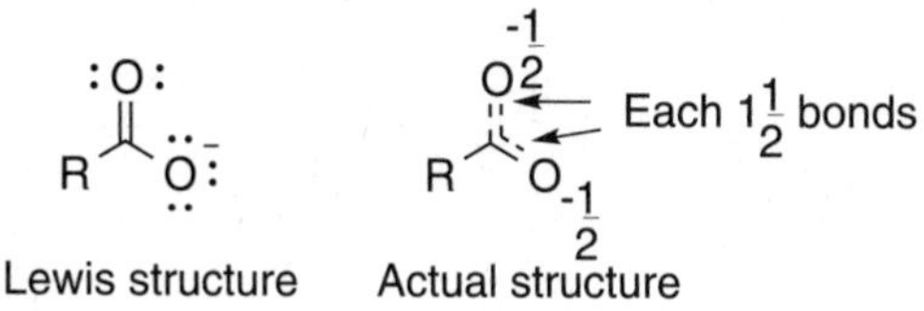

Figure 3-16:
The Lewis
structure of
a carboxyl-
ate ion
compared
to the actual
structure.

Lewis structure

Actual structure

Each $1\frac{1}{2}$ bonds

Because this Lewis structure fails to correctly account for the location of the electrons, organic chemists use resonance structures to correct for this flaw. But how do you know when to use resonance structures to correct for this flaw? In general, any time more than one valid Lewis structure can be drawn for a given molecule, each of these alternate structures is considered a resonance structure, and the actual structure of a molecule will look like a hybrid of all the different resonance structures. That leads to the next question: How do you find all the alternative resonance structures?

Rules for resonance structures

Chemists use arrow pushing to find all alternative resonance structures for a molecule. Here are the three basic rules for finding alternative resonance structures:

✔ **Atoms are fixed and cannot move.** Because the purpose of resonance structures is simply to account for alternative locations of electrons, all atoms stay in fixed positions. Only the distribution of the electrons in the molecule changes from one resonance structure to the next.

✔ **Only lone-pair electrons and pi electrons — which are only found in double and triple bonds — can move.** Single bonds stay put (to review pi bonds, refer to Chapter 2).

✔ **You can't break the octet rule.** This law still applies. To put this rule into practical terms, the sum of an atom's lone pairs and bonds cannot be greater than 4 for second-row elements.

For example, the carboxylate anion in Figure 3-17 can be represented by two different and equally valid Lewis structures, one in which the negative charge rests on the top oxygen, and one in which the charge rests on the bottom oxygen. So, which Lewis structure represents the actual molecule? The answer: Neither one does! In fact, no single resonance structure describes a molecule perfectly, but the correct structure lies somewhere in between; the correct structure is a hybrid of all the resonance structures. In the case of the carboxylate anion, each of the C-O bonds is neither a double bond nor a single bond, but a bond that is somewhere in between (one-and-a-half bonds).

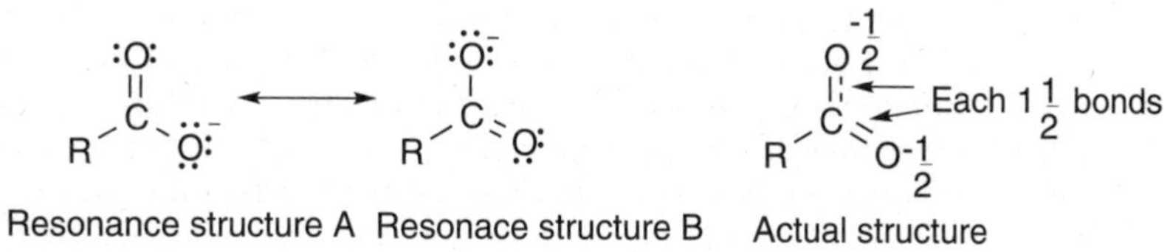

Figure 3-17:
Double-
headed
resonance
arrows
are used
to convert
between
resonance
forms.

Resonance structure A   Resonace structure B   Actual structure

You may ask why chemists don't simply draw the hybrid drawing (like the one shown in the right-hand portion of Figure 3-17), with the dotted line representing the partial (one-and-a-half) bonding character. One reason is that this drawing doesn't show the *number* of electrons in the molecule, so this drawing is an inconvenient representation for showing the reactions of that molecule. Instead, multiple resonance structures are typically used to represent the hybrid nature of the actual structure.

Note the use of the double-headed resonance arrow in Figure 3-17 to show the two equivalent resonance structures. This arrow serves as a reminder that resonance is not a reaction, not an equilibrium process, and not a change of any kind in a molecule. Instead, these resonance structures are simply a way of correcting the flaw in Lewis theory and accounting for the proper electron distribution of a *single molecule* on a piece of paper. A molecule does not flip between different resonance structures like Jekyll and Hyde, back and forth, back and forth. It exists instead as one structure all the time, one that simply looks like the hybrid of all the possible resonance structures.

For the carboxylate anion, two valid Lewis structures can be drawn, so each of these is considered a resonance structure.

Problem solving: Drawing resonance structures

So, how do you recognize when a structure has alternative resonance structures, and how do you go about drawing resonance structures for a given molecule? Any time a molecule can be represented by an alternative valid Lewis structure, all such alternatives are considered resonance structures. But how do you determine when a molecule can be represented by alternative Lewis structures?

One way that resonance structures are found is by pushing around the nonbonding and pi electrons with arrows. (Recall that pi electrons are electrons that contribute to double and triple bonds. Read more in Chapter 2.) Electron-pushing arrows are used to convert one resonance structure into the next. It's a bit unfortunate that usually the first "arrow pushing" you do in your organic chemistry course is for resonance structures, because the meaning behind the arrow pushing used for resonance structures is entirely different than the meaning behind the arrow pushing that you use again and again later in the course to show how a reaction occurs. They're fundamentally different, even if they seem similar.

How so? In the preceding section, I use a double-headed arrow in Figure 3-17 to show the protonation of water (the addition of H^+ to water). In that figure, the arrow is used to show how the reaction happens: The lone-pair electrons on the oxygen attack the proton (H^+), forming a new bond.

Resonance structures use arrows for a different purpose. They don't show how a reaction happens. Instead, the arrows are simply used as a convenient tool to determine all the alternative resonance structures. These arrows are not really showing the movement of electrons, because resonance structures are simply our way of modeling a single, unchanging molecule.

With that in mind, a few common patterns can be used to recognize when a structure can be better described using resonance structures. In the sections that follow, I take you through four kinds of structural features that are indicative of molecules that will have two or more resonance structures. In addition to these four patterns, keep in mind that resonance structures must adhere to the three basic rules listed earlier: Atoms are fixed and don't move, you must leave single bonds fixed, and you can't break the octet rule. Now, look at each of these patterns individually.

A lone pair next to a double bond or triple bond

You use arrows to convert one resonance structure into the other. Note that in cases where a lone pair is adjacent to a double bond, at least two arrows are drawn (see Figure 3-18). The first begins at the lone-pair electrons and moves these electrons in the direction of the double bond. The next arrow places the double bond electrons onto the adjacent carbon as a lone pair.

You can't stop after you've drawn just the first arrow because to do so would result in a structure that would violate the octet rule. To prove this to yourself, draw out all the hydrogens in the structure and then push some arrows!

Figure 3-18:
Lone-pair
resonance.

A common confusion among students involves the frequent symmetry of resonance structures. They see that both of the resonance structures look like the same thing — in other words, if you were to flip one of them over like a pancake, one would be superimposable on the other. So aren't they the same? The answer is no, because the first rule of resonance structures is that the atoms cannot move, so you could not perform such a flip without moving the atoms. The purpose of showing resonance structures is to indicate that half the negative charge resides on each of the end carbons and the two C-C bonds are neither a single bond nor a double bond but somewhere in between. Therefore, even though the resonance structures "look the same," each of them is a *different* resonance structure and must be included.

A cation next to a double bond, triple bond, or lone-pair of electrons

In a somewhat similar situation, a double bond that is situated next to a positive charge will also yield resonance structures. In this case, you can't draw the arrow from the positive charge because there are no lone-pair or pi electrons on that carbon. (Recall that with arrow pushing you never draw an arrow originating from a positive charge because arrows always originate from electrons.) Instead, you move the double-bond electrons onto the adjacent single bond, making a new double bond. This re-forms the double bond on the other side of the molecule and shifts the positive charge from the left side to the right, as shown in Figure 3-19. (Insert the hydrogens and calculate the charges for yourself if you don't believe me!) Note that the octet rule is not violated here.

Figure 3-19:
Pi bond
resonance.

Similarly, when a lone pair of electrons is situated next to a positive charge, an alternative resonance structure involves moving the lone pair of electrons in the direction of the cation to form a double bond. The positive charge then moves to the atom that originally held the lone pair, because that atom has become deficient in electrons (see Figure 3-20).

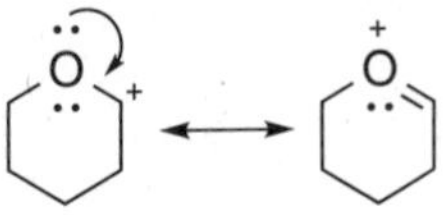

Figure 3-20:
Lone pair
resonance.

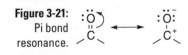

A double or triple bond containing an electronegative atom

When a double or triple bond includes an electronegative atom (like oxygen or nitrogen), an alternative resonance structure can be found by moving one of the double (or triple) bonds onto the electronegative element as a lone pair. In such situations, the electrons always go onto the more electronegative element, not the carbon. (Find out more about this rule in the "Assigning importance to resonance structures" section, later in this chapter.) An example of this kind of resonance is shown in Figure 3-21 for acetone (the chemical in nail polish remover).

Figure 3-21:
Pi bond
resonance.

Alternating double bonds around a ring

When double bonds alternate all the way around a ring, you can move each of the double bonds over to the next carbon to make an alternative resonance form. Benzene, for example, has two resonance structures, as shown in Figure 3-22. Here you would break the octet rule if you were to stop after drawing either one or two of the arrows. So, you have to go all the way around the ring.

Figure 3-22:
Benzene
resonance
structures.

Drawing more than two resonance structures

Sometimes a structure will have more than two resonance structures. Figure 3-23, for example, shows a molecule that has four resonance structures. If you were asked to draw all the resonance structures for this molecule, you would start by recognizing that the molecule had a double bond containing an electronegative atom, which conforms to pattern number three described in the preceding section.

Figure 3-23:
Four
resonance
structures
for 2-hexa-
dienone.

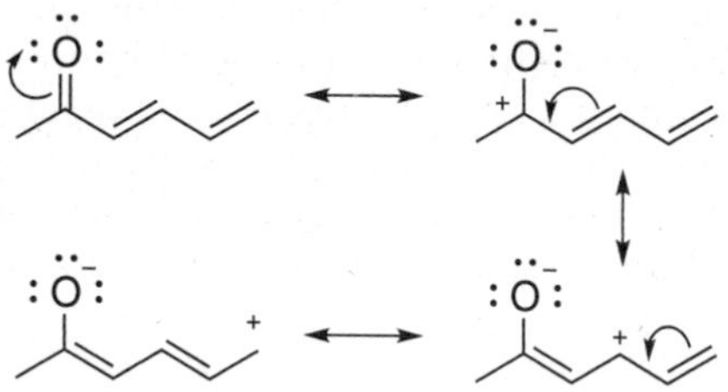

Drawing the resonance structure by moving the double-bond electrons onto the oxygen creates a species with a positive charge on carbon. But this is not the end of the resonance structures. This positive charge, now, is next to a double bond, which is another of the common patterns (pattern number two) discussed in the previous section. So, another resonance structure can be drawn by moving the double-bond electrons onto the single bond to the left, which moves the location of the positive charge. But there is still one more resonance structure that can be drawn. Moving the last double bond over to the left in the same fashion as you did previously puts the positive charge on the end carbon atom. All four of these structures are valid resonance structures and the actual structure looks like a hybrid of all of them.

Often, you can get to any of the resonance structures in a single step. Figure 3-24 illustrates this one-step process. Generally, though, drawing out all the resonance structures one at a time, as I've shown (at least until you get the hang of it), is helpful to make sure that you don't leave out any resonance structures.

Figure 3-24:
Converting
between
resonance
structures in
one step.

Assigning importance to resonance structures

Some resonance structures contribute more to the overall hybrid than others. As a rule, the more stable resonance structures will contribute more to the hybrid than unstable ones. Three major factors are used to determine the relative stability of resonance structures (and, therefore, their relative importance). In this section, I illustrate each of these three factors.

Fewest charges

The resonance structures that have the fewest charges within the molecule will contribute the most to the overall hybrid. This rule makes sense because keeping positive and negative charges separate from each other costs energy. For the molecule acetone, shown in Figure 3-25, the first resonance structure is uncharged, so it will contribute more to the overall hybrid than the resonance structure on the right that has both positive and negative charges.

Figure 3-25:
Acetone
resonance
structure.

Charges on the best atoms

Negative charges generally prefer to rest on electronegative elements (elements such as oxygen and nitrogen), while positive charges will prefer to rest on electropositive elements (such as carbon). An alternative resonance structure for acetone is one in which the negative charge rests on the carbon and the positive charge rests on the oxygen. This resonance structure is considered a bad resonance structure (see Figure 3-26). Because oxygen is an electronegative atom (an electron pig), it really does not want to have a positive charge; therefore, the resonance structure on the right would contribute an insignificant amount to the overall hybrid. Although technically it would be considered a valid resonance structure, this resonance form is never drawn because it's so unstable that it would not significantly contribute to the overall hybrid.

Figure 3-26:
Bad
resonance
structure.

In the example shown in Figure 3-27, both resonance structures *would* contribute to the overall hybrid. In this example, the negative charge rests on either oxygen (in the right-hand structure) or carbon (in the left-hand structure). Because oxygen is more electronegative, the oxygen is more comfortable holding the negative charge than carbon is, so the resonance structure on the right would contribute more to the overall hybrid.

Figure 3-27:
Comparing
charge
stability.

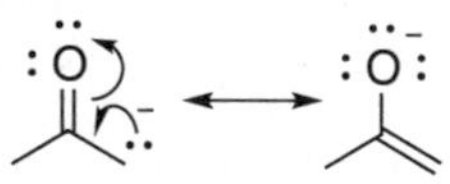

Filled octets

In the next example, shown in Figure 3-28, you may expect, based on the last argument about placing positive charges on electropositive atoms and negative charges on electronegative atoms, that the resonance structure on the left would contribute more to the overall hybrid, because carbon would be more willing to hold a positive charge than would nitrogen. However, this analysis neglects one key factor: a full octet of electrons.

Figure 3-28:
Octet con-
siderations
trumps
charge con-
siderations.

In fact, the resonance structure on the right contributes the most to the overall hybrid, because every atom in that resonance form has a full octet of electrons. In the resonance structure on the left, the carbon does not have a complete octet, owning only six electrons in its outermost shell. Generally, the desire of atoms to achieve a full octet trumps the desire to put the charges on the best atoms. This is why the resonance structure on the right contributes more to the overall hybrid than does the resonance structure on the left, even though the structure on the right has the charge on a less preferable atom.

Common mistakes in drawing resonance structures

Here are four common mistakes that students make in drawing resonance structures. Avoid these at all costs.

✔ **Forgetting charges:** After drawing out the arrows and determining what the alternative resonance structure will look like, leaving out a charge or two is an easy mistake to make (see Figure 3-29). One way to quickly check for missing charges is to see if all your charges balance. That is, the net charge that you begin with must equal the net charge that you end up with. If your starting resonance structure has a net negative charge, then all other resonance structures must have a net negative charge as well.

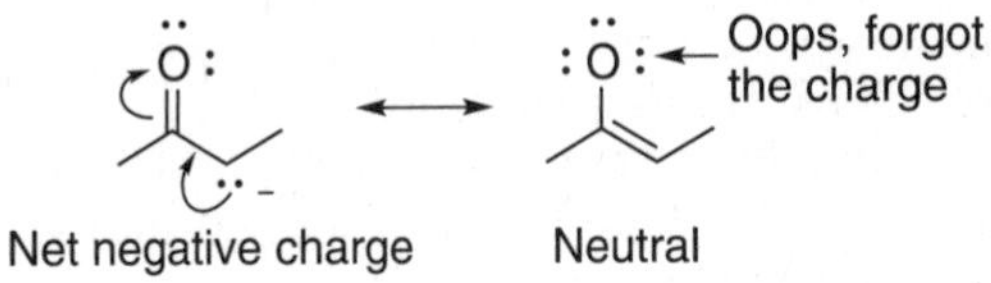

Figure 3-29:
Keeping
track of
charges.

Net negative charge   Neutral

Oops, forgot
the charge

That's not to say that the *number* of charges must be the same, just that the net charge must be the same. For example, if you start with no charges on a resonance structure, an alternative resonance structure can end up with one positive charge and one negative charge, because the net charge remains zero. But starting with a neutral molecule and ending up with one negative charge or one positive charge is not acceptable. If you see that your charges don't balance, this generally indicates that you simply neglected to place a charge on one of the atoms.

✔ **Breaking the octet rule:** Breaking the octet rule is a fairly common mistake, and, unfortunately, it's a big one. A general rule to remember is that an atom in the second row of the periodic table (like C, N, O, or F) will never have more than eight electrons around it. This means that the sum of all the bonds and lone pairs around one of these atoms cannot total more than four. I've known professors who have threatened to flunk students if on the final exam they drew a *Texas carbon* (a pentavalent carbon with five bonds, so called because everything's bigger in Texas) like the one shown in Figure 3-30. So, avoid making this mistake.

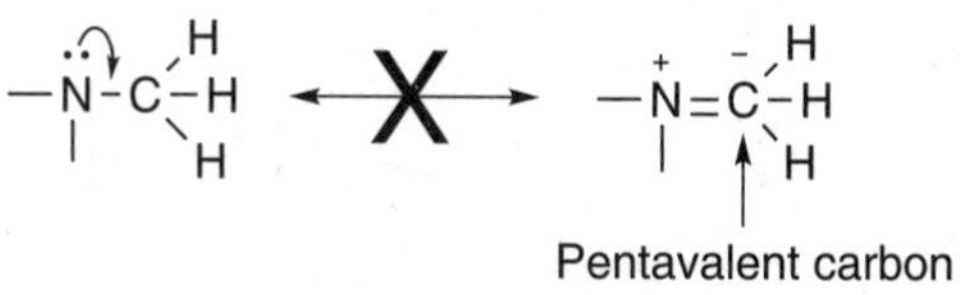

Figure 3-30:
A Texas (pentavalent) carbon. A major organic chemistry no-no.

Pentavalent carbon

✔ **Moving single bonds:** Resonance structures involve the movement of only lone-pair and pi electrons. Therefore, you can't move single bonds. To do so would break up a molecule into fragments, as shown in Figure 3-31.

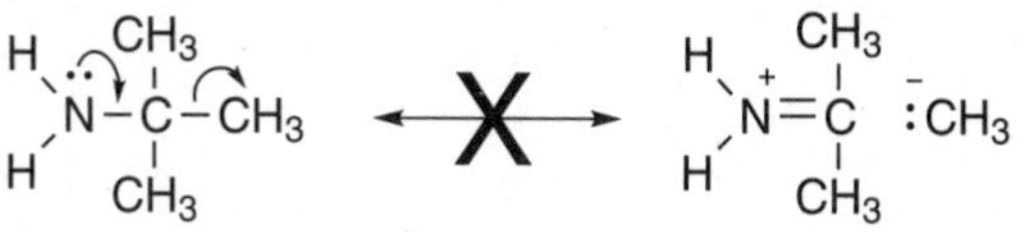

Figure 3-31:
Single bond resonance — also a major no-no.

✔ **Not following the electron flow:** Electrons are always drawn "flowing" in a single direction. So arrows won't start going one way and then double back in the opposite direction, as shown in Figure 3-32.

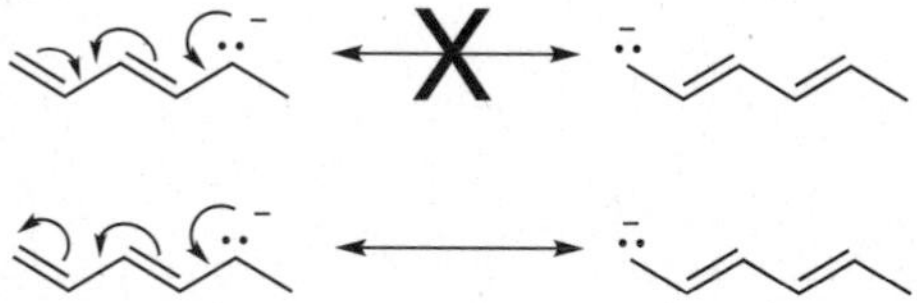

Figure 3-32:
Resonance going against the flow.

Chapter 4

Covering the Bases
(And the Acids)

You've surely done some acid-base chemistry in your lifetime, or at least observed some. Have you ever put lemon juice on fish to neutralize the fishy odors? Have you ever made a bottle rocket or fake volcano using baking soda and vinegar? Have you ever baked bread, cookies, or cakes? If so, then you've done acid-base chemistry. Certainly, you've dealt with acids and bases at some point. Foods such as tomatoes, oranges, lemons, sodas, and coffee are acidic, while household items such as bleach, ammonia, baking soda, and soaps are basic.

In fact, almost every reaction in organic chemistry involves acid-base chemistry — almost every one of them. Understanding how acids and bases work, therefore, is critical in understanding the reactions of organic molecules.

In this chapter, I define acids and bases using the three prominent definitions in use today. I show you how you can qualitatively predict the relative acidities of organic molecules by structural comparison, and I describe how organic chemists use a quantitative scale, called the pKa scale, to numerically define a molecule's acidity. Then I show you how you can use this pKa scale to predict the direction of an acid-base reaction at equilibrium.

A Defining Moment: Acid-Base Definitions

Before I get to the nitty-gritty of how acids and bases work, I need to define what they are. Three definitions are currently accepted for defining whether a molecule is an acid or a base: the Arrhenius definition, the Brønsted-Lowry definition, and the Lewis definition.

Arrhenius acids and bases: A little watery

Svante Arrhenius, a prominent chemist from the early 20th century (whose ideas on acids and bases later earned him the Nobel Prize), defined *acids* as molecules that dissociate in water to make the hydronium ion, H_3O^+. Strong Arrhenius acids are those that completely dissociate in water to make hydronium ions, while acids that only partially dissociate in water are said to be weak Arrhenius acids. Nitric acid (HNO_3), shown in Figure 4-1, is a strong acid because it completely dissociates in water to make hydronium ions; acetic acid (CH_3COOH) only partially dissociates in water and is a weak acid. (The direction of the equilibrium in Figure 4-1 is shown with the larger arrow.)

Figure 4-1:
The dissociation of strong and weak acids.

$$HNO_3 + H_2O \xrightarrow{\text{Complete dissociation}} NO_3^- + H_3O^+$$

Nitric acid

$$CH_3COOH + H_2O \rightleftharpoons CH_3COO^- + H_3O^+$$

Acetic acid — Partial dissociation

Arrhenius bases, on the other hand, are molecules that dissociate to make hydroxide ions, OH^-. As is the case with acids, bases that dissociate completely to generate hydroxide ions are strong bases, while bases that only partially dissociate to generate hydroxide ions are weak bases. Potassium hydroxide (KOH), shown in Figure 4-2, is a strong base because it completely dissociates in water to make hydroxide ions; beryllium hydroxide ($Be[OH]_2$) is a weak base because it only partially dissociates in water.

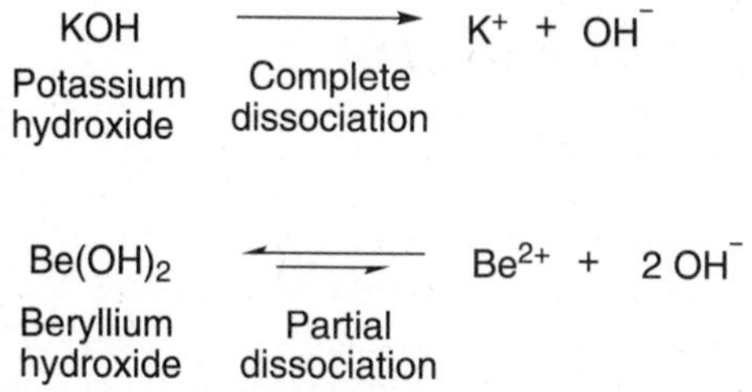

$$KOH \longrightarrow K^+ + OH^-$$

Potassium   Complete
hydroxide   dissociation

Figure 4-2:
The dissociation of strong and weak bases.

$$Be(OH)_2 \rightleftharpoons Be^{2+} + 2\,OH^-$$

Beryllium   Partial
hydroxide   dissociation

Pulling for protons: Brønsted-Lowry acids and bases

The Arrhenius definition, though useful, has some "basic" problems. First, the definition can't apply to all molecules, because many molecules aren't soluble in water. Second, not all bases dissociate to generate the hydroxide ion. Ammonia (NH_3), for example, creates hydroxide ion in solution but has no hydroxide ion in its formula, so it isn't a basic molecule by virtue of dissociation.

Therefore, other acid-base definitions that are more widely applicable are necessary. The most commonly used acid-base definition in organic chemistry is the Brønsted-Lowry definition of acids and bases. A *Brønsted-Lowry acid* is a molecule that donates a proton (H^+) to a base; a *Brønsted-Lowry base* is a molecule that accepts a proton from an acid.

An H^+ ion is called a proton because the hydrogen atom has no neutrons or electrons — just a single proton at the nucleus.

To keep the terminology straight, the deprotonated acid becomes what is known as the *conjugate base* (usually negatively charged, but not always), while the protonated base becomes the *conjugate acid,* as Figure 4-3 shows.

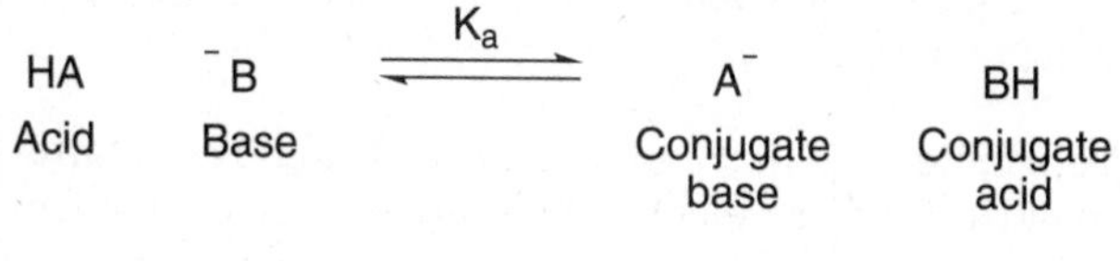

Figure 4-3:
An acid-base reaction.

$$HA \quad {}^-B \xrightleftharpoons{K_a} A^- \quad BH$$

Acid   Base   Conjugate   Conjugate
                        base   acid

Electron lovers and haters: Lewis acids and bases

While the Brønsted-Lowry definition of acids and bases is more general than the Arrhenius definition, it still isn't all-encompassing because it doesn't include molecules such as BF_3 or $AlCl_3$, which neither donate protons nor accept them. The most general method for classifying acids and bases is the Lewis acid and base definition. A *Lewis acid* is a molecule that accepts a pair of electrons to make a covalent bond, and a *Lewis base* is a molecule that donates electrons to make a covalent bond.

Borane (BH_3) is an example of a Lewis acid. Borane is a very unhappy molecule because it doesn't have a full octet of valence electrons (see Chapter 2). Because it doesn't have a full octet of electrons, it can accept a lone pair from a molecule like methylamine (CH_3NH_2), which has a lone pair of electrons, and use this electron pair to fill its octet (see Figure 4-4). Because BH_3 accepts a pair of electrons to make a covalent bond, it's said to be a Lewis acid; because methylamine donates electrons to make a bond, it's said to be a Lewis base.

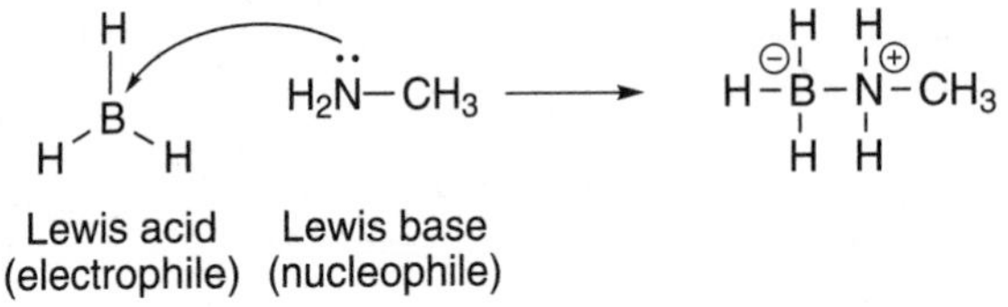

Figure 4-4:
Lewis acids (electron acceptors) and Lewis bases (electron donors).

Lewis acid (electrophile)   Lewis base (nucleophile)

Lewis acids are also called *electrophiles*, which means "electron lovers." Lewis bases are also called *nucleophiles*, which means "nucleus lovers." You see the terms *nucleophile* and *electrophile* used repeatedly throughout organic chemistry courses.

REMEMBER

Nucleophiles are molecules that can donate electrons (Lewis bases) to form bonds and are nucleus lovers; electrophiles are molecules that can accept electrons (Lewis acids) to make a new bond and are electron lovers.

Lewis acids and bases encompass Brønsted acids and bases as well. Any Brønsted acid will also be a Lewis acid, and any Brønsted base will also be a Lewis base (see Figure 4-5 for an example). You may reasonably ask, "If that's

---

A word about acids and bases

It's easy to call a molecule an acid or a base, and say, "That's all there is to it, folks." But the terms *acid* and *base* are a little more elusive. A molecule is an acid only in comparison to another molecule, and likewise for a base. When the terms *acid* or *base* are used when discussing a particular molecule, they're used in comparison to a reference molecule — water. Water is capable of acting both as an acid and as a base. Any molecule that's more acidic than water is generally considered an acid, and any molecule that's more basic than water is generally considered a base.

But keep in mind that these terms are general. Most people would agree that nitric acid is an acid; its name even includes the word *acid*. But even nitric acid can act as a base under the right conditions! In the presence of the more acidic sulfuric acid, nitric acid acts as a base (you see this reaction in the nitration of benzene in Chapter 15). This reaction is an extreme case, but I hope it makes you wary of rigidly classifying a molecule as an acid or a base, even though doing so is convenient. Whether a molecule acts as an acid or as a base really depends on what's thrown into the reaction pot along with it.

---

the case, why use Brønsted acids and bases at all?" The answer is that it's often more convenient for the organic chemist to think of acid-base reactions in terms of proton transfers rather than in terms of electron transfers. That's why when acids and bases are discussed in organic chemistry, with only a few exceptions, it's in terms of Brønsted acid and bases.

Figure 4-5:
Brønsted acids are also Lewis acids.

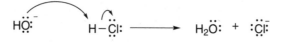

Lewis base / Brønsted base (donating electrons)   Lewis acid / Brønsted acid (accepting electrons)

Comparing Acidities of Organic Molecules

In general, the strength of an acid is directly proportional to the stability of the acid's conjugate base. In other words, an acid that has a more stable conjugate base will be more acidic than an acid that has a less stable conjugate base. Because the conjugate base is usually negatively charged (anionic), a convenient tactic when comparing the acidities of molecules is to think about what structural features contribute to stabilizing negative charges. The more stable these negative charges are in the conjugate base, the more acidic the acid. In the sections that follow, I talk about structural features that stabilize negative charges and lead to stronger acids.

Strong acids have stable conjugate bases.

In general, acidic molecules have structural features that allow the anion in the conjugate base to delocalize the charge over a larger space. Delocalization of the negative charge (such that one atom doesn't have to bear the full negative charge) makes the molecule more stable. The most important features that stabilize negative charges include the electronegativity, hybridization, and size of the atom upon which the negative charge is located, the electron-withdrawing effects of neighboring electronegative atoms, and resonance effects.

Comparing atoms

On which atom does the negative charge of the acid's conjugate base rest? A negative charge prefers to rest on electronegative (electron-loving) elements. Therefore, a negative charge is more stable on oxygen than it is on nitrogen; similarly, a negative charge is more stable on nitrogen than it is on carbon. For that reason, alcohols (R—OH) are more acidic than amines (R—NH$_2$), which in turn are more acidic than alkanes (R—CH$_3$), as shown in Figure 4-6.

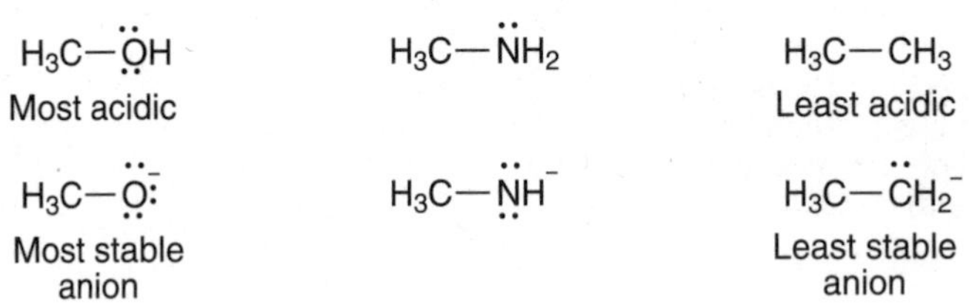

Figure 4-6:
Negative charges prefer to rest on the more electronegative atoms.

$H_3C-\ddot{O}H$
Most acidic

$H_3C-\ddot{N}H_2$

H_3C-CH_3
Least acidic

$H_3C-\ddot{O}\!\!:^-$
Most stable anion

$H_3C-\ddot{N}H^-$

$H_3C-\ddot{C}H_2^-$
Least stable anion

Electronegativity increases as you go up and to the right on the periodic table.

The size of the atom also plays a role in stabilizing the negative charge. Charges prefer to be on larger atoms than on smaller atoms. This preference results from large atoms allowing the negative charge to delocalize over a much larger region of space, instead of being concentrated in a small region (as it would on a small atom). As a rule, atom size trumps electronegativity considerations. So, even though fluorine is a more electronegative atom than iodine, HI is more acidic than HF. The much larger iodine atom allows the negative charge to delocalize over a larger space than does the much smaller fluorine atom, and thus makes hydrogen iodide more acidic. The effect that atom size has on acidity is shown in Figure 4-7.

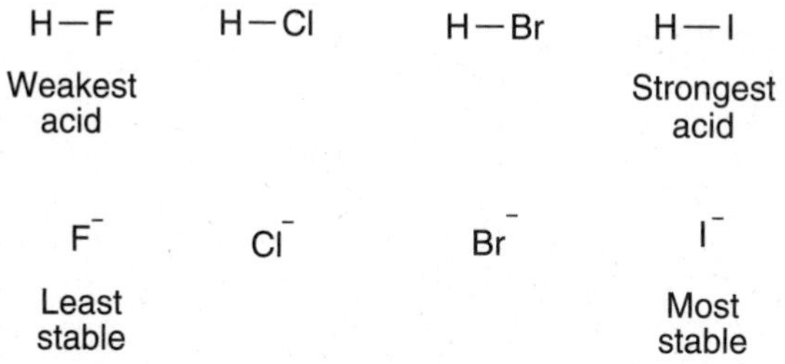

H—F   H—Cl   H—Br   H—I

Weakest   Strongest
acid   acid

Figure 4-7:
The size of
the atom
contributes   F⁻   Cl⁻   Br⁻   I⁻
to acidity.

Least   Most
stable   stable

Seeing atom hybridization

The orbital on which the lone-pair anion rests also affects the acidity. Lone-pair anions prefer to reside in orbitals that have more *s* character than *p* character, because *s* orbitals are closer to the atom's nucleus than *p* orbitals, and the electrons are stabilized by being closer to the nucleus. Orbitals that are *sp* hybridized have 50 percent *s* character, orbitals that are sp^2 hybridized have 33 percent *s* character, and orbitals that are sp^3 hybridized have 25 percent *s* character. Therefore, anions prefer to be in orbitals that are *sp* hybridized over those that are sp^2 hybridized orbitals, and they prefer to be in orbitals that are sp^2 hybridized over those that are sp^3 hybridized. The effect of orbital type on acidity is shown in Figure 4-8. (Refer to Chapter 2 to read more about orbital hybridization.)

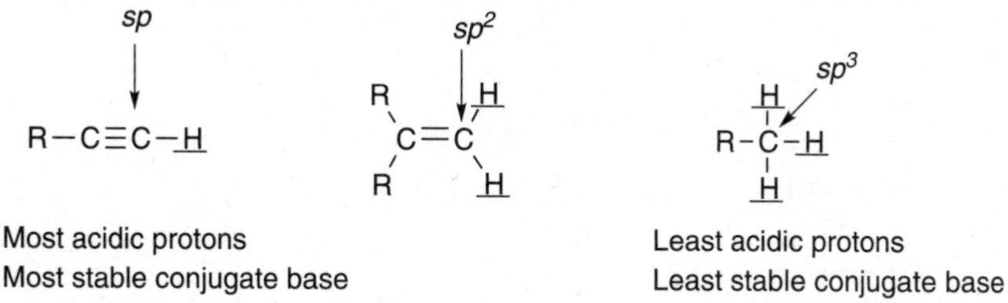

sp sp^2 sp^3

Figure 4-8:
The types
of orbitals
affect acidity.   R—C≡C—H

Most acidic protons   Least acidic protons
Most stable conjugate base   Least stable conjugate base

Seeing electronegativity effects

Electron-withdrawing groups on an acid also stabilize the conjugate base anion by allowing some of the charge on the anion to delocalize to other parts of the molecule. For example, trifluoroethanol (shown in Figure 4-9) is more acidic than ethanol. The highly electronegative fluorine atoms (which are electron pigs) on trifluoroethanol pull electron density away from the anion, taking away some of the negative charge from the oxygen, thereby stabilizing the molecule.

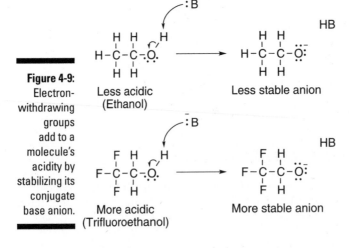

Figure 4-9: Electron-withdrawing groups add to a molecule's acidity by stabilizing its conjugate base anion.

Less acidic (Ethanol)

Less stable anion

More acidic (Trifluoroethanol)

More stable anion

Seeing resonance effects

Acids with conjugate bases that allow the negative charge to be delocalized through resonance are stronger acids than acids whose conjugate bases don't have resonance structures. For example, acetic acid, shown in Figure 4-10, is much more acidic than ethanol because the conjugate base anion of acetic acid can delocalize the negative charge through resonance.

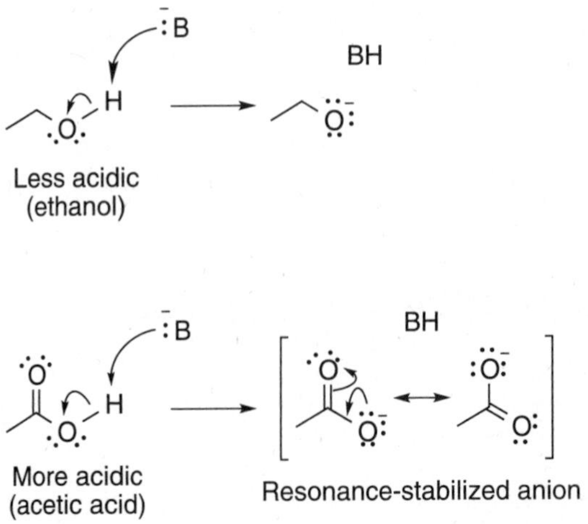

Less acidic (ethanol)

Figure 4-10: Resonance effects contribute to acidity.

More acidic (acetic acid)

Resonance-stabilized anion

REMEMBER

Resonance structures are a stabilizing feature of molecules and ions.

Defining pKa: A Quantitative Scale of Acidity

The pKa value of an acid is a quantitative measurement of a molecule's acidity. The pKa is derived from the equilibrium constant for the acid's dissociation reaction, Ka, and uses a logarithmic scale to allow the pKa values to span wide ranges.

$$pKa = -\log Ka$$

REMEMBER

The lower the pKa value of an acid, the stronger the acid. The higher the pKa value, the weaker the acid. Very strong acids have pKa values of less than zero, while weak acids generally have pKa values of between 0 and 9.

A brief pKa table of acids is shown in Table 4-1.

Table 4-1	Approximate pKa Values of Common Acids		
Acid	**Approximate pKa**	**Acid**	**Approximate pKa**
H_2SO_4	−7	H_3O^+	−2
HCN	9	O‖R–C–OH	5
R–OH	16	O O‖ ‖–C–CH–C– (H)	10
O‖R–C–CH₂–H	20	CH_4	50

Problem Solving: Predicting the Direction of Acid-Base Reactions at Equilibrium

With pKa table in hand (or in memory, if your professor insists upon it), you can predict the equilibrium direction of acid-base reactions. Weak acids and bases are lower in energy than strong acids and bases, and because equilibria favor the reaction side with the lowest-energy species, acid-base reactions will go to the side with the weakest acids and bases.

As a rule, the equilibrium of a reaction will favor the side with weaker acids and bases.

For example, you can predict the direction of the acid-base reaction between hydrogen cyanide (HCN) and acetate ($C_2H_3O_2-$) shown in Figure 4-11. Because hydrogen cyanide (pKa = 9) has a higher pKa value than acetic acid (pKa = 5), the equilibrium will lie to the left, in the direction of the weaker acid and base. That's really all there is to predicting the direction of an acid-base reaction. If you know the pKa values of the two acids on both sides of the equation, then you know in which direction the equilibrium lies, because equilibrium will favor the side with the acid that has the highest pKa.

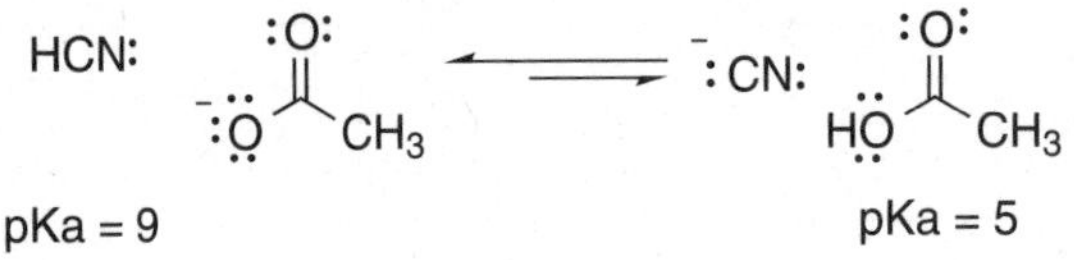

Figure 4-11:
The pKa values predict the direction of the acid-base equilibrium.

HCN: [structure: acetate ion, CH_3]   pKa = 9   ⇌   :CN: [structure: acetic acid, HO, CH_3]   pKa = 5

Chapter 5

Reactivity Centers: Functional Groups

Consider this: Millions of reactions of organic molecules are currently known, and that number is getting bigger by the day. That's probably a scary thought. Can you imagine trying to learn all of them? The good news is that you don't have to learn all the reactions of specific molecules because organic molecules often react in predictable ways based on what kinds of groups a particular molecule contains.

Alkanes (molecules containing just singly bonded hydrogen and carbon atoms) are pretty much inert under most conditions. Carbon, though, is unique among the elements in that it has the capability of forming stable compounds that have multiple bonds to other carbons, in addition to stable bonds to other non-carbon atoms, forming reactive centers. These reactive centers are called *functional groups* and are the reactive portions in an organic molecule. Chemists organize organic compounds based on what functional groups are present in a particular molecule.

One reason learning the functional groups is important is that if you learn the general reactions of a particular functional group, you have essentially learned the reactions of thousands of specific molecules that contain that functional group. The naming of organic molecules is based on what functional groups are present in a molecule because the functional groups dictate how a molecule will react. So, the sooner you learn these functional groups, the better.

In this chapter, I present the most important functional groups that you encounter throughout organic chemistry. I give examples of natural sources in which these molecules are found to show their relevance to biological systems. Additionally, I discuss their commercial uses, general nomenclature points, and interesting properties. This overview, I hope, will help you remember these functional groups and will give you a feel for the uses of particular functional groups.

Hydrocarbons

Hydrocarbons, as the name suggests, are molecules that contain just hydrogen and carbon. Simple hydrocarbons are generally cheap and commercially available, because they're found as components in crude oil. Hydrocarbons include alkanes (which contain only single bonds and are generally not considered a functional group), alkenes (molecules containing carbon-carbon double bonds), alkynes (molecules containing carbon-carbon triple bonds), and aromatics (double-bond-containing ring systems).

Double the fun: The alkenes

An *alkene* is a molecule that contains a carbon-carbon double bond (see Chapter 2 for more on double bonds). The general form of an alkene is shown in Figure 5-1. (See Chapter 3 for an overview of molecular drawings.)

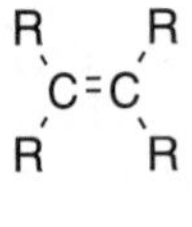

Figure 5-1:
The general form of an alkene.

$$\begin{array}{ccc} R & & R \\ & \diagdown \!\! C = C \!\! \diagup & \\ R & & R \end{array}$$

The *R* is an abbreviation for the rest of the molecule. An R group most often implies a hydrogen atom or a hydrocarbon group. It's used when a generality is being demonstrated or when the rest of the molecule isn't very important in understanding what's being discussed.

Alkenes are often found in natural products, compounds isolated from living organisms. In the simplest alkene, ethylene, each R group is a hydrogen atom (see Figure 5-2). Ethylene is a gaseous plant hormone that is released when the plant reaches maturity, signaling that it's time for the fruit to start ripening. Farmers often have special equipment that they use to spray ethylene on their crops when they want to force the fruit to ripen.

Figure 5-2:
The structures of some common alkenes.

Ethylene   2-methyl-2-butene   Cyclopentene   Cyclopropene

Alkenes are also used commercially. Ethylene *polymerizes* (combines many small units to make large molecules) to make polyethylene — a molecule formed from many ethylene molecules strung one after another to form a big chain. Polyethylene is a type of plastic used in containers such as grocery bags, milk bottles, and many different items that you see and use every day.

Alkenes can even be arranged in rings, like with cyclopropene, although they generally don't like to be in very small rings like cyclopropene (refer to Figure 5-2). I talk about alkenes in Chapters 9 and 10.

Alkenes are particularly important to organic chemists because they can be transformed into many different functional groups. They're easily made and converted into other things; this makes them particularly useful as go-betweens in the synthesis of complex molecules. In this book, I show how alkenes are converted into alkanes, cycloalkanes, cyclic ethers, alcohols, alkyl halides, aldehydes, and carboxylic acids. Talk about versatile!

The names of alkenes generally end with the suffix *–ene.* One particularly important alkene is vitamin A (also known as retinol). Vitamin A (see Figure 5-3) is an organic compound that contains five double bonds. It is important in vision, protects against sickness, and is used in skin-care products as an antioxidant that protects the skin from free radicals suspected of causing premature aging. (Read more about free radicals in Chapter 8.) I discuss the reactions, properties, and nomenclature of alkenes in detail in Chapters 9 and 10.

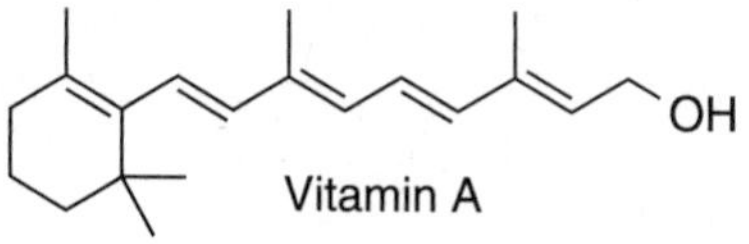

Figure 5-3:
The structure of vitamin A (retinol).

Vitamin A

Alkynes of fun

Alkynes are molecules that contain a carbon-carbon triple bond. See Figure 5-4. Many of their reactions and properties are similar to those of the alkenes, although the chemistry of alkenes and alkynes has some interesting differences.

Figure 5-4:
The general form of an alkyne.

$$R-C\equiv C-R$$

The simplest alkyne, acetylene, has the two R groups substituted with hydrogen (see Figure 5-5). Acetylene is a gas that is used as a welding fuel; it usually smells of garlic because of sulfurous impurities contained in it. Because acetylene burns cleanly and produces a very hot flame, acetylene torches reach temperatures of over 3,000°C, hot enough to weld metals, melt glass, and scorch off your pinkies if you aren't careful. Other common alkynes include methyl acetylene (or propyne) and dimethyl acetylene (or 2-butyne); see Figure 5-5 for the structures of these alkynes.

Figure 5-5:
The structures of some common alkynes.

$$H-C\equiv C-H$$
Ethyne
or acetylene

$$H-C\equiv C-CH_3$$
Propyne
or methyl acetylene

$$H_3C-C\equiv C-CH_3$$
2-butyne
or dimethyl acetylene

Alkynes are less common than alkenes in nature, but they do pop up occasionally. Calicheamicin (try saying that three times fast), for example, is a complicated organic molecule that is made by a bacterium that grows in a kind of sedimentary rock called caliche. Calicheamicin was recently found to selectively attack and kill the DNA of cancer cells and has, along with similar compounds, been used as a drug in anticancer therapies.

The biologically active portion of the calicheamicin molecule contains two triple bonds (see Figure 5-6). This portion is called an enediyne. The "diyne" part of the name means "two ynes," or "two triple bonds." Can you guess where the "ene" portion of the name comes from?

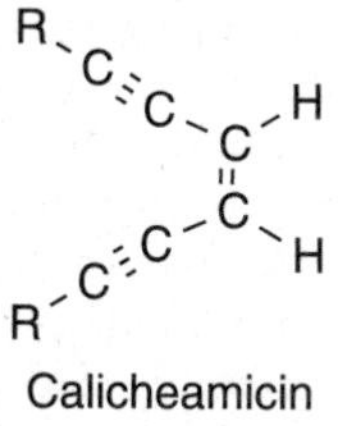

Figure 5-6:
The structure of calicheamicin.

Calicheamicin

Alkynes prefer to have bond angles of 180 degrees, where the triple bond and the two R groups on either side lie in a straight line. Triple bonds aren't particularly stretchy, and because of this, molecules with triple bonds are generally unstable in rings of fewer than eight carbons.

Alkyne names end with the suffix *–yne*. Sometimes, they're called by the common name that derives from the simplest alkyne, acetylene. I cover the specifics of alkyne properties and reactions in Chapter 11.

Smelly compounds: The aromatics

The *aromatics* (or *arenes,* as they're often called) consist of rings containing alternating double bonds. The principal aromatic compound is benzene, a six-carbon ring containing three alternating double bonds (see Figure 5-7). You might think that benzene would behave like a ring containing three alkenes, but this turns out not to be the case. Aromatic compounds have a special property that makes them significantly more stable, and less reactive, than alkenes. In Chapter 15, I discuss why benzene and other aromatics are so stable, and also talk about aromatic compounds other than benzene.

Figure 5-7:
The structure of benzene.

Benzene

These rings are called aromatics because the first of these compounds to be discovered (even before their structures were determined) had funky smells. Benzene itself is found as a component in crude oil; it has even better burning properties than does the octane fuel that you use to top off your gas tank. Unfortunately, it is a carcinogen and has other undesirable properties, so it isn't used as a fuel. However, it is often used by organic chemists (with careful handling) as a solvent and as a cheap starting material in multistep syntheses.

Many compounds made by living organisms contain aromatic rings. (See Figure 5-8 for the structures of some aromatic compounds; see Chapter 6 for more on 3-D structure.) Morphine, for example, contains a benzene ring that is vital to its pain-relieving properties. Auto exhaust, soot, and tobacco smoke contain fused rings of benzene, like benzopyrene, where the rings are squished together to make very large aromatic compounds. These compounds have been found to be carcinogenic. In fact, chimney sweeps (the guys who used to clean chimneys in the days when coal was used for home heating) had a remarkably high incidence of testicular cancer because of their high exposure to these aromatic compounds in the chimney soot.

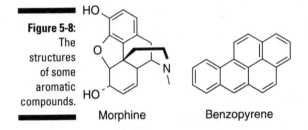

Figure 5-8: The structures of some aromatic compounds.

Morphine   Benzopyrene

Singly Bonded Heteroatoms

Heteroatoms are atoms other than carbon or hydrogen. They include such important atoms as the halogens, oxygen, and sulfur, and are the components of the halide, alcohol, ether, and thiol functional groups. Each of these functional groups is described in this section.

Happy halides

The *halides* are organic compounds that contain one or more halogens. (Halogens are those elements found in column 7A of the periodic table.) The four halogens that you frequently see in organic compounds are fluorine, chlorine, bromine, and iodine. The general form of a halide is shown in Figure 5-9.

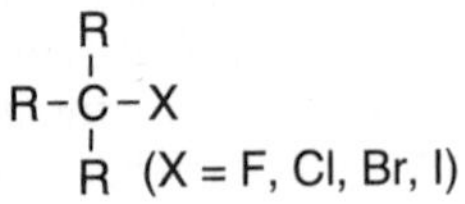

Figure 5-9: The structure of a simple halide.

$$R-\underset{\underset{R}{|}}{\overset{\overset{R}{|}}{C}}-X \quad (X = F, Cl, Br, I)$$

Halides, like alkynes, are seldom found in natural products, and when they are found, they're often in compounds that are toxins. Commercially, halides are used as propellants in aerosol cans such as hairspray and spray paint, solvents such as chloroform, and as refrigerants. Evidence that certain alkyl halides (that is, halides attached to alkanes) persist in the atmosphere and that these compounds contribute to depleting the ozone layer has led many countries to limit or outlaw their use as propellants. An example of a refrigerant halide is shown in Figure 5-10.

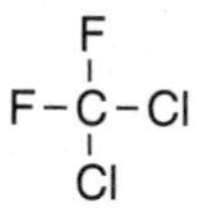

Figure 5-10:
The struc-
ture of a
refrigerant.

Halides are also used as insecticides. For example, DDT (dichlorodiphenyl-trichloroethane; see Figure 5-11) was widely used to protect crops from insects, until evidence suggested that this compound accumulated in the environment and caused unwanted side effects in wildlife, like the thinning of the eggshells of wild birds, and the death of bats and rodents. In large enough doses, DDT was even found to cause death to humans. Not surprisingly, this compound is now banned in the United States.

Other halides are still used in homes and businesses. For example, Teflon is a fluorine-containing polymer used in nonstick surfaces like pots and pans (see Figure 5-11). The brackets with the *n* subscript in the structure of Teflon indicate that that portion of the molecule is a repeating unit.

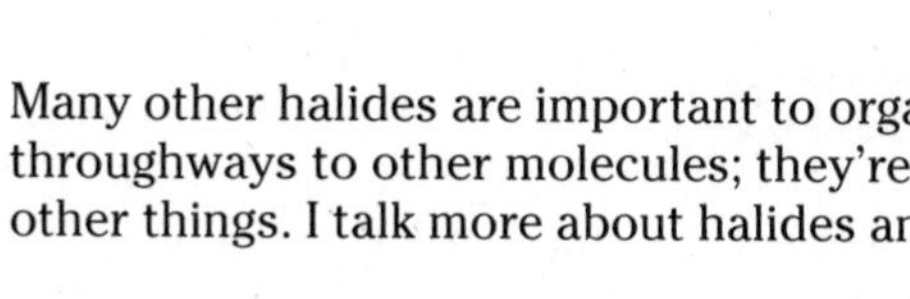

Figure 5-11:
The struc-
tures of
some
common
halides.

DDT

Teflon

Many other halides are important to organic chemists because they're good throughways to other molecules; they're easily made and converted into other things. I talk more about halides and their reactions in Chapter 12.

For rubbing and drinking: Alcohols

Alcohols are also a very common and important group of organic compounds. *Alcohols* consist of the general formula R-OH, and have names that end with the suffix –*ol*. Probably the alcohol you're most familiar with is ethanol (or ethyl alcohol) because it's the alcohol found in beer, wine, and liquor. Ethanol is toxic in large enough doses. Some of its effects include a decrease in your body's motor coordination, lowered inhibition, loud and off-key karaoke singing, and crank calls to your mother-in-law. Simple alcohols, such as isopropanol (rubbing alcohol), are used in cleaning solutions and as disinfectants, and an alcohol with two OH groups, ethylene glycol (which is also toxic, so don't let your dog drink it), is used in antifreeze. Figure 5-12 illustrates some common alcohols.

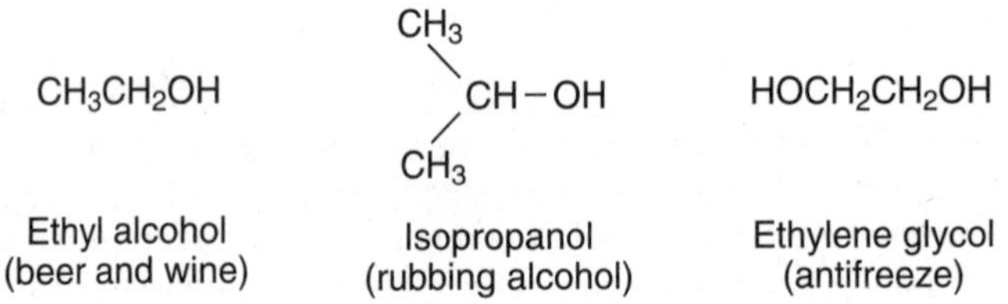

Figure 5-12:
The structures of some common alcohols.

CH_3CH_2OH

Ethyl alcohol (beer and wine)

CH_3
$\diagdown$
$CH-OH$
$\diagup$
CH_3

Isopropanol (rubbing alcohol)

$HOCH_2CH_2OH$

Ethylene glycol (antifreeze)

Alcohols are also commonly found in natural products. Sugars, for example, like the table sugar sucrose (shown in Figure 5-13), contain many OH groups.

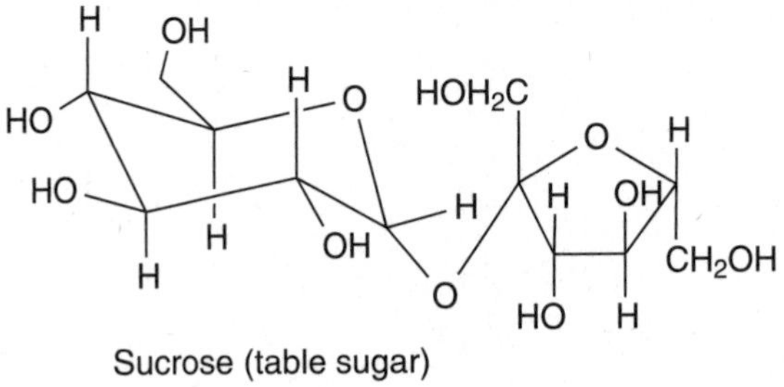

Figure 5-13:
The structure of table sugar (sucrose), showing the fructose and glucose components.

Sucrose (table sugar)

What stinks? Thiols

Very few organic chemists have a burning desire to work with thiols. That's because *thiols* are foul-smelling compounds, of the general formula R-SH. Thiols are the sulfur analog of alcohols and are often hideously unpleasant compounds. Certain thiols, for example, are found in skunk spray — that stinky stuff skunks use to fend off predators. Other thiols are responsible for the odors in flatulence, garlic, sewage, skunked beer, and rotten eggs. Commercially, very small amounts of thiols are added to methane gas so you can detect when a gas line has sprung a leak (methane itself is odorless).

Not all thiols are associated with such unpleasantries, however. Cysteine, for example, is one of the amino acids that the human body uses to produce proteins, and this amino acid plays a particularly important role in keratin, a protein found in hair. These cysteine amino acids (see Figure 5-14) are important because they can couple with each other and make disulfide bonds (S-S bonds) that hold the shape of your hair. If you've ever had a perm, you experienced thiol chemistry. A perm involves treating the hair with chemicals and heat to break the disulfide bonds, reshaping the hair in the desired way, and then resetting the disulfide bonds.

Generally, names of thiols end with the suffix *–thiol,* as is the case with two of the thiols found in skunk spray — 2,3-dimethyl-2-buten-1-thiol and 3-methyl-1-butanethiol, both of which are shown in Figure 5-14.

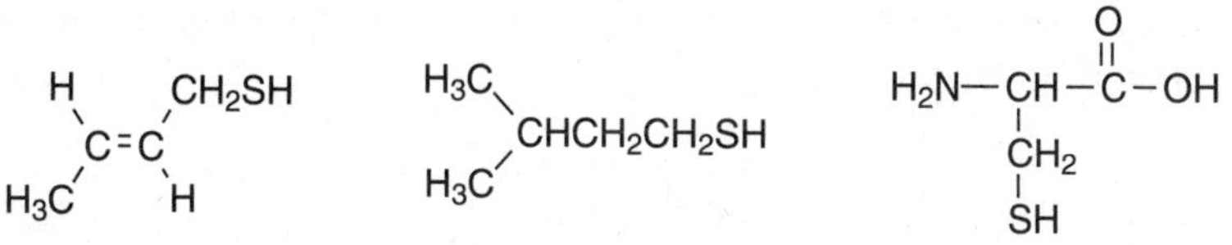

Figure 5-14: Cysteine and skunk-spray thiols.

2-buten-1-thiol   3-methyl-1-butanethiol   Cysteine

How ethereal

Ethers are molecules containing an oxygen sandwiched between two carbons (see Figure 5-15). Molecules containing ether functional groups are widely used as solvents in organic reactions. Additionally, before it was replaced by modern anesthetics that make recovery less painful, diethyl ether (commonly known as "ether") was used as an anesthetic to knock patients unconscious for surgery. Ethers found in three-membered rings are given the specific name of *epoxides* (the simplest epoxide shown in Figure 5-15 is ethylene oxide). Epoxides are found in epoxy resins and in epoxy glues, and are common as intermediates in multistep syntheses.

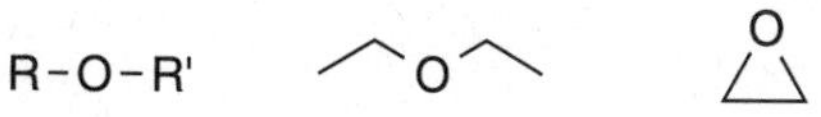

Figure 5-15: Ethers.

An ether   Diethyl ether   Ethylene oxide

Carbonyl Compounds

The chemistry of living things is largely the chemistry of carbonyl compounds. A *carbonyl group* is a C=O group — in other words, a carbon atom double-bonded to oxygen. A carbonyl group is not considered a functional group in itself; instead, it's considered a component in some of the most important functional groups, including the aldehydes, ketones, esters, amides, and carboxylic acids. If you're considering taking a course in biochemistry, you should have a good understanding of the reactions and properties of carbonyl compounds because most of the reactions in the body are carbonyl reactions (including the reactions in the Krebs Cycle, in glycolysis, and in the synthesis of fatty acids and polyketides).

Living on the edge: Aldehydes

Aldehydes are the simplest of the carbonyl compounds. In an *aldehyde* (see Figure 5-16), the carbonyl group is flanked by one hydrogen and one R group. It may be helpful to think of an aldehyde as a carbonyl group at the end of an organic molecule.

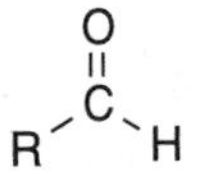

Figure 5-16:
An aldehyde.

The simplest aldehyde, formaldehyde — in which the R group is hydrogen — has many uses, including as a preservative. Often the smells that waft from a biology laboratory come from the formaldehyde used to preserve specimens (not that chemists have much right to complain, though, considering some of the smells emitting from their laboratories). Retinal is a large aldehyde; it's one of the pigments that traps light in the eyes of humans, making vision possible. Benzaldehyde is a wonderfully sweet-smelling compound that gives almonds its odor. Both retinal and benzaldehyde are shown in Figure 5-17.

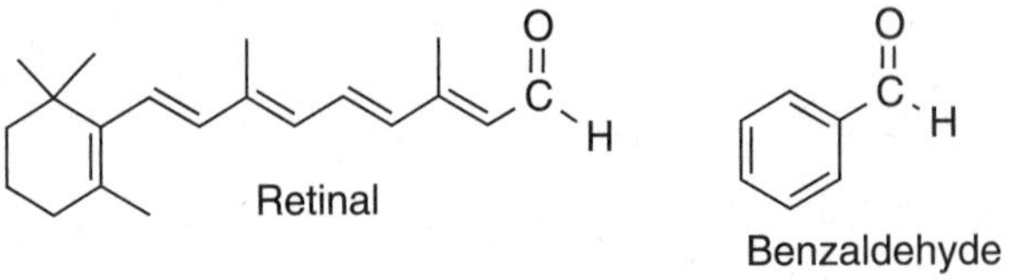

Figure 5-17:
Two important aldehydes.

Retinal

Benzaldehyde

You'll often see aldehydes represented as R-CHO. Don't confuse this condensed form with an alcohol, which is represented as R-OH.

Stuck in the middle: Ketones

Compounds that contain a carbonyl group sandwiched between two carbons are called *ketones*. If an aldehyde can be thought of as a carbonyl at the end of a molecule, then a ketone is a carbonyl somewhere in the middle of a molecule. The simplest ketone, in which both R groups are CH_3 units, is called acetone, a common laboratory solvent (see Figure 5-18). Acetone is the smelly stuff found in nail polish remover, and is the reason why polished nails don't stay polished for very long in organic laboratories. The names of ketones generally end with the suffix –*one*.

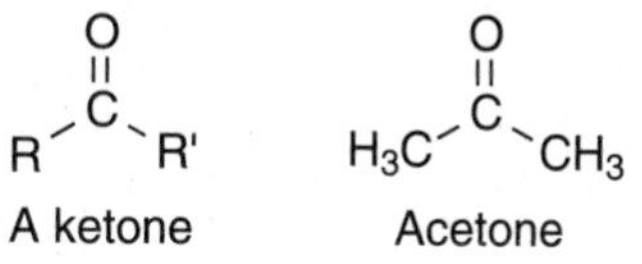

Figure 5-18:
The general structure of a ketone and the specific structure of acetone.

A ketone   Acetone

Carboxylic acids

The *carboxylic acid* functional group is made up of a carbonyl group attached to an OH group. The general form of carboxylic acids is shown in Figure 5-19.

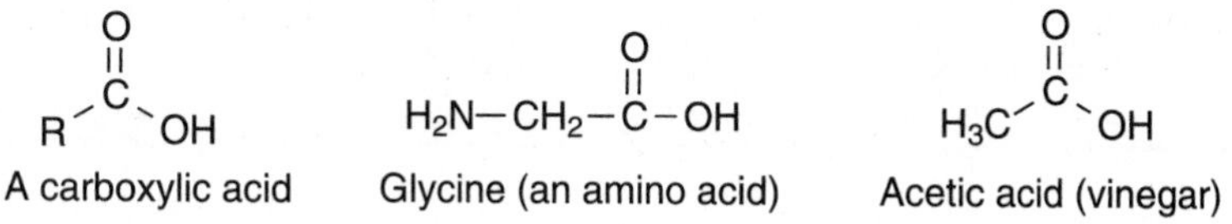

Figure 5-19:
Common carboxylic acids.

A carboxylic acid   Glycine (an amino acid)   Acetic acid (vinegar)

Don't confuse a carboxylic acid with a ketone or an alcohol. Carboxylic acids have entirely different properties and reactivities than either ketones or alcohols. In particular, the proton (H^+) on the oxygen in a carboxylic acid is unusually acidic (hence the name!), for reasons I talk about in Chapter 4.

Sneaky orchids, ketones, and chemical siren songs

In 2003, German and Australian scientists first reported their discovery of the *diketone* (a molecule with two ketone functional groups) chiloglottone (see the figure), produced by an Australian orchid flower. This orchid, they discovered, uses this diketone to reproduce, to get its pollen to another orchid for pollination. But how, you ask, can a chemical be responsible for transporting the orchid's pollen to another orchid, sometimes over several miles? By trickery, deception, and bamboozling, of course. This deception forms yet another interesting episode of the soap opera between plants and insects, a soap opera that mostly goes undetected by humans.

Chiloglottone

Chiloglottone, it turns out, not only is produced by the Australian orchid, but also is the sex pheromone of a species of wasp native to Australia. When a female wasp emits this pheromone, she signals to all eligible male wasps in the area that she's ready to mate. When a male wasp detects the pheromone in the air, he does the equivalent of a wasp "hubba hubba hubba," and wastes no time in tracing the chemical trail back to its source for a little wasp hanky-panky.

The orchid, however, has cracked the chemical code of communication between the wasps, and takes advantage of their instant-messaging system for its own devious purposes. The orchid makes its mischief by producing and transmitting the same sex pheromone that the wasps produce, and uses the chemical as a lure.

When they detect the pheromone produced by the orchid, the wasps — in typical male fashion — are so swooned by the possibility of wasp love that they become irrational, and in the height of their passion, they neglect to realize that the orchid is not a female wasp. Swooping down, the male wasp tries his best to mate with the orchid's flower head, and when he's finished his business, he flies away from the orchid with a thin dusting of orchid pollen on his legs. Coated with pollen, the wasp flies off until he's attracted by another orchid emitting the sex pheromone, and, feeling amorous yet again, he "mates" with the second orchid. When the pollen from the first orchid rubs off the wasp's legs, the second orchid becomes pollinated.

In this way, the orchid tricks the male wasp into becoming a postman for its pollen, free of charge, no stamps required. This kind of bamboozling is actually quite common in nature, and is called *sexual deception*. Though insects get revenge in some ways on plants — munching on them, for instance! — this round belongs to the plants.

Carboxylic acids are abundant in nature. In fact, each of the amino acids that our body uses as building blocks to make proteins contains a carboxylic acid (such as the amino acid glycine in Figure 5-19), as do all the fatty acids. The names of carboxylic acids generally end with –*oic acid*, like ethanoic acid (more commonly called acetic acid), which is the carboxylic acid responsible for the flavor and bitterness of vinegar.

Sweet-smelling compounds: Esters

Esters are very similar in structure to carboxylic acids. An *ester* is basically a carboxylic acid with the hydrogen snipped off, and an R group glued in its place. (In fact, esters are made from carboxylic acids.) Esters are generally sweet-smelling compounds, and many of the lovely smells from fruits are esters. (See Figure 5-20 for the structures of some esters.) Because of this, they're generally found in deodorants and used as artificial flavorings in food. The names of esters generally end with the suffix *–oate*.

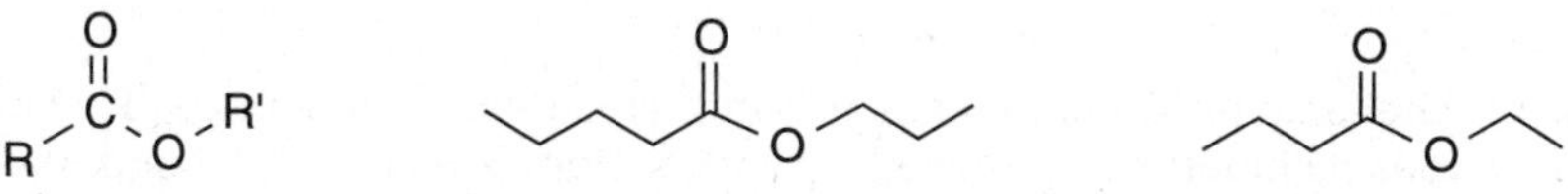

Figure 5-20:
Common
esters.

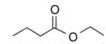

An ester   Propyl pentanoate (pineapples)   Ethyl butanoate (apples)

Nitrogen-containing functional groups

Just as the carbonyl compounds play an important role in nature, so do nitrogen-containing compounds. Illicit drugs and painkillers, which are often *alkaloids* (nitrogen-containing rings) often contain nitrogen as well.

Even though the primary component in air is nitrogen gas (N_2), most plants and animals cannot use this as a source of nitrogen because, in its gaseous state, nitrogen is so unreactive and stable. The N_2 gas needs to be converted to ammonia (NH_3) or nitrate (NO_3-) for animals and plants to be able to use it as a building block, a process called *nitrogen fixation*. This job is left up to microorganisms, which have special enzymes that can tackle this reaction. (Lightning strikes also account for some nitrogen fixation, although only a small amount.)

I am what I amide

Amides are close relatives of esters, except that amides have a nitrogen (rather than an oxygen) next to the carbonyl group. Amides are quite often found in nature. *Amide bonds* hold all our proteins together (the amide bond in proteins is called a *peptide bond*). See Figure 5-21 for the structures of some common amides.

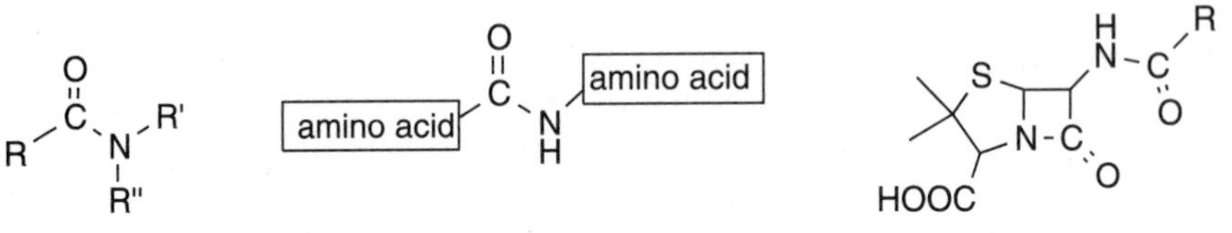

Figure 5-21:
The general structure of an amide and a peptide, and the specific structure of penicillin.

An amide

A peptide bond

Penicillin

The famous antibiotic penicillin contains two amide groups. The amide in the four-membered ring (called a β-*lactam ring*) is responsible for penicillin's bacteria-killing properties; this amide group reacts with enzymes responsible for building the cell walls of bacteria, which leads to bacterial death. (For more about penicillin, see Chapter 22.)

Be nice, don't be amine person

Amines are nitrogen atoms that take the place of a carbon atom in an alkane (the three forms of an amine are $R-NH_2$, R_2NH, or R_3N). Amines are not known for their pleasantness. For example, the smell of decaying animals comes from the *diamine* (a molecule with two amine functional groups) putrescine, for example. So, unless you're a vulture, you probably don't want to use these compounds as perfumes. Plants and animals and other organisms make many important amine compounds. Nicotine (see Figure 5-22), as well as many other familiar compounds like cocaine, mescaline, amphetamine, and morphine are all amine-containing compounds. Many drugs (both legal and illegal) contain amines that are essential to the drug's activity.

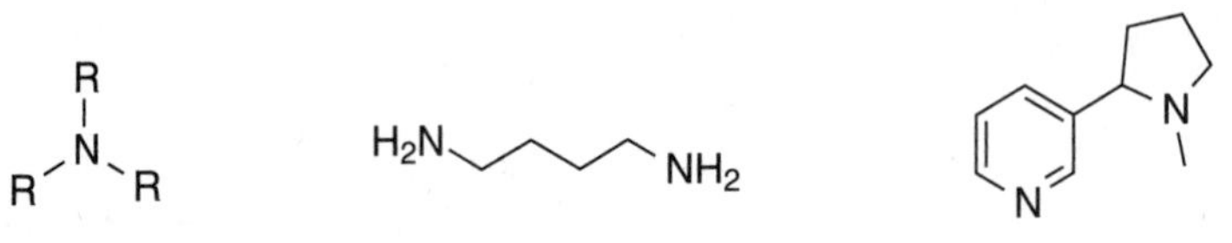

Figure 5-22:
Common amines.

An amine

Putrescine

Nicotine

Nitriles

Nitriles (see Figure 5-23) are compounds that contain a carbon triply bonded to nitrogen. Nitriles are often useful in organic synthesis. Nitriles can be converted into carboxylic acids and amines by well-known procedures. Acetonitrile, in which the R group is a methyl group (CH_3), is a common organic solvent. Nitriles generally end with the suffix *–nitrile*. As substituents on a molecule, they're referred to as *cyano groups*.

Figure 5-23:
Nitriles.

R−C≡N   H₃C−C≡N

A nitrile   Acetonitrile

Test Your Knowledge

Figure 5-24 gives you a made-up molecule that contains some of the functional groups that I mention in this chapter. How many can you identify? (**Hint:** Draw out all the condensed portions — like COOH, CN, and CHO — to help you identify which functional group it is.)

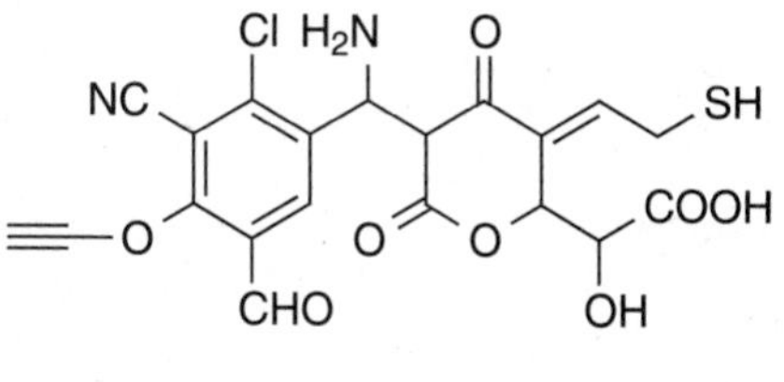

Figure 5-24:
A
hypothetical
molecule
with
different
functional
groups.

Chapter 6

Seeing in 3-D: Stereochemistry

*I*n this chapter, you look closely at molecules in three dimensions. Why should you care about the arrangements of atoms in three-dimensional space? Because nature does. Virtually all the compounds produced in nature — proteins, sugars, carbohydrates, DNA, and RNA, to name just a few — have atoms oriented in very specific ways in three-dimensional space.

Subtle differences in the way molecules are arranged in three-dimensions can be recognized, for instance, by your nose receptors. Limonene, for example, shown in Figure 6-1, can have two different three-dimensional configurations (solid wedges indicate a bond coming out of the plane of the paper; dashes indicate a bond going back into the paper). If the molecule is arranged in one configuration, it smells of oranges; arranged in the other configuration, it smells of lemons!

Figure 6-1:
Two config-
urations of
limonene.

Smells of oranges

Smells of lemons

This discrimination by your nose receptors (of molecules differing only in the orientation of atoms in three-dimensional space) is common in biological systems. Most drugs, in fact, are only active when they have a very specific three-dimensional configuration, while compounds with an identical structure but with different arrangements of the atoms in space are usually completely ineffective, or even harmful. (See the sidebar "The biological importance of stereochemistry: The thalidomide tragedy," later in this chapter for an example of such harmful drugs.) Because of this biological significance of the three-dimensionality of molecules, the study of *stereochemistry* (the way that molecules orient themselves in three-dimensional space) is very important, particularly in the pharmaceutical industry.

Drawing Molecules in 3-D

To be able to study stereochemistry, you need to be able to draw three-dimensional molecules on a two-dimensional piece of paper. When you draw molecules in three-dimensions, you typically use solid and dashed wedges. A solid wedge indicates a bond coming out of the plane of the piece of paper, and a dashed wedge indicates a bond going back into the paper. A regular line indicates a bond that's in the same plane of the paper. When you want to draw an sp^3 carbon atom (see Chapter 2) in three dimensions, you draw two bonds in the plane of the paper, one bond coming out of the plane of the paper, and one bond going back (behind the plane of the paper), in a tetrahedral arrangement (see Figure 6-2).

Figure 6-2: A tetrahedral arrangement.

Comparing Stereoisomers and Constitutional Isomers

The two different limonene molecules shown in Figure 6-1 are *stereoisomers* (molecules with the same atom connectivity but with different orientations of those atoms in space). Stereoisomers are different from constitutional isomers. *Constitutional isomers* are molecules with the same molecular formula but with atoms connected to each other in different ways (see Chapter 7).

Your hands, for example, are stereoisomers. On both hands, all your fingers are attached in the same way — thumb, index finger, middle finger, ring finger, and pinky, in that exact order. But although your hands have the same finger connectivity, your hands are not identical — they're stereoisomers of each other. Try putting a right-handed glove on your left hand if you don't believe me. If you happen to be an alien with one hand having a pinky where a thumb should be and a thumb where a pinky should be, then your hands would be constitutional isomers, because the connectivity of your fingers would be different.

Mirror Image Molecules: Enantiomers

Molecules that are nonsuperimposable mirror images of each other are called *enantiomers.* Your right hand is the enantiomer of your left hand because your hands are mirror images of each other (put your right hand up to a mirror, and you'll see what looks like a left hand). Enantiomers cannot be superimposed onto each other, just as your right hand cannot be superimposed on your left hand. To translate this feature to structures, the compound shown in Figure 6-3 cannot be superimposed on its mirror image; if you slide the compound over its mirror image, the various parts of the two molecules won't line up. You might try making the molecular models of the two enantiomers to prove to yourself that you can't superimpose these enantiomers over each other — the atoms simply don't align.

Mirror

Figure 6-3:
Enantiomers
(mirror
images).

Left hand   Right hand

If you really want to grasp stereochemistry, you need to mess around with molecular models. When you play with models, you get a feel for how stereochemistry works in your bones, not just in your head.

Molecules that are not superimposable on their mirror images (like the halogenated methane shown in Figure 6-3) are said to be *chiral* molecules (hard *k* sound, rhymes with *viral*). Conversely, molecules that *can* be superimposed on their mirror images are said to be *achiral*. Methane (shown in Figure 6-4) is an example of an achiral molecule. Methane is superimposable on its mirror image. That is, if you rotate the mirror image of methane, the atoms superimpose (or overlap) on the original molecule perfectly. Thus, methane doesn't have an enantiomer because its mirror image is identical to itself.

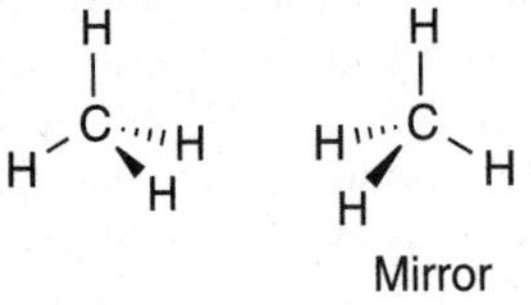

Figure 6-4:
Methane,
an achiral
molecule.

Mirror

Only chiral molecules have enantiomers.

Seeing Chiral Centers

How can you tell whether a molecule will be chiral or achiral without constructing the molecule's mirror image and seeing if the two can be superimposed? Most often, chiral molecules contain at least one carbon atom with four nonidentical substituents. Such a carbon atom is called a *chiral center* (or sometimes a *stereogenic center*), using organic-speak. Any molecule that contains a chiral center will be chiral (with one exception, discussed later). For example, the compound shown in Figure 6-5 contains a carbon atom with four nonidentical substituents; this carbon atom is a chiral center, and the molecule itself is chiral, because it's nonsuperimposable on its mirror image.

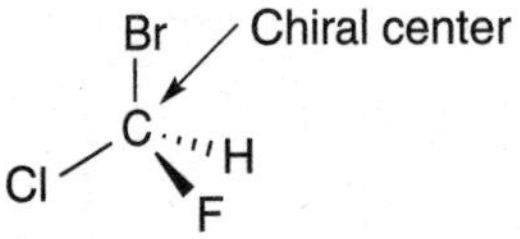

Figure 6-5:
A chiral
center.

Br Chiral center

You need to be able to quickly spot chiral centers in molecules. All straight-chain alkyl group carbons (CH_3 or CH_2 units) will *not* be chiral centers because these groups have two or more identical groups (the hydrogens) attached to the carbons. Neither will carbons on double or triple bonds be chiral centers because they can't have bonds to four different groups. When looking at a molecule, look for carbons that are substituted with four different groups. See, for example, if you can spot the two chiral centers in the molecule shown in Figure 6-6.

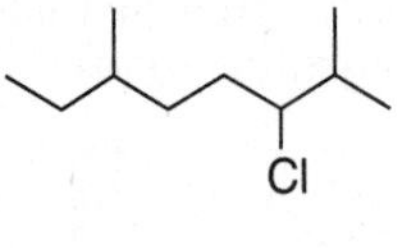

Figure 6-6:
A molecule
with two
chiral
centers.

Cl

Because not all CH_3 and CH_2 groups can be chiral centers, the molecule shown in Figure 6-5 has only three carbons that can be chiral centers. The two leftmost possibilities, identified in Figure 6-7, have four nonidentical groups and are chiral centers, but the one on the far right has two identical methyl (CH_3) groups and so is not a chiral center.

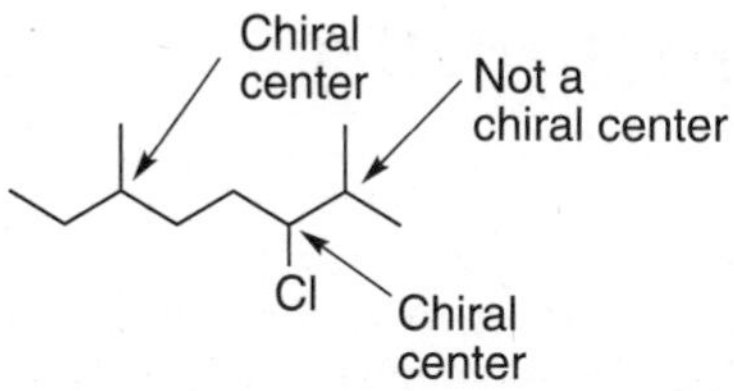

Figure 6-7:
The chiral centers in a long molecule.

Assigning Configurations to Chiral Centers: The R/S Nomenclature

Any chiral center can have two possible configurations (just as a hand can have two configurations, either right or left), and these configurations are designated either R or S by convention (the letters *R* and *S* come from the Latin words for right and left, *rectus* and *sinister*). If a molecule has a chiral center that is designated R, the chiral center will be S in the molecule's enantiomer.

You need to be able to assign whether a chiral center is R or S. To do so, you need to follow three steps:

1. **Number each of the substituents on the chiral center carbon using the Cahn–Ingold–Prelog system.**

   This numbering scheme is conveniently the same one used in determining E/Z stereochemistry on double bonds, as described (in more detail) in Chapter 11. According to the Cahn–Ingold–Prelog prioritizing scheme, the highest priority goes to the substituent whose first atom has the highest atomic number. (For example, Br would be higher priority than Cl, because Br has a larger atomic number.) If the first atoms on two substituents are the same, you keep going down the chain until you reach a larger atom and the tie is broken.

2. **After you've assigned priorities to each of the substituents, rotate the molecule so that the number-four priority substituent is oriented in the back.**

3. **Draw a curve from the first-priority substituent through the second-priority substituent and then through the third.**

If the curve goes clockwise, the chiral center is designated R; if the curve goes counterclockwise, the chiral center is designated S.

Problem Solving: Determining R/S Configuration

To get the hang of applying R/S nomenclature to chiral centers, you need lots of practice. Try to determine the R/S stereochemistry of the chiral center in the molecule shown in Figure 6-8.

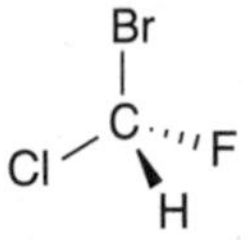

Figure 6-8:
A chiral
molecule.

$$Cl{-}\underset{H}{\overset{Br}{C}}{\cdots}F$$

Step 1: Prioritizing the substituents

The first step is to prioritize all the substituents from one to four. Bromine is the atom with the largest atomic number, so this substituent is given the highest priority; hydrogen has the smallest atomic number, so it's given the lowest priority. Chlorine gets the number-two priority because it's bigger than fluorine, which is given priority three. The priorities are shown in Figure 6-9.

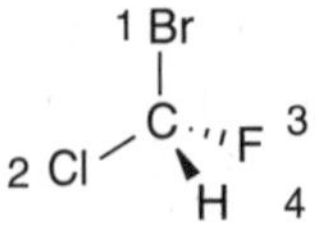

Figure 6-9:
Prioritizing
the
substituents
in a chiral
center.

$$2\ Cl{-}\underset{H\ \ 4}{\overset{1\ Br}{C}}{\cdots}F\ 3$$

Hydrogen will always be given the lowest (fourth) priority.

Step 2: Putting the number-four substituent in the back

The next step is to rotate the molecule so that the number-four substituent is pointed toward the back, as shown in Figure 6-10. For many people, this is the most difficult step because to rotate the molecule requires visualizing the molecule in three dimensions.

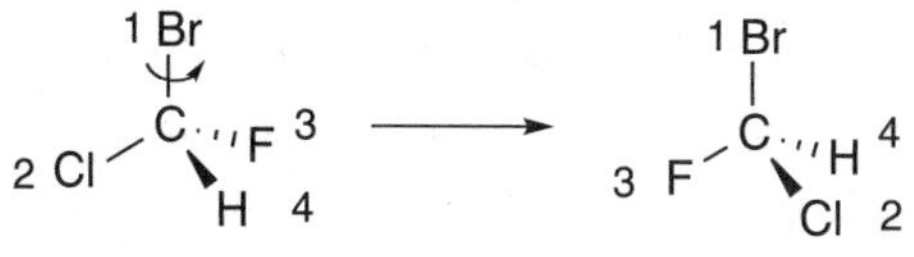

Figure 6-10:
Rotating the molecule to put the fourth priority in the back.

If you're not good at visualizing in three dimensions, you can use some tricks to put the number-four priority group in the back without having to mentally rotate the molecule in three-dimensional space.

First, swapping any two substituents changes the configuration — that is, if the chiral center was R before the swap, it becomes S after the swap (and vice versa). So, if you swap the number-four substituent with the substituent located in the back, as shown in Figure 6-11, the configuration of the chiral center switches. After you've made the switch to put the number-four priority substituent in the back, you could then go to Step 3 and determine the configuration, remembering that the actual configuration of that center is the opposite of the one determined.

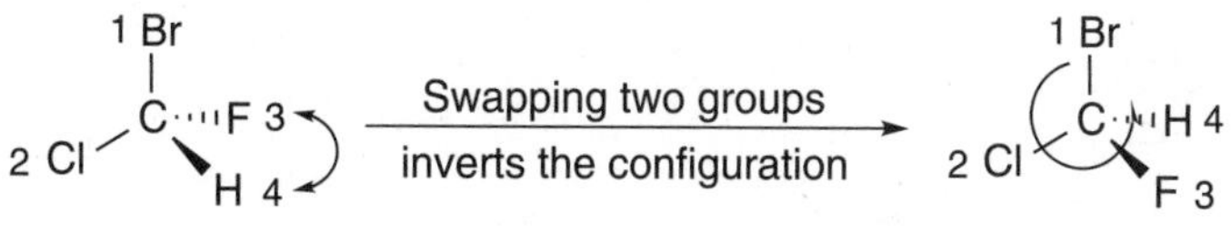

Figure 6-11:
Swapping two groups in a chiral center.

Swapping two groups inverts the configuration

Or you could keep going, swapping two more positions. Swapping the positions of the first two substituents inverts the configuration, but swapping the remaining two substituents after you perform this operation reverts the configuration back to the one you started with, as shown in Figure 6-12. So, if you first swap the number-four substituent with the group in the back, and then swap the remaining two substituents, you get two inversions of the configuration, which amounts to a net retention of the configuration. (For example, if the chiral center starts R, and you invert two substituents, it becomes S; if you invert two more substituents, the configuration goes back to R.) Doing this double swap is an easy way of getting the number-four priority substituent into the back without doing mental rotations of the atoms.

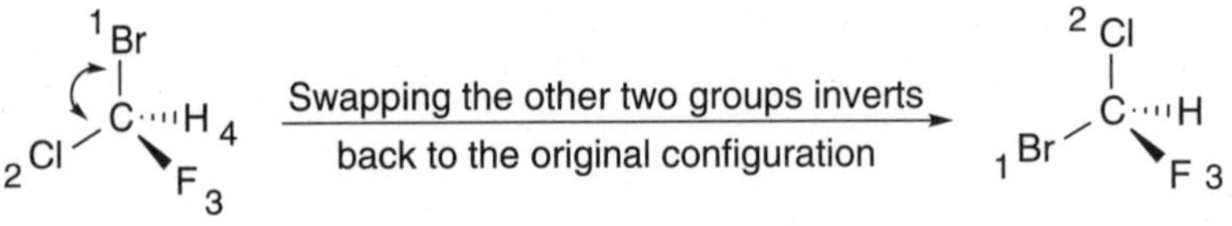

Figure 6-12:
Swapping two more groups in a chiral center.

Swapping the other two groups inverts back to the original configuration

REMEMBER

Swapping the positions of two substituents inverts the configuration.

Step 3: Drawing the curve

Now that the number-four priority substituent is in the back, you draw a curve from the first-priority substituent through the second-priority substituent, to the third-priority substituent. In this case, the curve goes clockwise, so the molecule has R stereochemistry, as shown in Figure 6-13.

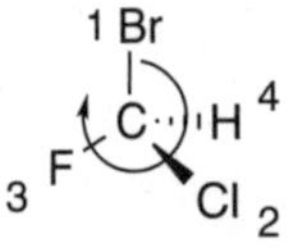

Figure 6-13:
The R stereo-chemistry.

TIP

If the fourth-priority substituent is in front, you can just reverse the rules of the last step and cut out Step 3. With the number-four substituent in front, a clockwise curve from the first- to the third-priority substituents results in an S configuration; a counterclockwise curve would indicate an R configuration.

All thumbs: An alternative technique for assigning R and S stereochemistry

A very convenient way to determine the R/S configuration of a chiral center is with the thumb technique. With this technique, shown in the accompanying figure, you point your thumb in the direction of the number-four priority substituent. If, with your thumb pointed toward the fourth-priority substituent, the fingers in your right hand can curl in the direction from the first- to second- to third-priority substituent in that order, the configuration is R; if you must use your left hand to get your fingers to curl from the first- to second- to third-priority substituent, then the configuration is S. The main advantage of this technique is that you don't have to put the number-four priority substituent into the back; you just point your thumb in whatever direction the number-four priority substituent is facing.

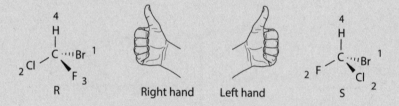

Right hand   Left hand

The Consequences of Symmetry: Meso Compounds

Any molecule that contains a chiral center will be chiral, with one exception: Those molecules that contain a plane of symmetry will be achiral, regardless of whether the molecule has a chiral center. A plane of symmetry is a plane that cuts a molecule in half, yielding two halves that are mirror reflections of each other.

By definition, a molecule that's not superimposable on its mirror image is a chiral molecule. Compounds that contain chiral centers are generally chiral, whereas molecules that have planes of symmetry are achiral and have structures that are identical to their mirror images. Molecules that contain chiral centers but are achiral as a result of having a plane of symmetry are called *meso compounds,* using organic-speak. For example, cis-dibromo cyclopentane (shown in Figure 6-14) is meso because a plane cuts the molecule into two halves that are reflections of each other. Trans-dibromo cyclopentane, however, is chiral because no plane splits the molecule into two mirror-image halves.

Figure 6-14:
The plane
of symmetry
in meso
compounds.

Plane of symmetry
meso

No plane of symmetry
chiral

Look at the mirror images of these two molecules (shown in Figure 6-15) to
prove this generality to yourself. Even though the cis compound has two
chiral centers (indicated with asterisks), the molecule is achiral because the
mirror image is identical to the original molecule (and is, therefore, superimposable on the original molecule). Molecules with planes of symmetry will
always have superimposable mirror images and will be achiral. On the other
hand, the trans stereoisomer has no plane of symmetry and is chiral.

Mirror   Mirror

Figure 6-15:
The mirror
images
of achiral
(meso)
and chiral
molecules.

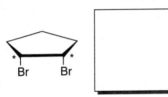 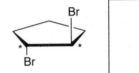

Superimposable   Non-superimposable

Meso compounds are achiral.

In organic chemistry, you need to be able to spot planes of symmetry in molecules so you can determine whether a molecule with chiral centers will be
chiral or meso. For example, can you spot the planes of symmetry in each of
the meso compounds shown in Figure 6-16?

Figure 6-16:
Some meso
compounds.

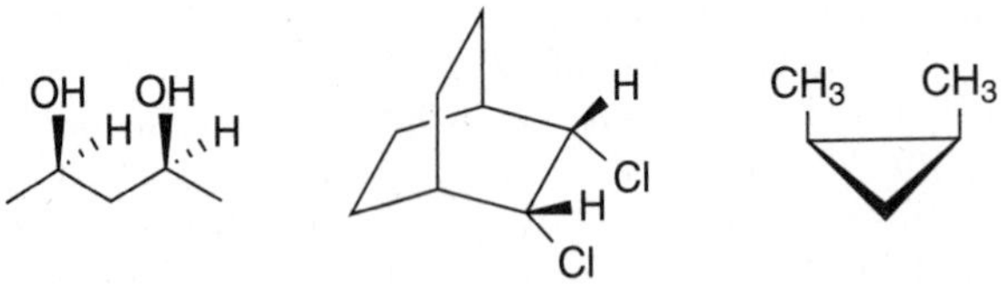

Rotating Plane-Polarized Light

Enantiomers share nearly all physical properties with each other — they have the same boiling points, melting points, dipole moments, and all other properties, save one. Enantiomers rotate plane-polarized light in opposite directions. Light waves propagate in all directions, and plane-polarized light is simply light that has been filtered to allow only light propagating in one direction to pass through the filter (this is the way that polarized sunglasses work to block out light). When plane-polarized light is passed through a sample containing a chiral molecule, the light is rotated either to the right or to the left by a certain number of degrees (labeled α), as shown in Figure 6-17.

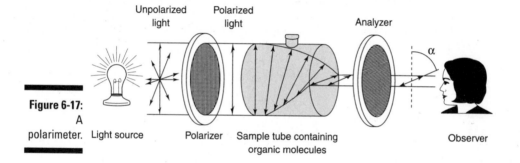

Figure 6-17:
A
polarimeter.

Unpolarized light · Polarized light · Analyzer · α

Light source · Polarizer · Sample tube containing organic molecules · Observer

Only chiral molecules rotate plane-polarized light. The number of degrees that a chiral molecule rotates the light must be determined by experiment and is not something that you can figure out on your own. The direction of light rotation by a molecule is not something you can determine from looking at the structure of a molecule either (R or S designation does not tell you which way the light will be rotated). However, if you know that a chiral molecule rotates light by 30 degrees, then you know its enantiomer will rotate plane-polarized light by exactly –30 degrees (that is, 30 degrees in the opposite direction).

Enantiomers rotate plane-polarized light in equal and opposite directions.

If you mix two enantiomers in equal proportions, you get a *racemic mixture* — a 50/50 mix of enantiomers. Racemic mixtures do not rotate plane-polarized light. This is because half of the molecules in the mix rotate the light one direction, and the other half of the molecules rotate the light the other direction, resulting in a net rotation of zero. Such racemic mixtures are, therefore, said to be *optically inactive,* which means that they don't rotate plane-polarized light.

Multiple Chiral Centers: Diastereomers

When only one chiral center exists in a molecule, the only possible stereo-isomer is the enantiomer, the molecule's mirror image. But when more than one chiral center is present in a molecule, you have the possibility of having stereoisomers that are not mirror images of each other. Such stereoisomers that are not mirror images are called *diastereomers*.

Typically, you can only have diastereomers when the molecule has two or more chiral centers.

The number of possible stereoisomers that a molecule can have is a function of 2^n, where n is the number of chiral centers in the molecule. Therefore, a molecule with five chiral centers can have 2^5 or 32 possible stereoisomers! As the number of chiral centers increases, the number of possible stereoisomers for that compound increases rapidly.

For example, the molecule shown in Figure 6-18 has two chiral centers.

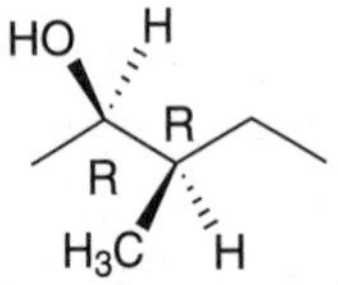

Figure 6-18:
A molecule
with two
chiral
centers.

Because this molecule has two chiral centers, this molecule can have a total of 2^2, or 4, possible stereoisomers, of which only one will be the enantiomer of the original molecule. Because both chiral centers in this molecule are of R configuration, the enantiomer of this molecule would have the S configuration for both chiral centers. All the stereoisomers of this molecule are shown in Figure 6-19. Those molecules that are not enantiomers of each other are diastereomers of each other.

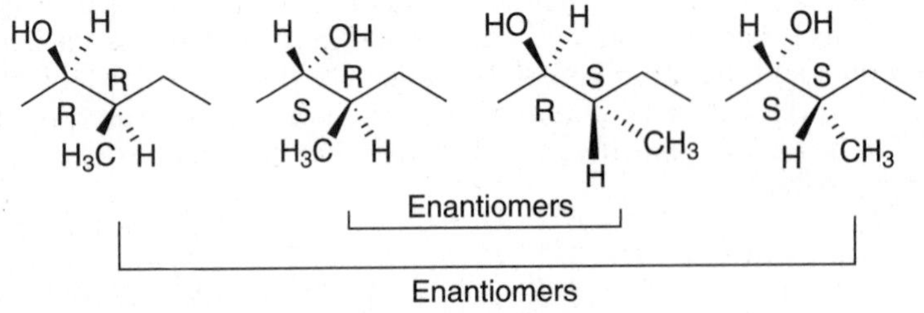

Figure 6-19:
The four
stereoiso-
mers of a
molecule
with two
chiral
centers.

Enantiomers

Enantiomers

Representing 3-D Structures on Paper: Fischer Projections

The most convenient way of viewing molecules with more than one chiral center is with Fischer projections. A Fischer projection is a convenient two-dimensional drawing that represents a three-dimensional molecule. To make a Fischer projection, you view a chiral center so that two substituents are coming out of the plane at you, and two substituents are going back into the plane, as Figure 6-20 shows. Then the chiral center becomes a cross on the Fischer projection. Every cross on a Fischer projection is a chiral center.

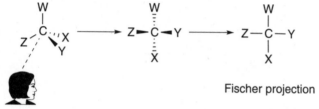

Figure 6-20: Creating a Fischer projection.

Fischer projection

Rules for using Fischer projections

Fischer projections are convenient for comparing the stereochemistries of molecules that have many chiral centers. But these projections have their own sets of rules and conventions for how you can rotate and move them. The two main ways to rotate a Fischer projection are as follows:

- ✔ You can rotate a Fischer projection 180 degrees and retain the stereochemical configuration, but you cannot rotate a Fischer projection 90 degrees (see Figure 6-21).

- ✔ You can rotate any three substituents on a Fischer projection while holding one substituent fixed and retain the stereochemical configuration.

Determining R/S configuration from a Fischer projection

Here's how to determine the configuration of a chiral center drawn in a Fischer projection. First, you prioritize each of the substituents using the Cahn–Ingold–Prelog prioritizing scheme. (This is the same prioritizing scheme used in a previous section of this chapter and discussed in detail in Chapter 9.) Then, you put the fourth priority substituent on the top, and draw a curve from the first- to the second- to the third-priority substituent.

If the curve goes clockwise, the configuration is R, if the curve goes counter-clockwise, the configuration is S. To get the number-four priority substituent at the top of the Fischer projection, you have to use one of the two allowed moves diagramed in Figure 6-21. (You can make a 180-degree rotation, or you can hold one substituent fixed and rotate the other three.) Two examples of the determination of the configuration from Fischer projections are shown in Figure 6-22.

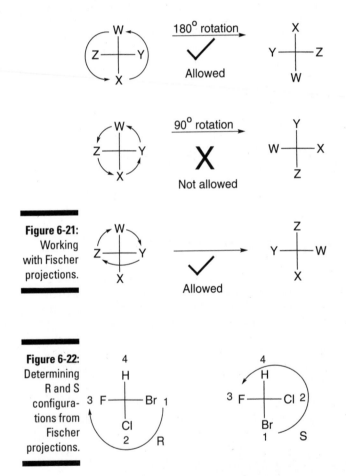

Figure 6-21:
Working with Fischer projections.

Figure 6-22:
Determining R and S configurations from Fischer projections.

Seeing stereoisomerism with Fischer projections

One of the main advantages of Fischer projections is that you can easily see the relationships between stereoisomers. In other words, with a Fischer projection, you can quickly determine whether two stereoisomers are diastereomers or enantiomers. For example, the relationships between some stereoisomers of a molecule are shown in Figure 6-23. When you compare stereoisomers, you want to make sure that the tops and bottoms are lined up correctly. In other words, when comparing two Fischer projections, make sure that both projections have the same top and the same bottom, and that one isn't upside-down (if one is upside-down, do an allowed 180-degree rotation). After the tops and bottoms are oriented correctly, you compare the directions of the substituents between the two molecules at every cross on the Fischer projection. A substituent that points to the left in a Fischer projection will point to the right in its enantiomer.

Figure 6-23:
Seeing the relationships between stereoisomers in Fischer projections.

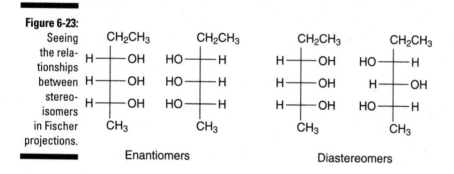

diastereomers or enantiomers.

Enantiomers   Diastereomers

Spotting meso compounds with Fischer projections

Another convenient feature of Fischer projections is that you can often spot meso compounds more easily with Fischer projections than you can with molecules drawn out with dashes and wedges. With the Fischer projection, the symmetry of the molecule is easier to spot. An example of symmetry in a Fischer projection is shown in Figure 6-24.

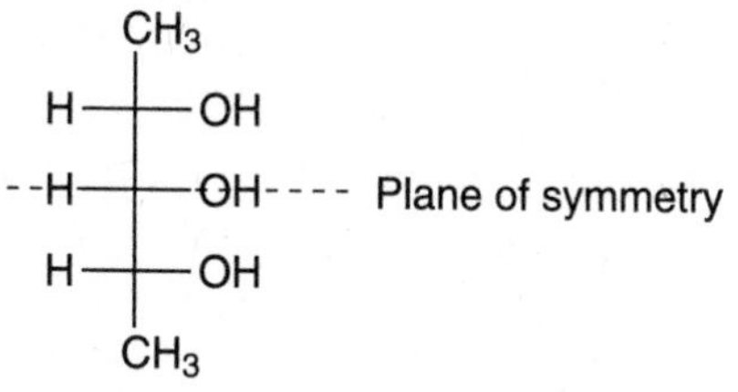

Figure 6-24:
The Fischer projection of a meso compound.

Keeping the Jargon Straight

One of the most confusing aspects of stereochemistry is the massive amount of new terminology that you need to remember. So here's a mini glossary of stereochemical terminology to keep you straight:

- **Achiral molecule:** An achiral molecule is one that can be superimposed over its mirror image; an achiral molecule is identical to its mirror image. Achiral molecules do not rotate plane-polarized light.

- **Chiral center:** Typically, for a molecule to be chiral, it must have at least one chiral center — a carbon with four nonequivalent groups. These chiral centers are assigned either R or S configuration. Any molecule that contains a chiral center will be chiral, unless the molecule has a plane of symmetry, in which case the molecule is achiral (and meso).

- **Chiral molecule:** A chiral molecule is one that cannot be superimposed over its mirror image. Chiral molecules rotate plane-polarized light.

- **Diastereomer:** Diastereomers are stereoisomers that are not mirror images of each other. For a molecule to have a diastereomer, it must typically have more than one chiral center.

- **Enantiomer:** Molecules that are mirror images of each other are called enantiomers.

- **Meso compound:** A meso compound is a molecule that contains a chiral center but is achiral as a result of having a plane of symmetry in the molecule.

- **Optically active compound:** An optically active compound (such as a pure solution of a single enantiomer of a molecule) has the ability to rotate plane-polarized light. The direction of light rotation is experimentally determined and is not predictable simply by looking at a structure.

- **Racemic mixtures:** A racemic mixture is a 50/50 mixture of two enantiomers. Racemic mixtures do not rotate plane-polarized light.

The biological importance of stereochemistry: The thalidomide tragedy

Thalidomide, a chiral compound whose structure is shown in the accompanying figure, was a drug prescribed to pregnant women in Europe to counter the symptoms of morning sickness. Even though only one of the enantiomers of thalidomide actively worked against preventing morning sickness, the drug was prescribed as the racemic mixture (a 50/50 mix of both enantiomers).

Why was the drug dispensed as the racemic mixture rather than as the single active enantiomer? For practical reasons. Usually, selectively synthesizing one enantiomer over another is very difficult, and with most reactions that produce compounds with chiral centers, you get products that are racemic mixtures. Racemic mixtures are fiendishly difficult to separate because enantiomers share important physical properties. They have the same boiling points, so they can't be separated by distillation; they have the same dipole moments and solubilities so they can't be separated by recrystallization. Separating enantiomers is possible, but the task is often tedious and costly.

Unfortunately, while one enantiomer of thalidomide controlled the symptoms of morning sickness, the other enantiomer caused terrible side effects, including severe defects (like stunted limb growth) in the unborn children. The drug was pulled from the market when scientists discovered that these birth defects were caused by the drug.

Part II

Hydrocarbons

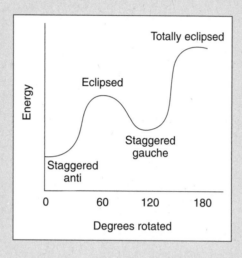

In this part . . .

✔ See the structures of hydrocarbons.

✔ Name simple hydrocarbons.

✔ Work with organic reactions using arrow pushing.

✔ Synthesize alkenes and alkynes.

Chapter 7

What's in a Name? Alkane Nomenclature

In This Chapter

▶ Seeing how straight-chain alkanes are named

▶ Naming a branched alkane

*A*lkanes are compounds that contain only carbon-carbon single bonds. Because they have the maximum possible number of hydrogens, alkanes are said to be *saturated hydrocarbons*. Alkanes have the molecular formula of C_nH_{2n+2}, where *n* is the number of carbons in the molecule. An alkane that has eight carbons, for example, would have a molecular formula of C_8H_{18}.

Naming molecules is an important skill to have. Unlike biology, where organisms are given seemingly arbitrary names, organic molecules are named systematically. When you understand the nomenclature rules and how to apply them, you can name most molecules you come across. And thank goodness for that, as there are *millions* of known organic compounds. Can you imagine the amount of memorization that would be required if each one were given some arbitrary name? I don't even want to *think* about it.

Furthermore, every unique molecule in chemistry *must* have a different name. George Foreman may be able to name each of his five sons George, but in chemistry, if two molecules are different, then they must have different names to distinguish them.

All in a Line: Straight-Chain Alkanes

The names of alkanes all end with the suffix *–ane*. The prefixes for alkanes with one, two, three, and four carbons come from historical roots, while alkanes with more than five carbons have prefixes that derive from Greek (see Table 7-1). These prefixes simply must be memorized. The mnemonic, "**M**ary **e**ats **p**eanut **b**utter" may help you remember the prefixes for the first four alkanes that have historical roots — **m**ethane, **e**thane, **p**ropane, and **b**utane. After that, knowing the Greek numbers (for example, knowing that *hex* refers to six, and *oct* refers to eight) makes the names more familiar.

Table 7-1	The Names of the Straight-Chain Alkanes	
Number of Carbons	*Name*	*Structure*
1	Methane	CH_4
2	Ethane	CH_3CH_3
3	Propane	$CH_3CH_2CH_3$
4	Butane	$CH_3(CH_2)_2CH_3$
5	Pentane	$CH_3(CH_2)_3CH_3$
6	Hexane	$CH_3(CH_2)_4CH_3$
7	Heptane	$CH_3(CH_2)_5CH_3$
8	Octane	$CH_3(CH_2)_6CH_3$
9	Nonane	$CH_3(CH_2)_7CH_3$
10	Decane	$CH_3(CH_2)_8CH_3$
11	Undecane	$CH_3(CH_2)_9CH_3$
12	Dodecane	$CH_3(CH_2)_{10}CH_3$

Reaching Out: Branching Alkanes

Oh, if life were only that simple! All the alkanes in Table 7-1 have carbon atoms that are joined in a line, so they're called *straight-chain alkanes*. Alkanes, though, are not limited to staying in a line — they can have structures that branch. For example, the alkane with the formula C_4H_{10} can have two possible structures (see Figure 7-1) — one alkane that's a straight chain (butane), and one alkane that's branched (isobutane).

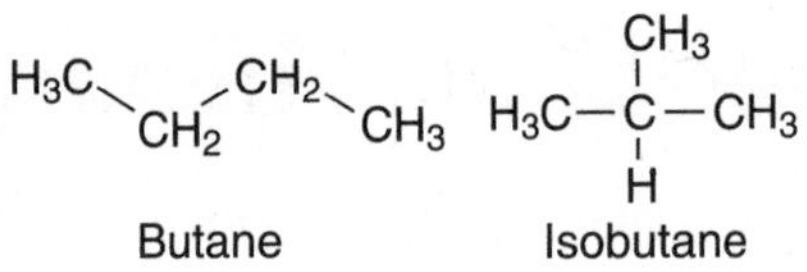

Figure 7-1:
The isomers of C_4H_{10}.

Butane   Isobutane

Molecules that have the same molecular formula but different chemical structures are referred to as *isomers,* using organic-speak. Branching makes the nomenclature of these molecules a bit more difficult, but these branching molecules can still be named systematically. With a bit of practice, you'll soon get the hang of naming all alkanes — including branched ones — and nomenclature questions on exams will become gimmes.

Several steps need to be followed in order to name a branched alkane. In the following sections, I detail how to perform each step, using an example to clarify what I mean as I go along.

Finding the longest chain

The first, and potentially the trickiest, step is to find the longest chain of carbon atoms in the molecule. This task can be tricky because, as a reader of English, you're so used to reading from left to right. Often, however, the longest chain of carbons is not the chain that follows simply from left to right, but one that snakes around the molecule in different directions. Organic professors like to make the parent chain one that curves around the molecule and doesn't necessarily flow from right to left, so you have to keep on your toes to make sure that you've spotted the longest carbon chain.

The longest carbon chain for the molecule shown in Figure 7-2 is seven carbons long, so the parent name for this alkane is heptane (refer back to Table 7-1).

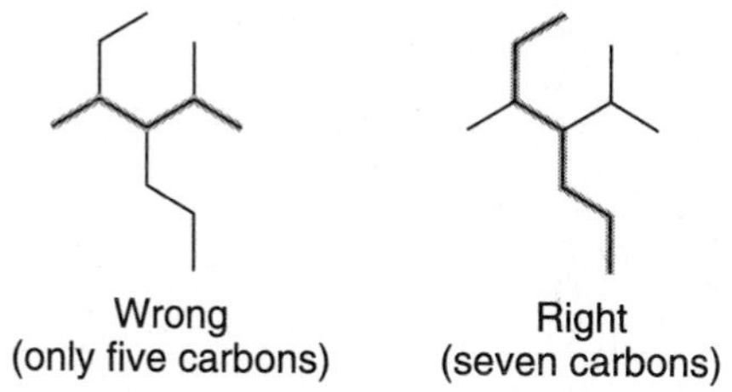

Figure 7-2:
The right and wrong ways to count the parent chain.

Wrong
(only five carbons)

Right
(seven carbons)

Numbering the chain

Number the parent chain starting with the end that reaches a substituent first. A chain can always be numbered in two ways. In the case of the molecule in Figure 7-2, the numbering could start at the top and go down, or it could start at the bottom and go up. The correct way to number the parent chain is to start with the end that reaches the first substituent sooner. A *substituent* is organic-speak for a fragment that comes off of the parent chain. In this example, if you number from the top down, the first substituent comes at carbon number three (see Figure 7-3); if you number from the bottom up, the first substituent comes at carbon number four. The correct numbering in this case, then, starts at the top and goes down.

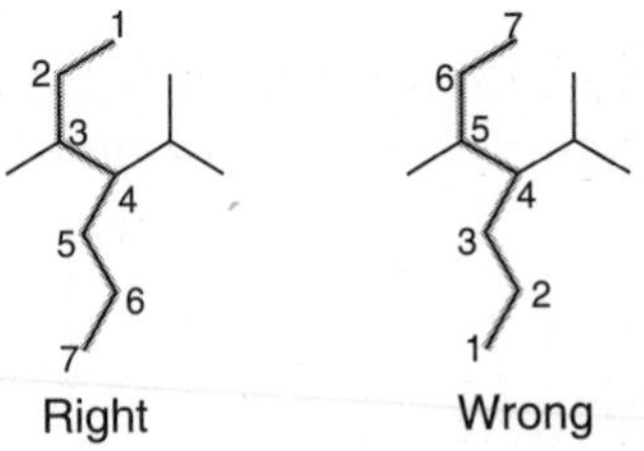

Figure 7-3:
The right and wrong ways to number the carbons of the parent chain.

Right   Wrong

Seeing the substituents

After numbering the parent chain, the next step is to determine the names of all the substituents that stick off of the parent chain. On this molecule, two substituents come off the parent chain — one substituent at carbon number three, and one substituent at carbon number four (see Figure 7-4). Substituents are named in the same manner as the parent chains, except that instead of ending with the suffix –*ane* they end with the suffix –*yl*, which indicates that the group is a substituent off the main chain. For example, the one-carbon substituent at carbon number three is a *methyl* substituent (not a *methane* substituent). A two-carbon substituent would be *ethyl*, a three-carbon substituent would be *propyl*, a four-carbon substituent would be *butyl*, and so on (refer to Table 7-1).

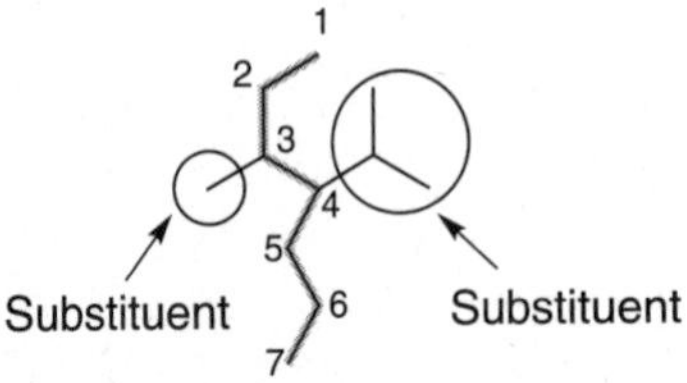

Figure 7-4:
The locations of substituents on a parent chain.

Substituent   Substituent

Some complex substituents have common names rather than systematic ones. These simply must be memorized. The most important common substituents are the isopropyl group (a three-carbon group that looks like a snake's tongue), the *tert*-butyl (or *t*-butyl) group and the *sec*-butyl group, shown in Figure 7-5.

Figure 7-5:
The common names of some substituents.

Isopropyl

Tert-butyl or t-butyl

Sec-butyl

In the example shown in Figure 7-4, the substituent at carbon number four is, in fact, an isopropyl group. Now you have the name of the parent chain and the names of all the substituents. The one-carbon substituent at carbon three is a methyl group, and the three-carbon substituent at carbon four is an isopropyl group. Now you just have to put it all together!

Ordering the substituents

The next step is to order the substituents alphabetically in front of the parent name, using numbers to indicate the location of the substituents. Because *i* comes before *m* in the alphabet, the isopropyl group is placed in front of the methyl group in the name of the molecule: 4-isopropyl-3-methylheptane. Note that dashes are used to separate the numbers from the substituents, and that there is no space between the last substituent and the name of the parent chain.

Of course, there's always a stick to throw into the spokes. One quirk involving the common names of *tert*-butyl and *sec*-butyl substituents comes when placing them in alphabetical order, as the *tert* and *sec* portions of the name are ignored. In other words, *tert*-butyl would be ordered as if it started with the letter *b*, the same as with *sec*-butyl. Isopropyl, however, is alphabetized normally, under the letter *i*. (There always has to be weird exceptions like this, doesn't there?)

More than one of a kind

What if the molecule contains more than one of the same substituent? For example, what if the compound has two methyl-group substituents on the molecule? In such a scenario, you don't name each of the methyl groups individually. Instead, you put a prefix in front of a single substituent name to indicate the number of these substituents that the molecule contains. (See Table 7-2 for a list of prefixes.)

Table 7-2	Prefixes for Identical Groups
Number of Identical Groups	*Prefix*
2	di
3	tri
4	tetra
5	penta
6	hexa
7	hepta
8	octa
9	nona
10	deca

For example, if the molecule has two separate methyl groups, the prefix *di–* goes in front of the name *methyl,* making the substituent name dimethyl. In addition, two numbers are placed in front of the substituent name to show the locations of the two methyl groups on the parent chain. Students often forget that for every substituent you must explicitly list a number, even if the substituents are attached to the same carbon. If the molecule has three methyl groups coming off the parent chain, the substituent name is trimethyl, if it has four methyls, the name is tetramethyl, if it has five methyls, the name is pentamethyl, and so on. These prefixes, like *tert* and *sec* in the previous examples, are ignored when placing the substituents in alphabetical order (dimethyl would be alphabetized under *m,* for example). For two examples of molecules that have identical substituents, see Figure 7-6. Note the use of commas to separate the numbers preceding the substituent name.

Figure 7-6:
Examples
of multiple
identical
substituents.

2,2-dimethylpropane   2,3,4,5,6-pentamethylheptane

Naming complex substituents

Sometimes, substituents will branch and will have no common name, like the substituent shown in Figure 7-7. In such a case, you name the substituent just as if it were a separate alkane, but with the parent name of the substituent ending with the suffix *–yl* instead of *–ane,* indicating that this portion of the molecule is a substituent off the main chain.

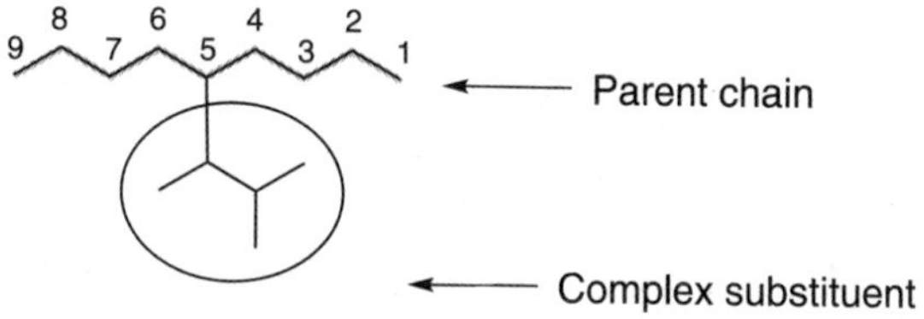

Figure 7-7:
An alkane
with a
complex
substituent.

←——— Parent chain

←——— Complex substituent

The main catch in naming complex substituents, however, is that the carbon that attaches the substituent to the parent chain must be the number-one carbon; thus, when you number the carbons in the substituent chain, you force the carbon that attaches to the main chain to be carbon number one. (See Figure 7-8 for an example of how this works.)

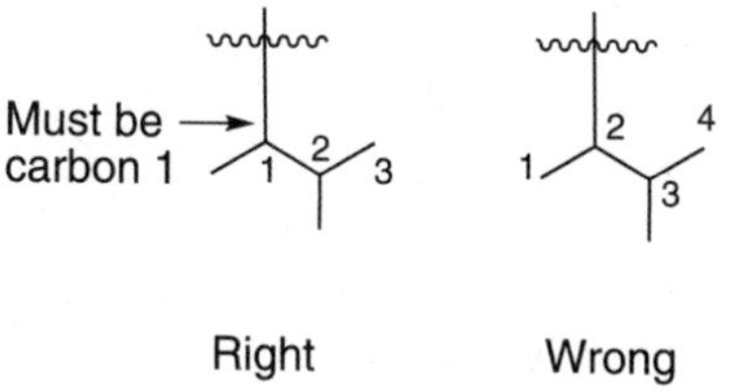

Figure 7-8:
The right
and wrong
ways to
number a
complex
substituent.

Must be →
carbon 1

Right   Wrong

Because the complex substituent shown in Figure 7-8 is three carbons long and has methyl groups at carbons one and three, the complex substituent is named 1,2-dimethylpropyl.

Because you're naming a substituent off the parent chain, the name must end with the suffix –*yl*, not –*ane*.

Now you can put all the parts of the name together. By convention, when a complex substituent is included in a name, the name of the complex substituent is set off with parentheses. So the name of the molecule in Figure 7-7 is 5-(1,2-dimethylpropyl)-nonane.

Getting lots of practice with naming alkanes is essential in order to master nomenclature of organic molecules.

Chapter 8

Drawing Alkanes

In This Chapter

- Seeing the structures of alkanes
- Analyzing the conformations of alkanes
- Seeing alkane reactions

In this chapter, I cover the skeletons of organic molecules — the alkanes. Alkanes are the simplest organic molecules. Because they're the simplest of the organic molecules, they're useful for introducing several key principles in organic chemistry, such as *isomerism* (molecules with the same molecular formula but different structures), *conformation* (the different ways atoms can flex and bend), and *stereochemistry* (the orientation of atoms in three-dimensional space). Also, because alkanes make up the simplest organic molecules, they're the best place to start honing your skills in *nomenclature* (the rules for naming organic molecules).

Converting a Name to a Structure

It's just as important to be able to determine the structure from a name as it is to be able to determine the name of a structure. Although English reads from left to right, the best way to read a structure's name is *from right to left*. For example, if you want to draw the structure for 4-*t*-butyl-2,3,5-trimethylheptane, you would start by first drawing the parent chain — heptane. Heptane is seven carbons long, so you draw a seven-carbon chain (see Figure 8-1).

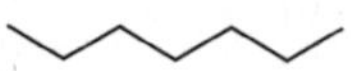

Figure 8-1: The parent chain.

A word about drawing and interpreting structures

It's easy to become confused and misled by structures written on paper, because drawn structures are simply an organic chemist's way of conveniently representing a three-dimensional molecule on two-dimensional paper. For example, look at the following two structures and ask yourself if they refer to the same molecule or if they refer to different molecules.

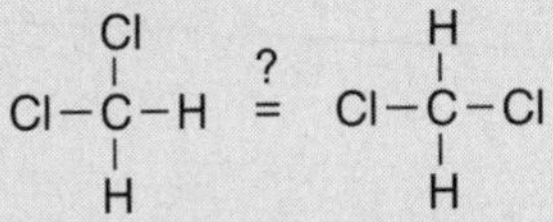

$$
\begin{array}{ccc}
\text{Cl} & & \text{H} \\
| & ? & | \\
\text{Cl}-\text{C}-\text{H} & = & \text{Cl}-\text{C}-\text{Cl} \\
| & & | \\
\text{H} & & \text{H}
\end{array}
$$

Many people would say that they're different because they look different on paper. But both drawings are of the same molecule. The confusion results from the fact that a three-dimensional molecule is represented with a two-dimensional drawing. It looks on paper as if these molecules are flat and that each bond juts from the carbon at a 90-degree angle; in fact, the bonds are 109.5 degrees apart from each other, and the molecule has a tetrahedral geometry (refer to Chapter 2).

Likewise, it's easy to get confused by drawings that look slightly different when they really represent the same thing. Whether you draw

a straight chain with the first carbon up or the first carbon down is simply a matter of personal preference, because both represent the same molecule.

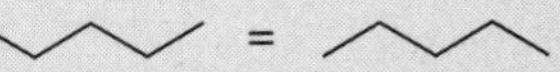

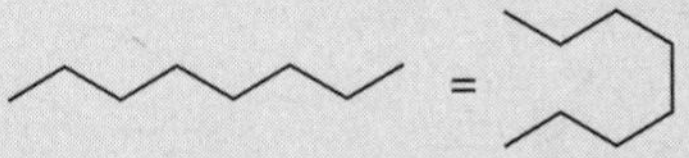

The main point I want to make is that you should be mindful of the assumptions made in these structural drawings. I also want to encourage you to use molecular models (those plastic balls and sticks that represent atoms and bonds). Especially at the beginning of the course, molecular models come in handy, allowing you to get the feel of how two-dimensional drawings relate to the actual three-dimensional structures. After you get the feel of how the drawn structures relate to the molecular models, you'll be much less likely to be confused by the different ways that structures are drawn.

Then you number the carbon chain as shown in Figure 8-2.

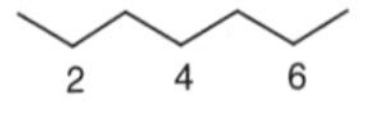

Figure 8-2: Numbering the chain.

Then you stick on the substituents. Substituents can be added in any order. A *t*-butyl group goes on the number-four carbon (see Figure 8-3).

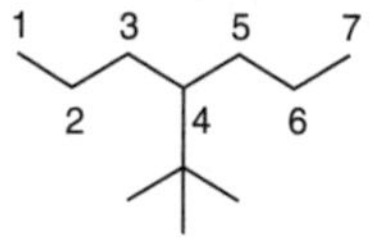

Figure 8-3:
Adding
a *t*-butyl
group.

Then, methyl groups need to be plugged onto the carbons numbered two, three, and five, as shown in Figure 8-4.

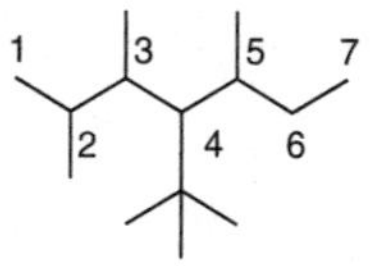

Figure 8-4:
Adding
methyls.

That's it! By reading the molecule from right to left, the structure is easily drawn from the name.

Conformation of Straight-Chain Alkanes

Carbon-carbon single bonds are capable of rotating. This allows alkanes to exist in different *conformations*. What are conformations? Think of a molecule in terms of your body — you can flex it and bend it. You snuggle up in bed to read a book, you sit in class to hear a lecture, and you bend to do calisthenics to get your blood flowing in the morning. All these activities put your body into different conformations. But these conformations are not of equal energy — lying down to read a book requires very little energy, sitting in class (and paying attention!) takes a bit more energy, and doing calisthenics requires a lot more energy. Now, which of these activities would you rather do? I'm guessing you'd rather be in the lower-energy conformation reading a book than in the higher-energy conformation doing calisthenics.

Similarly, alkanes can exist in different conformations — different spatial arrangements of atoms — and not all the conformations are of equal energy. And molecules — like this author — prefer to be in low-energy conformations rather than high-energy conformations.

In the following sections, I talk about the different conformations of alkanes, how you can draw alkanes in different conformations using Newman projections, and how you can predict the relative energies of conformations.

Newman! Conformational analysis and Newman projections

One of the best ways to look at different conformations of molecules (individual conformations are called *conformers*) is to use Newman projections. A *Newman projection* is a convenient way of sighting down a particular carbon-carbon bond. Figure 8-5 shows a perspective drawing of ethane and the Newman projection of the same molecule. (See Chapter 6 for more on perspective drawings.)

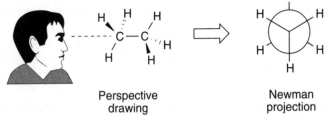

Figure 8-5:
Making a
Newman
projection.

Perspective
drawing

Newman
projection

The solid wedges in Lewis structures indicate a bond that's coming out of the paper, while dashed wedges indicate a bond that's going back into the paper.

In a Newman projection, the three lines in the shape of a Y represent the three bonds of the first carbon that you're sighting down; where the three lines connect is where the front carbon is. A circle (drawn "behind" the front carbon) represents the back carbon; the three lines coming out of the circle represent the three bonds that come off that carbon. (Note that the fourth bond for each of these carbons is the carbon-carbon bond that you're looking down.) A Newman projection can help you analyze the rotations around a particular carbon-carbon bond.

Using Newman projections, rotating around a specific bond to reach other conformers is a fairly straightforward task. The best way to reach other conformers is to rotate just one carbon at a time — either the front carbon or the back carbon. For consistency, in this book I always hold the front of the Newman projection fixed and rotate the back carbon.

Although an infinite number of conformations exist (just rotate one of the carbons by a fraction of a degree and — *voilà!* — you have a different conformation), the two most important ones are the eclipsed and staggered conformations (see Figure 8-6). An *eclipsed conformation* results when the bonds from the front carbon and the bonds from the back carbon align with each other, and the angle between the bonds (called the *dihedral angle*) is 0 degrees. On the Newman projection, the eclipsed bonds are drawn a little ways apart so that the substituents on the back carbon can be seen.

A *staggered conformation* results when the bonds from the front carbon and the bonds from the back carbon have a dihedral angle of 60 degrees. In the staggered conformation, the bonds coming off the front and back carbons are as far apart from each other as possible.

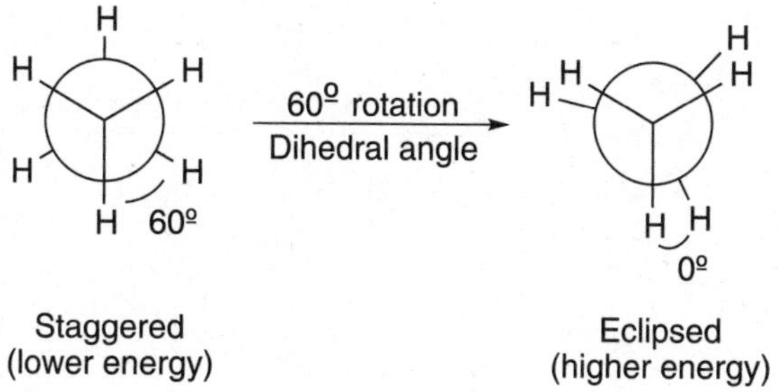

Figure 8-6:
The staggered and eclipsed conformations in Newman projections.

Staggered
(lower energy)

60° rotation
Dihedral angle

Eclipsed
(higher energy)

 When the bonds from the front carbon and back carbon are aligned in an eclipsed conformation, the electron repulsion between the bonds is higher than when the bonds are staggered and farther apart. This electron-electron repulsion between the bonds is called *torsional strain*. Because staggered conformations have less torsional strain than eclipsed conformations, staggered conformations are, as a general rule, more stable (that is, lower in energy) than eclipsed conformations.

Conformations of butane

The situation with butane is a bit more complicated than the situation with ethane. Figure 8-7 shows the Newman projection of butane, sighting down the bond between the second and third carbons (also known as the C2-C3 bond).

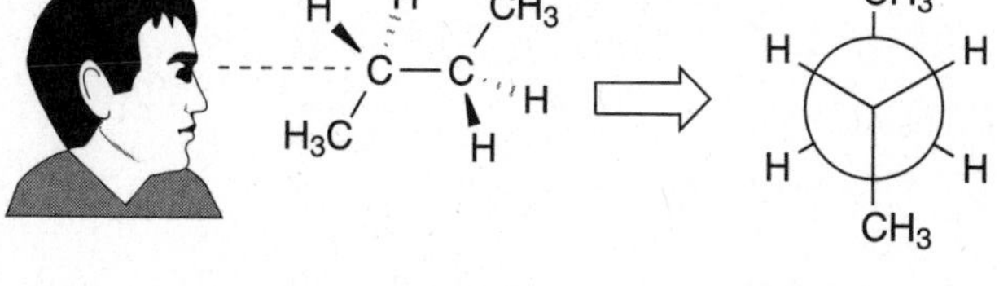

Figure 8-7:
Sighting down the C2-C3 bond of butane.

With butane, not all eclipsed conformations are of the same energy, and not all staggered conformations are of the same energy. In Figure 8-8, I show the rotation around the C2-C3 bond in butane. In the first staggered conformation, the two methyl groups are 180 degrees apart from each other. This conformer is called the *anti conformer*. In the other staggered conformer, the two methyl groups flank each other. The interaction between two adjacent large groups (like methyl groups) in a staggered conformation is called a *gauche interaction*. Gauche interactions act to increase the energy of a conformer. Because the staggered anti conformer has no gauche interactions, it's lower in energy than the staggered gauche conformer.

Figure 8-8:
The conformations of butane formed by the rotation around the C2-C3 bond.

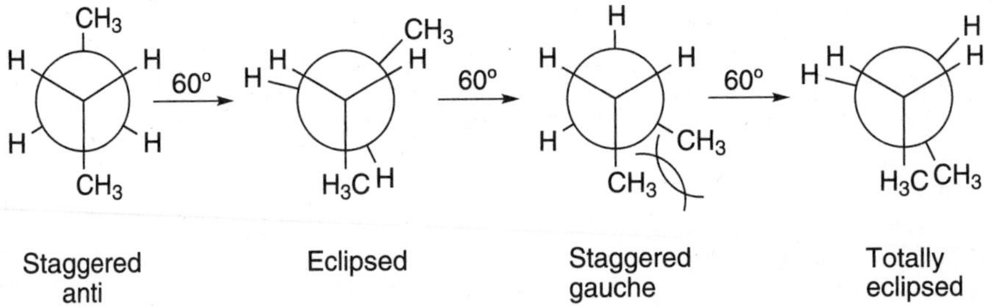

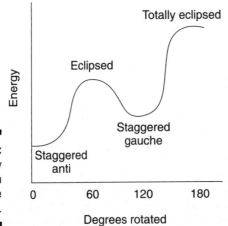

Staggered anti   Eclipsed   Staggered gauche   Totally eclipsed

Generally, the following principle holds true: Big groups (like methyl groups) want to be as far away from each other as possible (so they don't invade each other's space), and conformers with the big groups far apart will be lower in energy than conformers where the big groups are near each other.

Likewise, the two eclipsed conformations are not of equal energy. In the first eclipsed conformation, the methyl groups are 120 degrees apart; in the last eclipsed conformation, the methyl groups are 0 degrees apart. This last conformer is called *totally eclipsed* to indicate that the two big groups are eclipsed with each other. Because the two big groups are closer together in the totally eclipsed conformer, this conformer is higher in energy than the other eclipsed conformer. The energy diagram for converting the staggered anti into the totally eclipsed conformer is given in Figure 8-9, with the *x*-axis representing the rotation (in degrees) and the *y*-axis indicating the energy (in arbitrary units).

Figure 8-9:
The energy diagram for butane conformers.

Full Circle: Cycloalkanes

Alkanes also commonly form into rings, called *cycloalkanes*. Cycloalkanes are named by counting the number of carbons in the ring, giving the same suffix as you would to a straight-chain alkane, and then giving the molecule the prefix *cyclo–*. For example, a three-carbon ring is called cyclopropane. The smallest and most common rings are shown in Figure 8-10.

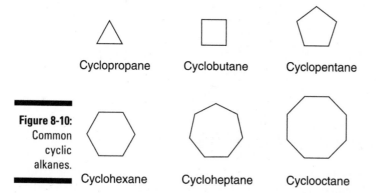

Cyclopropane   Cyclobutane   Cyclopentane

Figure 8-10:
Common
cyclic
alkanes.

Cyclohexane   Cycloheptane   Cyclooctane

The stereochemistry of cycloalkanes

One interesting feature of cycloalkanes is that each ring has two distinct faces. If two substituents are attached to a cycloalkane ring, they can both point off the same face, or they can point off opposite faces of the ring. Consider 1,2-dimethylcyclopentane, shown in Figure 8-11 in both the traditional perspective view and in a slanted side view — called a *Haworth projection* — that's convenient for looking at the stereochemistry of rings. The two methyl groups could either point off the same face (forming an isomer called a *cis stereoisomer*) or point off opposite faces (forming an isomer called a *trans stereoisomer*). *Stereoisomers* are molecules that have the same connectivity of atoms but differ in the arrangement of those atoms in space.

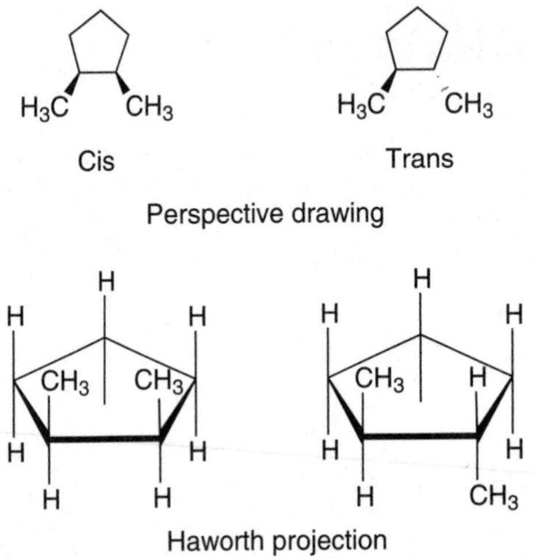

H3C   CH3

Cis

H3C   CH3

Trans

Perspective drawing

Figure 8-11:
The *cis* and *trans* stereo-isomers of 1,2-dimethylcy-clopentane.

H   H   H   H

CH3   CH3

H   H

H   H

H   H   H   H

CH3   H

H   H

H   CH3

Haworth projection

Conformations of cyclohexane

Although cyclohexane is typically drawn as if it were flat, in reality the structure is not flat at all. Most of the time, the structure exists in what is called the *chair conformation*. This conformation is called the chair because it looks (sort of) like a reclining lounge chair (see Figure 8-12). Other conformations for cyclohexane do exist — and they include the boat, half-chair, and twist-boat conformations — but the most important conformation is the chair because this conformation is typically the most stable conformation.

Figure 8-12:
The chair conforma-tion of cyclohexane.

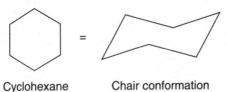

Cyclohexane   Chair conformation

Drawing chairs

Many students find that drawing chairs can be somewhat difficult at first, but mastering the correct drawing of this conformation is essential for your success in orgo. To begin, start by drawing two lines that are parallel to each other but not perfectly horizontal, as shown in Figure 8-13. Next, add a downward-pointing *V* tip to one end (this is the *tail* of the chair). Finally, add an upward-pointing *V* tip to the other end (this is the *nose* of the chair).

Figure 8-13:
The steps involved in drawing the chair con- formation of cyclohexane.

Adding hydrogens to a chair

A cyclohexane chair contains two kinds of hydrogens — axial hydrogens and equatorial hydrogens. *Axial hydrogens* are those hydrogens that stick straight up or straight down parallel to an imaginary axis through the chair; *equatorial hydrogens* are hydrogens that stick out along the equator of the chair (see Figure 8-14).

Figure 8-14:
The axial and equato- rial hydro- gens on cyclohexane.

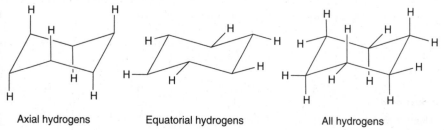

Axial hydrogens   Equatorial hydrogens   All hydrogens

In drawing the hydrogens on a chair cyclohexane, it's often easiest to draw the axial hydrogens first. At any point on the chair that sticks up, put the axial hydrogen sticking straight up; at any point on the chair that sticks down, draw the axial hydrogen straight down. After the axial hydrogens are drawn, adding in the equatorial hydrogens around the equator of the chair is a fairly straightforward task.

Flipping the chair

At room temperature, cyclohexane doesn't stay in one-chair conformation, but rapidly interconverts into an alternative chair conformation with a *ring flip* (shown in Figure 8-15). With the chair on the left of the figure, the nose of the cyclohexane chair goes down, and the tail of the cyclohexane goes up to make the new chair cyclohexane conformer. With a ring flip, all hydrogens that were originally axial become equatorial, and all hydrogens that were equatorial become axial. It's a good idea to get out your molecular models and try this ring flip yourself, so you can see it in three dimensions.

Figure 8-15:
The ring flip of cyclo-hexane.

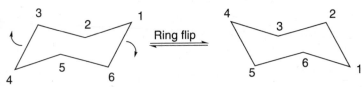

Ring flip

With *unsubstituted cyclohexane* (a cyclohexane that has only hydrogens attached to it), undergoing a ring flip doesn't change the molecule. With substituted cyclohexanes, however, the two chair conformers may not be identical. For example, isopropylcyclohexane is shown in Figure 8-16. One chair conformer puts the isopropyl group in the axial position. After undergoing a ring flip, the isopropyl becomes equatorial.

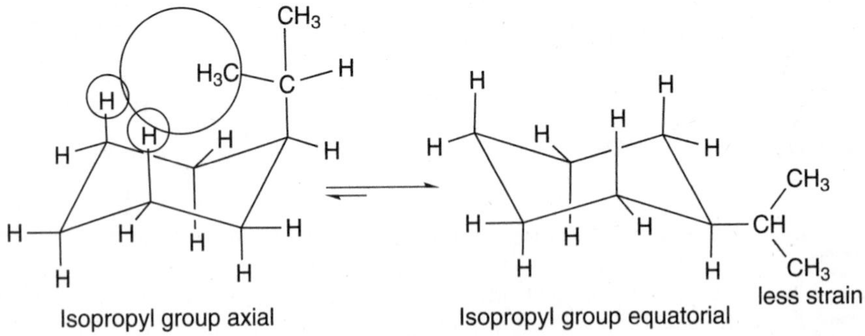

Figure 8-16:
The chair conformers of isopropyl-cyclo-hexane.

Isopropyl group axial

Isopropyl group equatorial

less strain

REMEMBER

Ring flips change all axial positions to equatorial and all equatorial positions to axial.

These two conformers are not identical, and they don't have the same energy. When a large group is axial, the large group invades the space of the hydrogens on carbons two positions away, introducing *1,3-diaxial strain.*

This interaction increases the energy of the axial conformer. Therefore, as a rule, large groups prefer to be equatorial because this conformation has no 1,3-diaxial strain — the right-pointing arrow is longer, indicating the direction of the equilibrium favors the more stable chair conformation (see Figure 8-16).

Don't confuse conformation with configuration. A *cis* configuration (a molecule with substituents pointing off the same face of the ring) or *trans* configuration (a molecule with substituents pointing off opposite faces of the ring) doesn't change by changing the conformation. For example, a ring flip of a cyclohexane will not change a *cis* configuration to a *trans* configuration, because after the flip the substituents will still be pointing off the same face (for a *cis* configuration) or off opposite faces (for a *trans* configuration). The only way to change the configuration is by chemical reactions that make and break bonds.

Problem Solving: Drawing the Most Stable Chair Conformation

The first step in drawing the most stable conformation of cyclohexane is to determine — based on whether the substituents are *cis* or *trans* to one another, and based on where they're located on the ring — what the choices of axial and equatorial positions are for the substituents. A handy way of determining the substitution alternatives is to use the Haworth projection in Figure 8-17. This projection is easy to make — simply start at some position on the ring and alternate axial (a) and equatorial (e) from carbon to carbon, and from top to bottom. This projection will tell you what the two options for the two-chair conformations are.

Figure 8-17:
The
Haworth
projection.

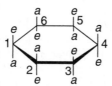

For example, if you were asked to draw the most stable conformation of *cis*-1,3-dimethylcyclohexane, both substituents could be on the top of the ring or both could be on the bottom of the ring, as shown in Figure 8-18.

(Recall that *cis* means that both substituents are on the same side of the ring.) As the figure shows, to get the *cis* stereochemistry, either both of the substituents could be equatorial (both e) or both could be axial (both a).

Figure 8-18:
The possible positions of *cis* substituents in positions 1 and 3 on cyclohexane.

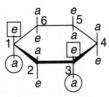

Putting large groups in the equatorial position to eliminate 1,3-diaxial strain is energetically favorable.

Because large groups prefer to be equatorial, the most stable conformer for *cis*-1,3-dimethylcyclohexane is the diequatorial conformer, shown in Figure 8-19. The diaxial conformer would be higher in energy.

Figure 8-19:
The diequatorial conformation of *cis*-1,3-dimethyl-cyclohexane.

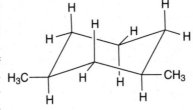

If cyclohexane has two substituents and one has to be placed axial and one equatorial (as is the case in trans-1,2-disubstituted cyclohexanes), the lowest-energy conformation will be the one in which the bigger group goes in the equatorial position and the smaller group goes in the axial position.

Reacting Alkanes: Free-Radical Halogenation

Because alkanes are essentially inert under most conditions, virtually the only reaction of alkanes that you see is the free-radical halogenation reaction. This reaction is often the first reaction that you encounter in organic chemistry.

The chlorination of methane is shown in Figure 8-20. In this reaction, a chlorine atom is substituted for a methane hydrogen.

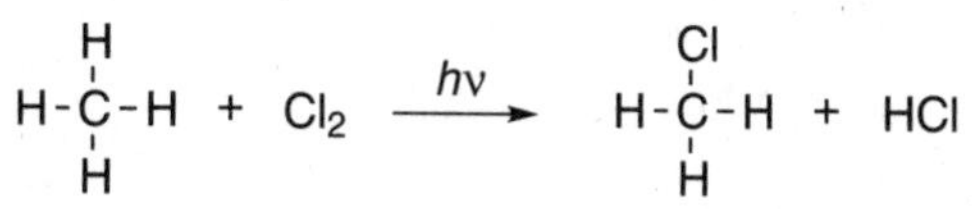

Figure 8-20:
The chlorination of methane.

$$\underset{H}{\overset{H}{H-\overset{|}{\underset{|}{C}}-H}} + Cl_2 \xrightarrow{h\nu} \underset{H}{\overset{Cl}{H-\overset{|}{\underset{|}{C}}-H}} + HCl$$

One feature that's interesting about this chlorination reaction is that the reaction is *photochemical:* Instead of using heat to start the reaction, the reaction uses light (abbreviated $h\nu$)! The reaction proceeds in three stages — initiation, propagation, and termination.

Getting things started: Initiation

In the initiation step, light is shone on the reaction and the radiation is absorbed by the chlorine (Cl_2). The light provides enough energy for the married chlorines to divorce — that is, for the chlorine-chlorine bond to break apart to form two chloride radicals, as shown in Figure 8-21. (Recall that free radicals are compounds that contain unpaired electrons.) This kind of bond dissociation is called *homolytic cleavage,* because the bond breaks symmetrically — one electron from the bond goes to one side, and the other electron goes to the other side, just as half of the shared property goes to each person in a divorce (theoretically). Note that you use one-headed fishhook arrows to show the movement of only one electron. See Chapter 3 for more on using arrows in organic chemistry.

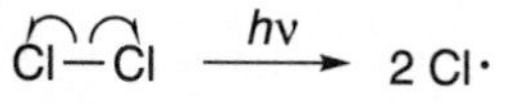

Figure 8-21:
The initiation step.

$$Cl-Cl \xrightarrow{h\nu} 2\ Cl\cdot$$

Keeping the reaction going: Propagation

After the reaction has been initiated by forming the chlorine radicals, the reaction proceeds to the propagation steps (see Figure 8-22). A chlorine radical is unstable because the chlorine atom only has seven valence electrons, one electron short of having its valence octet completely full. To fill its valence octet, a chlorine radical then plucks a hydrogen atom (not a proton) from the methane to make hydrochloric acid plus the methyl radical. Now, however, the methyl radical is one electron short of completing its octet. So, the methyl radical then attacks another molecule of chlorine to make chloromethane plus another chlorine radical.

Figure 8-22:
The propagation steps.

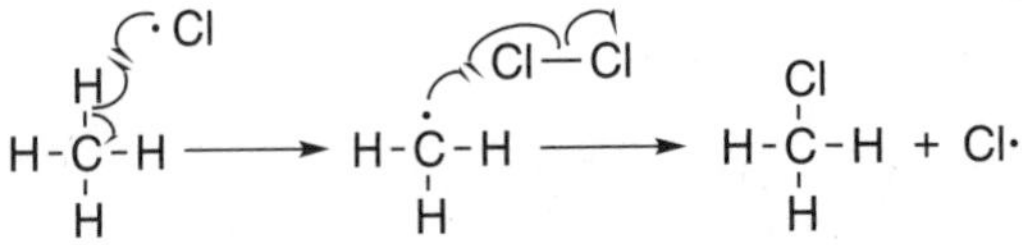

You're fired: Termination steps

Because this reaction generates chlorine radicals as a byproduct, the reaction is called a *chain reaction*. In a chain reaction, the reactive species (in this case, the chlorine radical) is regenerated by the reaction. If not for the termination steps, this reaction could theoretically continue until all the starting materials were consumed. Termination steps are reactions that remove the reactive species without generating new ones. Any of the radical couplings shown in Figure 8-23 are considered termination steps because they remove the reactive species (the free radicals) from the reaction without replacing them.

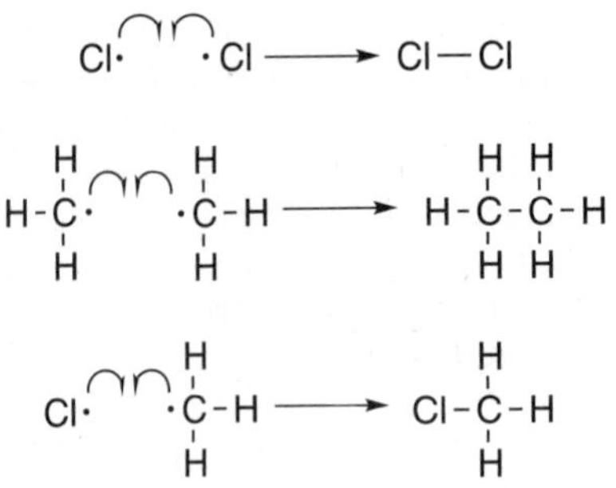

Figure 8-23:
The termination steps for the chlorination of methane.

When solving multistep synthesis problems (see Appendix A), you're often given an alkane-starting material. One of the best ways to get your foot in the door (often the only way) as far as putting a functional group on the molecule is to brominate (or chlorinate) the alkane using this free-radical reaction.

What about the chlorination of larger molecules that have different kinds of hydrogens? In methane only one kind of hydrogen is available to be pulled off — and so only one possible product can be made — but in larger molecules, several products can be formed. For example, butane (see Figure 8-24) has two types of hydrogen. Hydrogens are classified according to the substitution of the carbon to which they're attached. Hydrogens attached to primary carbons (or carbons bonded to only one other carbon) are called *primary hydrogens,* hydrogens attached to secondary carbons (or carbons bonded to two other carbons) are called *secondary hydrogens,* and so on. Butane has two types of hydrogens — primary hydrogens (1 degree) and secondary hydrogens (2 degrees).

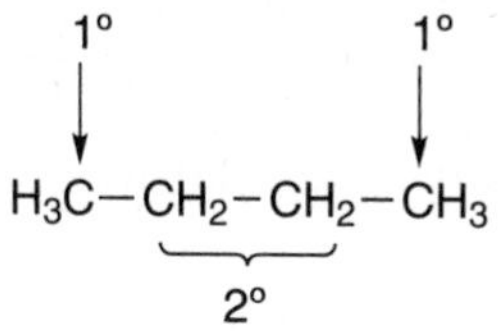

Figure 8-24:
The pri-
mary and
secondary $1°$ $1°$
hydrogens $H_3C-CH_2-CH_2-CH_3$
on butane. $2°$

The chlorination of butane selectively forms the product that results from the chloride radical abstracting a secondary hydrogen to make the secondary radical, as shown in Figure 8-25.

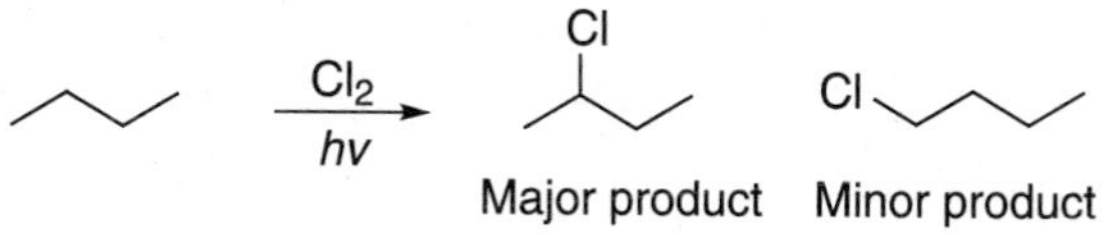

Figure 8-25:
The chlo-
rination of
butane.

$$\xrightarrow[hv]{Cl_2}$$

Major product   Minor product

To see why this is so, you need to consider the stabilities of free radicals. Radicals are more stable when they rest on more highly substituted carbons, as Figure 8-26 shows. Thus, you preferentially get chlorine substitution on the more highly substituted carbon atom.

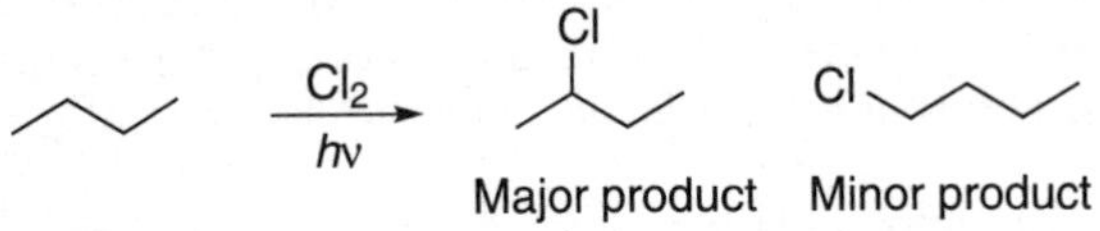

Figure 8-26:
The rela-
tive stabil-
ity of free
radicals.

Major product   Minor product

Selectivity of chlorination and bromination

The bromination of alkanes occurs in the same fashion as the chlorination of alkanes, except that Br_2 is used in the reaction instead of Cl_2. One difference between the chlorination and bromination of alkanes is that bromide radicals are more selective for hydrogen on more substituted carbons than chloride radicals are.

Chlorine radicals are less stable than bromine radicals and, thus, have only modest selectivity for reacting with hydrogens on more substituted carbons, reacting many times with any hydrogen it bumps into. Bromine radicals, however, are more stable than chlorine radicals and, thus, are happier to wait until they bump into a hydrogen on a more highly substituted carbon before reacting.

Chapter 9

Seeing Double: The Alkenes

*A*t this point, I assume you know the basics of organic chemistry. I assume you understand the difference between covalent and ionic bonds (Chapter 2), how to draw and interpret organic structures (Chapter 3), and how to name alkanes (Chapter 7). In construction terms, I assume you have the foundations of the house of organic chemistry. Now that the foundation is there, I get to the fun stuff — the chemical reactions, themselves — which build up the walls of the house of organic chemistry (that Jack built).

Specific chemical reactions are the meat and potatoes of most organic chemistry courses. And as the number of reactions start to pile up like dirty socks on the floors of college dormitories, organizing all the many reactions in as succinct and convenient a way as possible becomes essential. With the alkene functional group, numerous reactions are available to form alkenes and convert them into other functional groups.

In this chapter, I introduce the alkene functional group. I show you how to add the nomenclature of alkenes into your repertoire, and I discuss the many reactions that both make alkenes and transform them into new kinds of compounds. Along the way, I give you practical advice on how you can best learn the swarms of reactions that you see throughout this chapter and in your organic course.

Defining Alkenes

Alkenes are compounds that contain carbon-carbon double bonds. Because alkenes are so often found in valuable compounds (like pharmaceuticals), they're one of the most important functional groups in organic chemistry. Alkenes are also very versatile, because they're easy to make and convert into other molecules, as Figure 9-1 shows. Therefore, alkenes become useful as waypoints on the road to synthesizing other molecules.

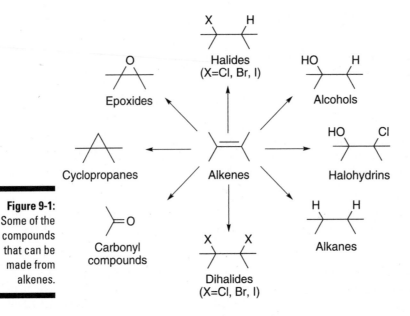

Figure 9-1: Some of the compounds that can be made from alkenes.

What do I mean by *waypoints?* Think of air travel as an analogy. You may not be able to get a flight directly from Tampa to Timbuktu, but you may be able to get a connection flight from Tampa to Chicago, and from Chicago you could get a flight to Timbuktu. Similarly, you may not be able to convert one functional group directly into the functional group that you want, but you may be able to convert the functional group into a connection functional group (like an alkene), which could then be converted into the desired product. So, even if a desired product contains no carbon-carbon double bonds, alkenes may still play a role in the synthesis of that molecule.

Mastering organic reactions

Many students find that making reaction note cards (like the one shown in the accompanying figure) is a helpful way to learn the reactions (I recommend this practice). Other students find that reaction schemes that organize functional group conversions visually (like the one shown in Figure 9-1) are very helpful for learning the reactions.

No matter how you choose to learn the reactions, you should work many problems that test your ability to recall reactions and apply these reactions to new situations. Because most of organic chemistry involves reactions of organic molecules, you need to become an "expert" in synthetic organic chemistry. And to become an expert at anything you need to practice. To become an expert pianist, you need to

practice the piano; to become an expert painter, you need to practice painting; and to become an expert whistler, you need to practice whistling. In the same way, to become an expert in organic reactions, you need to practice, practice, practice.

In a practical sense, you should work many problems that test both your ability to recall the reactions (including which reagents to use), and your ability to apply what you know to new circumstances. At first, you'll just deal with one-step reactions — what reagents convert compound A into compound B, for example. But eventually you'll need to be able to work multi-step synthesis problems — that is, syntheses involving more than one step, which I discuss in more detail in Appendix A.

Mechanism:

Bromonium ion   Anti-Stereochemistry

Taking Away Hydrogens: Degrees of Unsaturation

Alkanes are said to be saturated hydrocarbons because these molecules are saturated with hydrogens, holding the maximum number of hydrogens while still obeying all the rules of valence (having no more than four bonds for second-row atoms). Because adding a carbon-carbon double bond in a molecule requires the loss of two hydrogens, alkenes are said to add a *degree of unsaturation* into a molecule (sometimes called a *double-bond equivalent* or *index of hydrogen deficiency*). So, although alkanes have the general formula C_nH_{2n+2}, where *n* is the number of carbon atoms in the molecule, alkenes have the general formula of C_nH_{2n}, and contain two fewer hydrogens than an alkane with the same number of carbons.

Determining degrees of unsaturation from a structure

Knowing the number of degrees of unsaturation in a molecule is useful because this number gives you an indication of how many double bonds are present in an unknown compound. (This morsel of information becomes very useful when you want to determine the structure of an unknown compound. See Part IV for how to apply this information to spectroscopy and spectrometry problems.)

The degrees of unsaturation in a molecule are additive — a molecule with one double bond has one degree of unsaturation, a molecule with two double bonds has two degrees of unsaturation, and so forth. Just as the formation of a double bond causes two hydrogens to be lost, the formation of a ring also results in the loss of two hydrogens, so every ring in the molecule also adds one degree of unsaturation. For every triple bond, two degrees of unsaturation are added to a molecule, because a molecule must lose four hydrogens to make a triple bond. Some examples of three-carbon molecules with different numbers of degrees of unsaturation are shown in Figure 9-2.

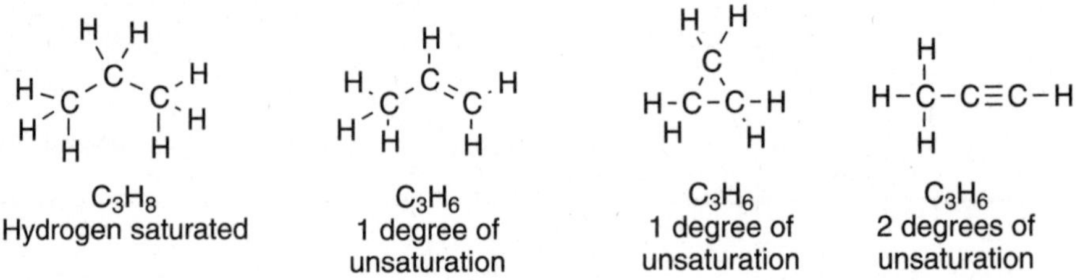

Figure 9-2: The degrees of unsaturation for three-carbon molecules.

C_3H_8
Hydrogen saturated

C_3H_6
1 degree of unsaturation

C_3H_6
1 degree of unsaturation

C_3H_6
2 degrees of unsaturation

To determine the number of degrees of unsaturation for any arbitrary structure, you sum all the individual elements of unsaturation in the molecule. Figure 9-3 shows a molecule that consists of one ring, one double bond, and one triple bond. This molecule, therefore, has four degrees of unsaturation because the double bond and the ring each add one degree of unsaturation, and the triple bond adds two degrees, for a total of four.

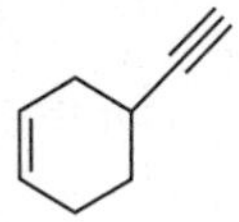

Figure 9-3:
A molecule
with four
degrees of
unsaturation.

Problem solving: Determining degrees of unsaturation from a molecular formula

More important than determining the number of degrees of unsaturation from a molecular structure is being able to determine the number of degrees of unsaturation from a molecular formula. The number of degrees of unsaturation can be determined from the molecular formula using the following equation.

$$\text{Degrees of Unsaturation} = \frac{\left[(\text{Number of Carbons} \times 2) + 2\right] - \text{Number of Hydrogens}}{2}$$

With this equation, the number of degrees of unsaturation can be determined for any hydrocarbon whose molecular formula is known. (For compounds whose structure and formula are not known, chemists use an instrumental technique called *mass spectrometry* to determine the molecular formula of the compound. See Chapter 17 for more on mass spec.)

But what about molecules that contain atoms other than hydrogen and carbon? In such cases, you need to convert these multi-atom molecular formulas into equivalent formulas that contain just carbon and hydrogen so they can be plugged into the preceding equation. To do so, you use the following conversion factors:

- ✔ **Halogens (F, Cl, Br, I):** Add one hydrogen to the molecular formula for each halogen present.

- ✔ **Nitrogen:** Subtract one hydrogen for each nitrogen present.

- ✔ **Oxygen or sulfur:** Ignore.

For example, to determine the number of degrees of unsaturation in the formula $C_8H_6F_3NO_2$, you first make the proper substitutions for all atoms that are not hydrogen and carbon. Fluorine is a halogen, so you add three hydrogen atoms to the molecular formula (one for each F). The molecule contains one nitrogen, so you subtract one hydrogen from the molecular formula. The two oxygens in the molecule you ignore. This gives a reduced equation of $C_8H_{6+3-1} = C_8H_8$. In other words, both the formula $C_8H_6F_3NO_2$ and the formula C_8H_8 have identical numbers of degrees of unsaturation. Plugging this reduced formula into the preceding equation gives five degrees of unsaturation for the molecular formula $C_8H_6F_3NO_2$.

The Nomenclature of Alkenes

If you know how to name alkanes (see Chapter 7), adding alkene nomenclature to your repertoire is a fairly straightforward task. Whereas the names of alkanes end with the suffix *–ane,* alkenes end with the suffix *–ene.* A two-carbon alkene, therefore, is named ethene; a three-carbon alkene is named propene, and an alkene in a five-membered ring is named cyclopentene (see Figure 9-4).

Figure 9-4:
The structures of some alkenes.

$H_2C{=}CH_2$ $H_2C{=}CH-CH_3$

Ethene   Propene   Cyclopentene

Numbering the parent chain

For molecules in which the double bond can be put in more than one position, you must put a number in front of the parent name to indicate the position of the double bond. For example, two alkene locations in pentene are possible (see Figure 9-5), so a number must be placed in front of the name to show the location of the double bond.

Figure 9-5:
The two possible locations.

1-pentene   2-pentene

The parent chain is numbered so that the alkene is given the lowest possible number. For example, you wouldn't name the pentene 3-pentene because that would be an identical structure to 2-pentene. Figure 9-6 shows the correct and incorrect numbering in a seven-carbon chain with a methyl substituent. Notice that giving the smaller number to the double bond takes precedence over assigning the smaller number to the substituent.

Figure 9-6:
The correct and incorrect numberings of a long-chain alkene.

Correct numbering   Incorrect numbering

Additionally, the alkene *must* be included in the parent chain, even if a longer chain of carbon atoms can be found that doesn't include the alkene. Figure 9-7 shows how this works.

Figure 9-7:
The correct and incorrect numberings of an alkene.

Correct numbering   Incorrect numbering

To number an alkene in a ring, number around the ring so that the double bond has the lowest possible number, as shown in Figure 9-8.

Figure 9-8:
The correct and incorrect numberings of a ringed alkene.

Correct numbering   Incorrect numbering

Adding multiple double bonds

The naming of alkenes with more than one double bond requires the use of a prefix (such as *di–, tri–, tetra–,* and so forth) to indicate the number of double bonds in the molecule. Also, the position of all double bonds is indicated with numbers at the beginning of the name. An example of how this is done is shown in Figure 9-9.

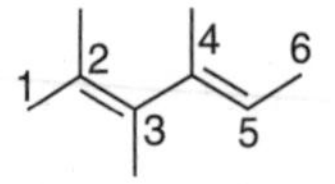

Figure 9-9:
The num-
bering and
naming of
a dialkene
containing
two methyl
substituents.

2,3,4-trimethyl-2,4-hexadiene

Common names of alkenes

Although alkenes can be named systematically using the IUPAC nomencla-ture (the official nomenclature system of chemists) as outlined earlier, some alkenes have common names that come from the olden days. These older names (some of which are shown in Figure 9-10) are still often seen in the literature and need to be memorized. For some smaller alkenes, the suffix *–ylene* is used instead of simply *–ene*. Ethene, for example, is usually called ethylene, and propene is called propylene. Styrene — the compound used to make the plastic polystyrene — is the common name for a molecule that consists of a double bond jutting off a benzene ring (see Chapter 15 for a discussion of benzene and other ringed compounds).

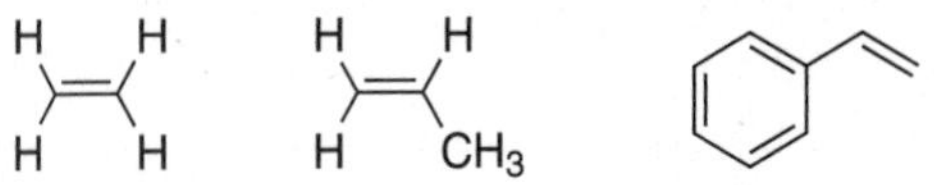

Figure 9-10:
The
common
names
of some
alkenes.

Ethylene   Propylene   Styrene

The Stereochemistry of Alkenes

Unlike carbon-carbon single bonds, which are free to rotate (see Chapter 8), double bonds are fixed and rigid. In other words, rotation around carbon-carbon double bonds is not possible at reasonable temperatures. Therefore, molecules containing double bonds have the possibility of having stereoisomers, just as ring systems do.

Molecules that have the same connectivity of atoms, but different orientations of those atoms in space are called stereoisomers (see Chapter 6).

You on my side or their side? Cis and trans stereochemistry

Consider the case of 2-pentene, shown in Figure 9-11. In this alkene, two stereoisomers are possible. One stereoisomer, called the *cis* stereoisomer, has both of the double-bond hydrogens on the same side of the double bond, while the other stereoisomer, called the *trans* stereoisomer, has the two hydrogens on opposite sides of the double bond. In general, when two identical groups are on the same side of the double bond, the molecule is said to possess *cis* stereochemistry; when two identical groups are on opposite sides of the double bond, the molecule is said to possess *trans* stereochemistry.

Figure 9-11:
Cis and *trans* 2-pentene.

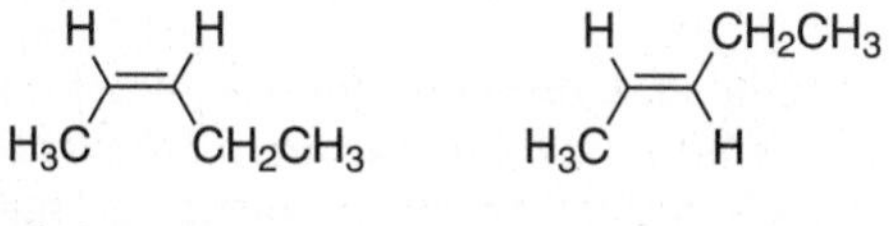

cis-2-pentene   *trans*-2-pentene

You can only have *cis-trans* stereochemistry in rings and on double bonds. You can't have *cis-trans* isomers on single bonds due to the rapid free rotation of these bonds at room temperature. (Take a peek at the section on Newman projections and conformations in Chapter 8.)

Playing a game of high-low: E/Z stereochemistry

What if a double bond holds four nonidentical groups? In such a case, the *cis-trans* nomenclature doesn't apply because *cis-trans* nomenclature can be used only when two *identical* groups are attached to a double bond or ring (in many cases, these identical groups are simply hydrogens). When four nonidentical groups are attached to a double bond, you must use the E/Z system of nomenclature (rather than the *cis-trans* nomenclature) to assign the stereochemistry of the double bond. (*E* stands for the German word *entgegen*, which means "opposite," while *Z* stands for the German word *zusammen*, which means "together.")

To use the E/Z system of nomenclature, you have to play a game of high-low. First, you must decide which substituent on each carbon is given higher priority and which is given lower priority using the Cahn–Ingold–Prelog prioritizing scheme on both sides of the alkene. Figure 9-12 helps you visualize this process

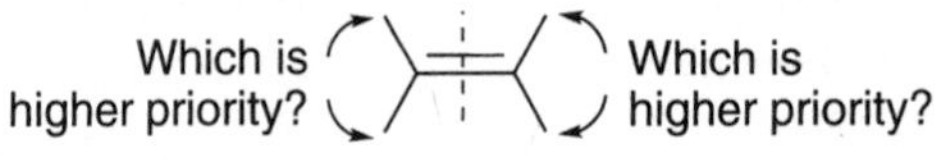

Figure 9-12: Assigning E/Z stereochemistry.

Which is higher priority? ... Which is higher priority?

If the two high-priority substituents are on the same side of the double bond, the alkene is given the Z nomenclature; if the highs are on opposite sides of the double bond, the alkene is given the E nomenclature. (See Figure 9-13 for an illustration of these nomenclature rules.)

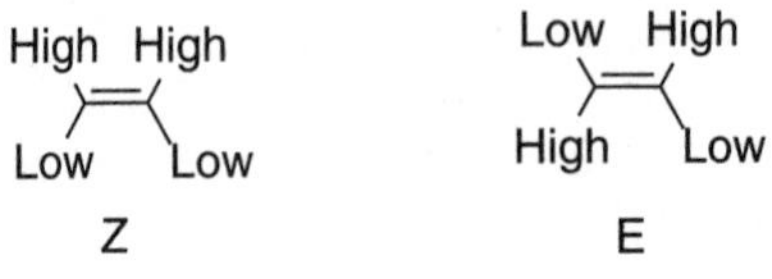

Figure 9-13: Playing high-low.

High High / Low Low — Z

Low High / High Low — E

Here's how to prioritize the double-bond substituents using the Cahn–Ingold–Prelog prioritizing scheme.

✔ **Individual substituents:** The substituent whose first atom has the highest atomic number gets the highest priority. For example, iodine would be higher priority than bromine, which would be higher than fluorine.

✔ **Ties:** In the case of a tie (both first atoms are carbon, for example), proceed to the next atom and decide which atom has the higher atomic number. If you get another tie, keep going until the tie is broken. (See Figure 9-14 for an example.)

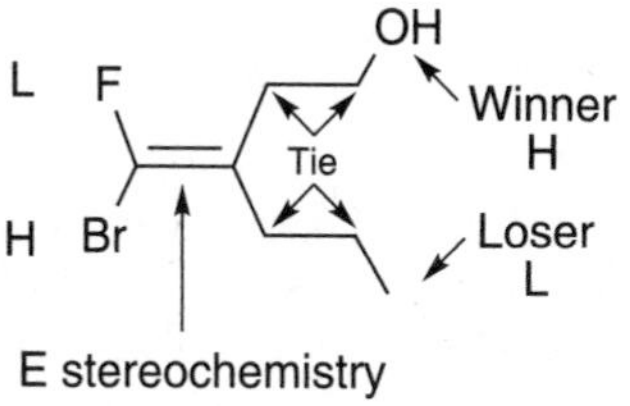

Figure 9-14:
Determining the priorities of double-bond substituents in cases of ties.

✔ **Multiple bonds:** You treat multiple bonds in a somewhat strange fashion using the Cahn–Ingold–Prelog prioritizing scheme. Using this prioritizing scheme, you treat multiple bonds as if they were multiple single bonds. A carbon double-bonded to oxygen, for example, would be treated as if the carbon had two carbon-oxygen single bonds (and as if each oxygen were bonded back to carbon). Which is higher priority, the carbon double-bonded to oxygen or the carbon triple-bonded to nitrogen, as shown in Figure 9-15? The carbon double-bonded to oxygen is higher priority because oxygen has a higher atomic number than nitrogen. (The fact that one group has three C-N bonds and the other has only two C-O bonds is irrelevant; the number of bonds only matters when comparing identical bonds, so C=NH would be higher priority than C-NH$_2$.)

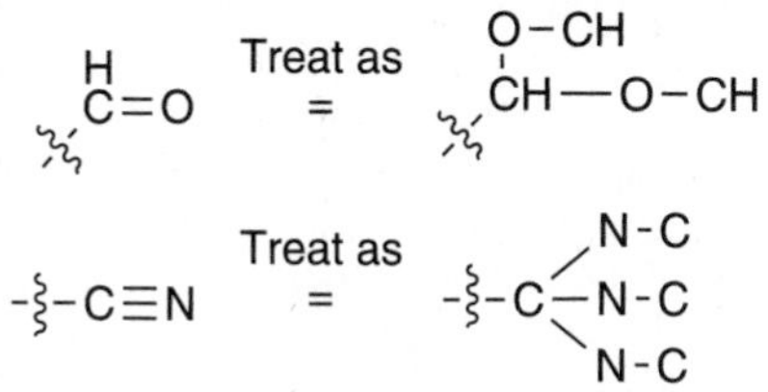

Figure 9-15:
Treating multiple bonds using the Cahn–Ingold–Prelog prioritizing scheme.

Stabilities of Alkenes

Reactions yielding alkenes often preferentially form certain alkene isomers over others. In many cases, this preference for one isomer over another is the result of one isomer being more stable than another, with the more stable isomer being formed preferentially. Understanding what structural features lead to this greater alkene stability helps you to justify why certain alkenes will be formed in particular reactions.

Alkene substitution

Perhaps the biggest factor affecting the stability of alkenes is the number of non-hydrogen substituents that come off the carbon-carbon double bond. In general, the more non-hydrogen substituents that come off a double bond, the more stable the alkene is, as shown in Figure 9-16.

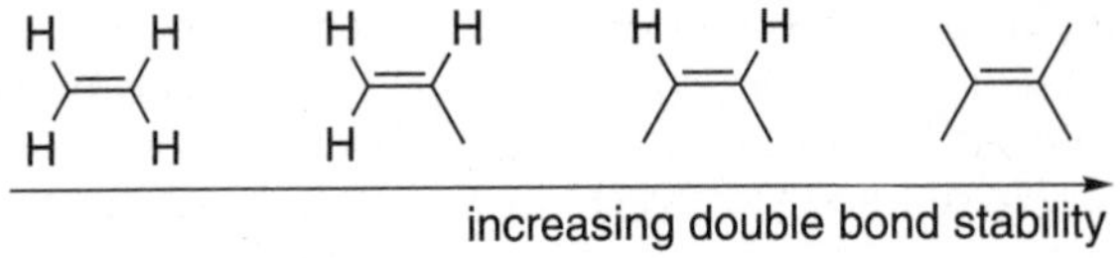

Figure 9-16: The relative stabilities of substituted alkenes.

increasing double bond stability

Stability of cis and trans isomers

Also affecting the stability of alkenes is *cis-trans* isomerism. Generally, *trans* isomers of alkenes are more stable than *cis* isomers. This preference of alkenes for *trans* isomers comes about because atoms (like people) want a certain amount of personal space and don't like to have their personal space invaded. Technically, this desire for adequate space results from atoms repelling each other because of the electron-electron repulsion of their electron clouds.

Alkenes in the *trans* configuration have substituents that are on opposite sides of the double bond, and the bulky substituents are, thus, farther away from each other; alkenes in the *cis* configuration, on the other hand, have substituents that are on the same side of the double bond, and the substituents are close to each other, invading each other's space. The electron-electron repulsion that results from atoms being forced into the personal space of other atoms is called *steric repulsion,* as shown in Figure 9-17. Therefore, alkenes prefer to be in *trans* configuration rather than *cis* configuration.

Figure 9-17:
The steric
repulsion of
cis alkenes
and the
relative sta-
bilities of *cis*
and *trans*
alkenes.

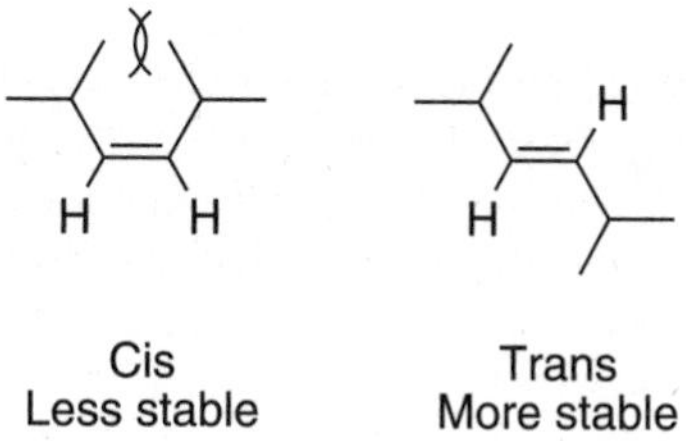

Cis
Less stable

Trans
More stable

> **TIP**
>
> To say that one molecule is more stable than another is generally synonymous with saying that one is lower in energy than another. In other words, stable molecules are generally lower in energy than unstable molecules.

Formation of Alkenes

The three primary ways to form alkenes is by dehydrohalogenation, dehydration, and the Wittig reaction.

Elimination of acid: Dehydrohalogenation

One of the most common methods of making alkenes is from alkyl halides. Halogens are often abbreviated *X*, which stands for any of the halogens chlorine (Cl), bromine (Br), or iodine (I). When a strong base (abbreviated B:⁻) is added to an alkyl halide, an elimination reaction occurs to make the alkene. This process is called *dehydrohalogenation* and is shown in Figure 9-18, indicating the loss of a hydrogen atom and a halide from the starting material in the reaction. The proton that's eliminated is a proton adjacent to the halide, not a proton on the same carbon as the halide. The two possible mechanisms for this elimination are discussed in detail in Chapter 12.

Figure 9-18:
The dehy-
drohaloge-
nation of an
alkyl halide.

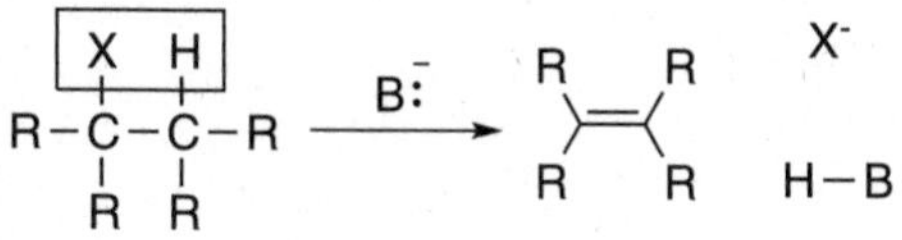

Losing water: Dehydration of alcohols

Similar to the process of dehydrohalogenation of alkyl halides is the process of *dehydration,* or the loss of water to make alkenes. In the presence of a strong acid and heat (abbreviated Δ), alcohols lose water to make alkenes, as shown in Figure 9-19. These dehydration reactions typically follow the E1 mechanism discussed in Chapter 12.

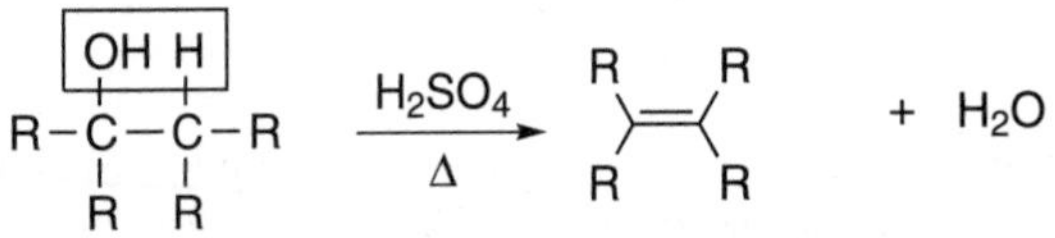

Figure 9-19: The dehydration of an alcohol.

Alkenes from coupling: The Wittig reaction

A very useful reaction for making alkenes is the Wittig reaction (pronounced *vit*-ig), a reaction for which Georg Wittig, the chemist who discovered it, received the Nobel Prize in chemistry.

If you want to win a Nobel Prize in chemistry, a good way to start is to discover a general carbon-carbon bond-making reaction, because such reactions are extremely valuable. Chemists who have won Nobel Prizes for doing just that include Otto Diels, Kurt Alder, Georg Wittig, and Victor Grignard.

In the Wittig reaction, an aldehyde or ketone is reacted with a phosphorane (phosphorous double bonded to carbon, $Ph_3P=CR_2$) to make an alkene, as shown in Figure 9-20. (Remember that Ph stands for a phenyl ring, C_6H_5.) From a conceptual standpoint, you can imagine snipping off the carbonyl oxygen and the triphenylphosphine (PPh_3) portion of the phosphorane, and then smooshing the two parts together to make the alkene.

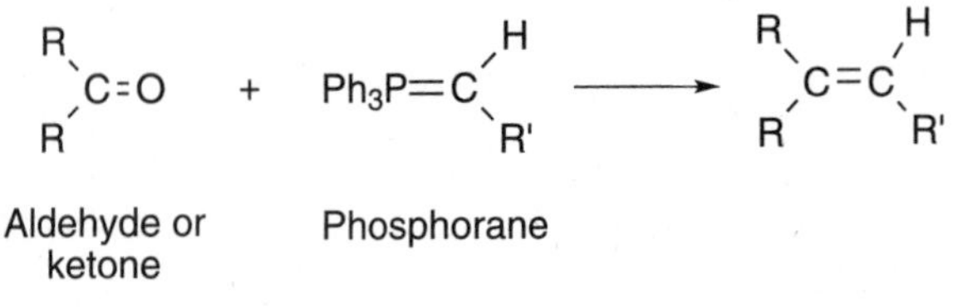

Figure 9-20: The Wittig reaction.

Aldehyde or ketone   Phosphorane

Making a phosphorane requires two steps. First, triphenylphosphine (Ph₃P) is reacted with a primary alkyl halide to make a phosphonium ion, as shown in Figure 9-21. Then, a strong base (B:⁻) is added to deprotonate the carbon adjacent to the phosphorous, making the phosphorane; this phosphorane has two major resonance structures.

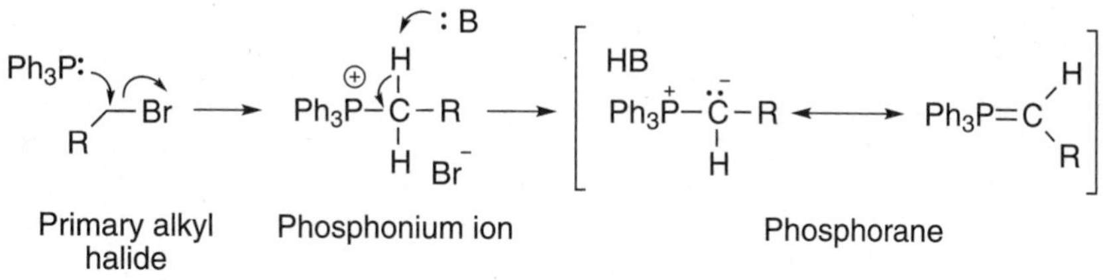

Figure 9-21:
Making
phosphorane.

Primary alkyl
halide

Phosphonium ion

Phosphorane

The mechanism of the addition of the phosphorane to the carbonyl compound (usually an aldehyde or ketone) is still disputed, but chemists think that the attack of the phosphorane on the carbonyl compound generates a charged species called a *betaine,* as shown in Figure 9-22. The *oxyanion* (negatively charged oxygen) of the betaine then attacks the positively charged phosphorous to generate a four-membered neutral species called an *oxaphosphetane.* The oxaphosphetane collapses to generate a very stable triphenylphosphine oxide and the alkene. (Whew!)

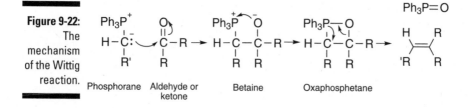

Figure 9-22:
The
mechanism
of the Wittig
reaction.

Phosphorane   Aldehyde or
                ketone

Betaine

Oxaphosphetane

Chapter 10

Reactions of Alkenes

. . . .

In This Chapter

▶ Adding a hydrohalic acid to a double bond

▶ Clueing in on carbocations

▶ Making alcohol by adding water across a double bond

▶ Cleaving carbon-carbon double bonds into two fragments

▶ Converting alkenes into cyclopropane rings

▶ Sussing out the Simmons–Smith reaction

▶ Making epoxides by reacting an alkene with a peroxy acid

▶ Converting alkenes into alkanes by hydrogenation

. . . .

*O*ne of the most useful aspects of alkenes is that they are incredibly versatile, as they can be easily made into such a variety of different compounds. In this chapter, I show you how you can convert alkenes into a variety of different functional groups like alkyl halides, alcohols, aldehydes and ketones, epoxides, and cyclopropane rings.

Adding Hydrohalic Acids across Double Bonds

Adding a hydrohalic acid (usually HCl or HBr) to a double bond converts the alkene into an alkyl halide. This reaction, shown in Figure 10-1, is just the reverse reaction of the elimination reaction that creates double bonds.

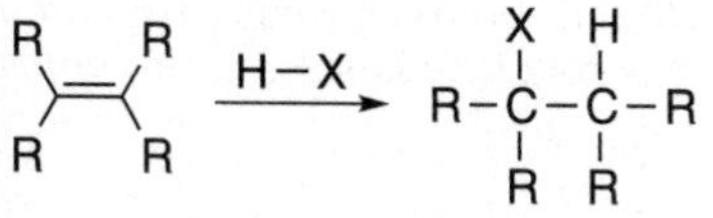

Figure 10-1:
The Markovnikov addition to alkenes.

The mechanism of the reaction is shown in Figure 10-2. The first step of the reaction is protonation of the double bond by the acid. This leads to a short-lived *carbocation* intermediate (cations on carbon atoms are called carbocations). The carbocation is then attacked by the free halide anion to generate the alkyl halide.

A double bond, however, can be protonated to generate two different cations, as shown in Figure 10-2. But which carbon in the double bond will receive the proton? (Or, alternatively phrased, which side of the double bond will become positively charged?)

The Russian chemist Vladimir Markovnikov observed that alkenes are protonated on the least substituted carbon in the double bond, generating the carbocation on the most highly substituted carbon atom. In other words, *tertiary carbocations* (carbocations substituted by three alkyl groups and abbreviated 3°) are preferred over *secondary carbocations* (cations substituted by two alkyl groups and abbreviated 2°). Secondary carbocations in turn are favored over *primary carbocations* (1°), those cations substituted by just one alkyl group.

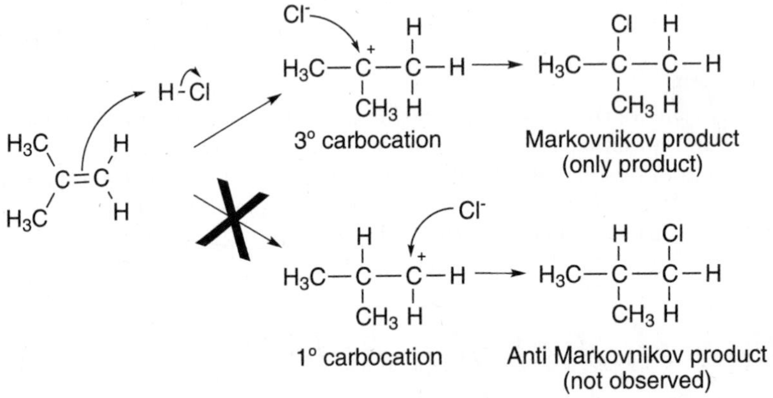

Figure 10-2:
The formation of the carbocation on the most highly substituted carbon (the Markovnikov product).

3° carbocation

Markovnikov product
(only product)

1° carbocation

Anti Markovnikov product
(not observed)

Because of this preference for more highly substituted carbocations, *halides add to the carbon that's most substituted with alkyl groups.* When addition occurs on the most substituted carbon, as in the reaction shown in Figure 10-2, the product is called the *Markovnikov product,* in honor of the discoverer of this phenomenon.

Because this reaction favors one of two products — halide addition to the more substituted side of the double bond as opposed to halide addition to the less substituted side — this reaction is said to be *regioselective.*

Regioselective reactions are those that prefer one constitutional isomer in a reaction to another. (Recall that constitutional isomers are molecules with the same molecular formula, but the atoms bonded in different ways.)

But what accounts for this preference? Why are the more highly substituted carbocations more stable (and therefore preferable) to those cations with fewer alkyl substitutions? To understand this preference you need to look at the structure of the carbocation.

I'm Positive: Carbocations

Carbocations are an unstable species because they leave the cationic carbon two electrons short of filling the atom's valence octet (and you know how much atoms hate that). In addition, the cationic carbon bears a full positive charge (which is something else that atoms don't like). The carbocation carbon is, therefore, highly electron deficient.

Helping a neighbor: Hyperconjugation

Alkyl substituents, however, can donate some electrons to the carbocation, thereby stabilizing it. For carbocations substituted with alkyl substituents, instead of the molecule localizing the full positive charge on just the one carbon atom, some of the positive charge is shared by the neighboring alkyl substituents. This sharing of charge is energetically preferable to having all the charge localized on a single atom. Figure 10-3 shows the relative stability of primary, secondary, and tertiary carbocations.

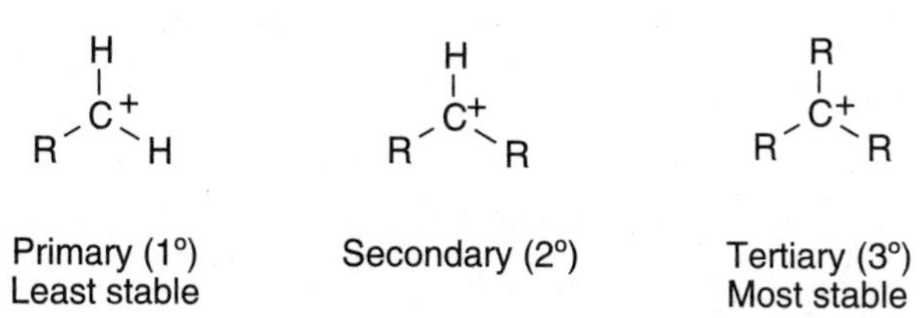

Figure 10-3: The relative stability of substituted carbocations.

Primary (1°) Least stable

Secondary (2°)

Tertiary (3°) Most stable

Alkyl groups are able to donate electron density to the carbocation center through *hyperconjugation.* Hyperconjugation is organic-speak for the weak overlap that occurs between the empty *p* orbital on the carbocation and the σ bond of an adjacent alkyl C-H bond (or C-C bond), as shown in Figure 10-4.

(Turn to Chapter 2 for a review of orbitals and bonds.) The more alkyl groups a carbocation center has, the more hyperconjugation takes place and the more the charge is shared; the more the charge is shared, the more stable the cation species is.

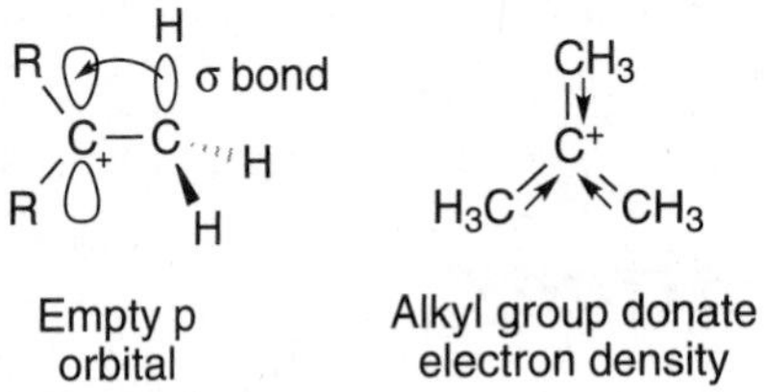

Figure 10-4:
Hyper-
conjugation
of neigh-
boring alkyl
groups.

σ bond

Empty p
orbital

Alkyl group donate
electron density

Resonance stabilization of carbocations

Carbocations are also stabilized by resonance. (See Chapter 3 for how to draw resonance structures.) Carbocations with resonance structures are more stable than carbocations without resonance structures, all else being equal. Thus, *benzylic cations* (cations next to a benzene ring), and *allylic cations* (cations adjacent to a double bond) are stabilized cations because of resonance delocalization of the positive charge onto other atoms, as shown in Figure 10-5.

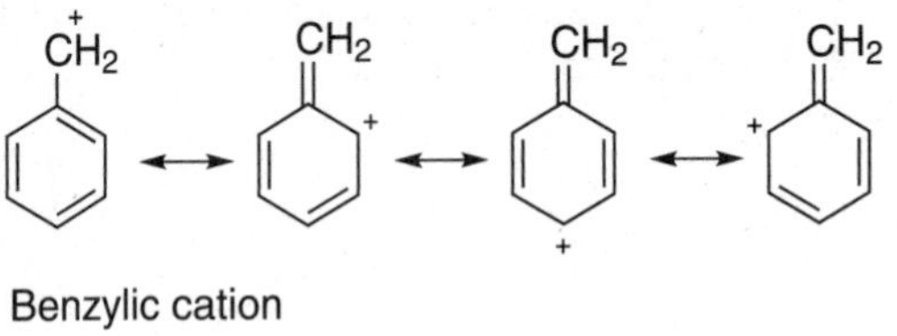

Figure 10-5:
The
resonance
stabilization
of the
benzylic
cation.

Benzylic cation

Resonance structures stabilize molecules.

Allylic cation

The stabilities of carbocations can be approximated as follows (starting with the least stable and going to the most stable cation): primary cations < secondary cations ≈ allylic cations < tertiary cations ≈ benzylic cations.

Carbocation mischief: Rearrangements

Carbocations, like unsupervised children, have the capability of making a great deal of mischief. Take the reaction shown in Figure 10-6, for example. Reacting the alkene with hydrochloric acid yields the expected product only in small amounts; the major product is a rearranged alkyl halide.

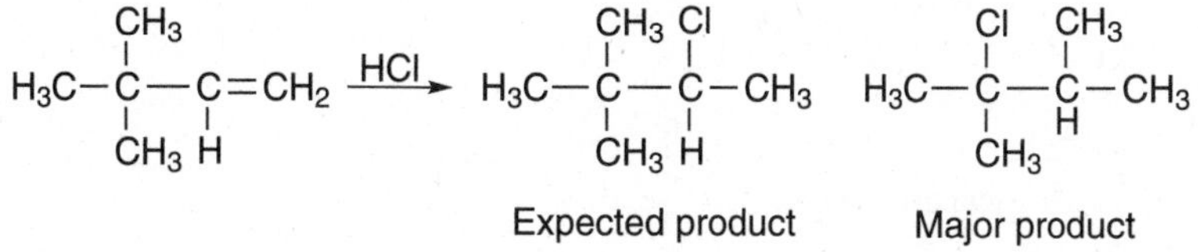

Figure 10-6:
The addition of HCl to an alkene.

H₃C—C—C=CH₂ $\xrightarrow{\text{HCl}}$ H₃C—C—C—CH₃   H₃C—C—C—CH₃

Expected product   Major product

How did this reaction occur? Through a carbocation rearrangement. Alkyl substituents or hydrogens can (and will) shift from a neighboring carbon to the carbocation center whenever doing so generates a more stable carbocation. In this example, protonating the double bond as shown in Figure 10-7 makes the secondary carbocation (which is more stable than the primary cation, the alternative in this reaction). Shifting a methyl group (along with the two electrons in the bond) from the adjacent carbon to the secondary carbocation shifts the cation to the more stable tertiary carbon. The halide then attacks the tertiary cation to make the rearranged product.

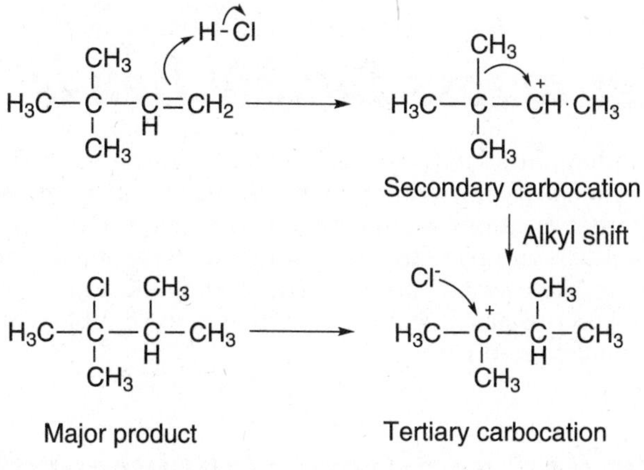

Figure 10-7:
The mechanism of the carbocation rearrangement.

Secondary carbocation

Alkyl shift

Major product   Tertiary carbocation

Spotting when these carbocation rearrangements may occur takes practice. When carbons adjacent to the cation center are highly substituted with alkyl groups, watch out for rearrangements.

Alkyl shifts can also open up small rings into larger ones, as shown in the reaction in Figure 10-8. Small rings (three- and four-membered rings) are less stable than medium-size rings (those containing five to six carbons) because of the ring strain from the constricted bond angles created by the small rings (recall that sp^3 carbons prefer to have bonds at 109.5-degree angles to each other).

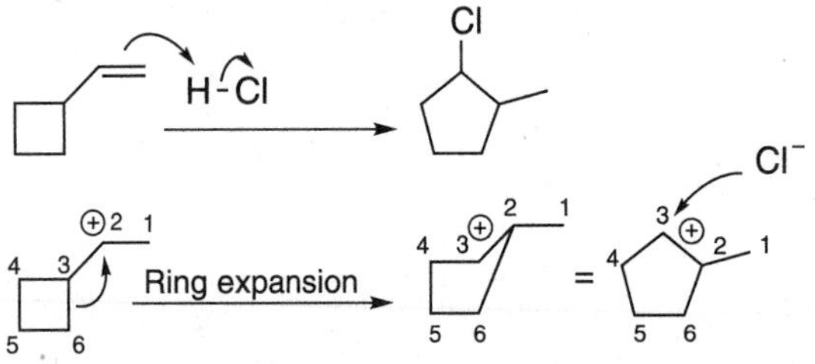

Figure 10-8:
The carbocation rearrangement in a small ring.

TIP

Number all the atoms when working on ring expansion problems so you can keep track of all the atoms and keep the charges on the correct atoms in the rearranged product.

REMEMBER

Alkyl rearrangements occur for two primary reasons:

- ✔ To make a more stable carbocation (for example to change a secondary carbocation to a tertiary carbocation)
- ✔ To relieve ring strain (for cations next to small rings)

Adding Water across Double Bonds

Hydration, or adding water across a double bond to make an alcohol, is a reaction that's similar to the addition of a hydrohalic acid across a double bond. Two different reactions accomplish the hydration. The first reaction adds the alcohol (OH group) to the most substituted carbon on the double bond to make the Markovnikov product, and the complementary reaction puts the alcohol on the least substituted carbon in the double bond to make the anti-Markovnikov product.

Markovnikov addition: Oxymercuration-demercuration

To make the Markovnikov product where the alcohol adds to the most substituted carbon, you react the alkene with mercuric acetate, $Hg(OAc)_2$ and water, followed by addition of sodium borohydride, $NaBH_4$ (see Figure 10-9).

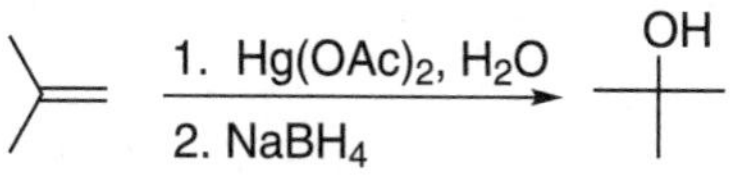

Figure 10-9:
The oxymer-
curation-
demer-
curation of
an alkene.

$$\text{alkene} \xrightarrow[\text{2. NaBH}_4]{\text{1. Hg(OAc)}_2,\ \text{H}_2\text{O}} \text{product with OH}$$

 The numbers over (or under) the reaction arrow indicate separate steps. In the case of oxymercuration-demercuration, the numbers specify that mercuric acetate is added first, followed by sodium borohydride. When no numbers are present over (or under) the arrow, this indicates that reagents are all added together in the same pot.

The mechanism for the oxymercuration-demercuration involves an attack of the double bond on the mercuric acetate to make a three-membered ring intermediate (called a *mercurinium ion*), as shown in Figure 10-10. Water then attacks the most highly substituted carbon to make the mercurial alcohol (after the loss of a proton). In the second step (when NaBH$_4$ is added), sodium borohydride replaces the mercuric portion with hydrogen.

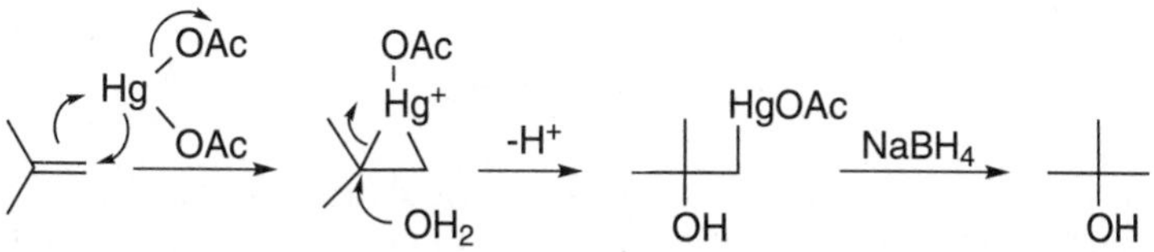

Figure 10-10:
Oxymer-
curation-
demer-
curation of
an alkene.

Anti-Markovnikov addition: Hydroboration

With oxymercuration-demercuration, you have a reaction that converts alkenes into Markovnikov-product alcohols. To make the alcohol on the least-substituted carbon (called the anti-Markovnikov product) you use hydroboration, as shown in Figure 10-11. The addition of borane (BH$_3$) in tetrahydrofuran solvent (THF) to the alkene, followed by the addition of hydrogen peroxide (H$_2$O$_2$) and sodium hydroxide (NaOH), make the anti-Markovnikov alcohol.

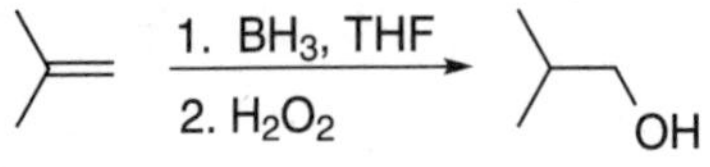

Figure 10-11:
The hydroboration and oxidation of an alkene.

1. BH_3, THF
2. H_2O_2

⟶ OH

The mechanism for hydroboration involves the cyclic transition state shown in Figure 10-12. Borane adds to the least substituted side of the double bond to make the alkyl borane. Because the addition is concerted (both the hydrogen and BH_2 are added simultaneously), the borane and hydrogen must add to the same face of the carbon-carbon bond (two groups adding to the same face of a double bond is called *syn addition*). In the second step, hydrogen peroxide (H_2O_2) in the presence of sodium hydroxide (NaOH) substitutes a hydroxyl group (OH) for the boryl unit (BH_2) to make the anti-Markovnikov alcohol.

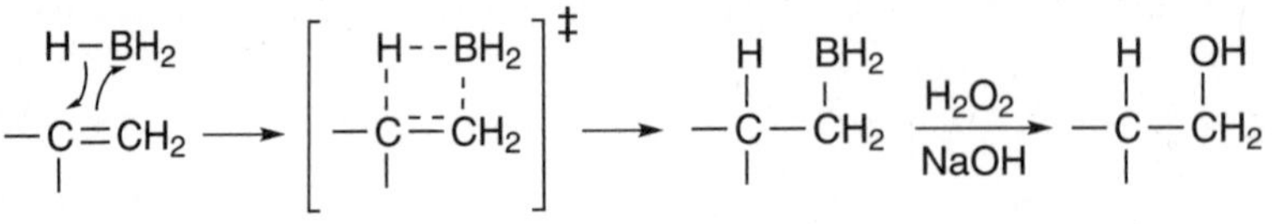

Figure 10-12:
Mechanism of hydroboration and oxidation.

$$H-BH_2 \qquad \left[H--BH_2 \right]^{\ddagger} \qquad H \quad BH_2 \qquad H \quad OH$$
$$-C=CH_2 \longrightarrow -C=CH_2 \longrightarrow -C-CH_2 \xrightarrow[\text{NaOH}]{H_2O_2} -C-CH_2$$

TIP

Syn addition is when two groups add to the same face of a double bond, and *anti addition* is where two groups add to opposite faces of a double bond. Therefore, *syn* addition to cycloalkenes results in the two added groups being *cis* to each other, and *anti* addition to cycloalkenes results in the two added groups being *trans* to each other.

A double shot: Dihydroxylation

You can add one hydroxyl group to a double bond using either oxymercuration-demercuration or hydroboration, but you can also add two hydroxy groups across a double bond using osmium tetroxide (OsO_4) in the presence of hydrogen peroxide (H_2O_2), as shown in Figure 10-13. This reaction is called a *dihydroxylation*.

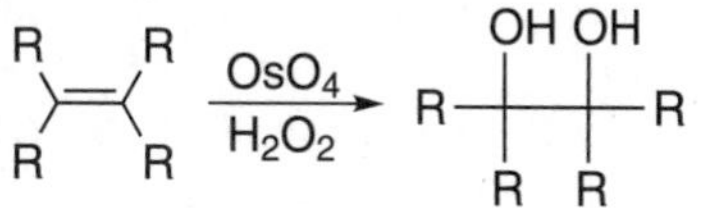

Figure 10-13:
Dihydroxyl-
ation of an
alkane.

The mechanism for dihydroxylation involves the reaction of the alkene and
the osmium tetroxide to generate a five-membered cyclic osmate ester, as
shown in Figure 10-14. Water regenerates the osmium tetroxide catalyst and
makes the diol (a *diol* is a molecule with two hydroxyl [OH] groups). Because
both oxygens in the osmium tetroxide are added to the same face of the
double bond, both hydroxyl groups end up on the same side of the carbon-
carbon bond (that is, they're added via syn addition).

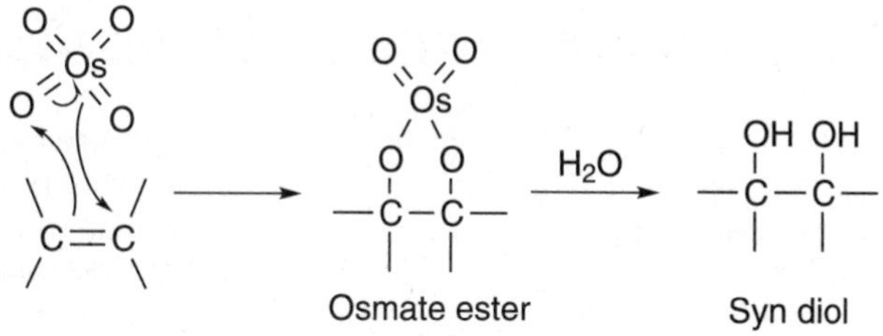

Figure 10-14:
The mecha-
nism for the
hydroxyl-
ation of an
alkene.

Osmate ester   Syn diol

Double the fun: Bromination

Alkenes react rapidly with bromine (Br_2) or chlorine (Cl_2) in carbon tetrachlo-
ride (CCl_4) solvent to make a dihalide, as shown in Figure 10-15. This reaction
is, in fact, used as a test to see whether an unknown compound contains an
alkene, because the blood-colored bromine (Br_2) turns clear when it reacts
with alkenes.

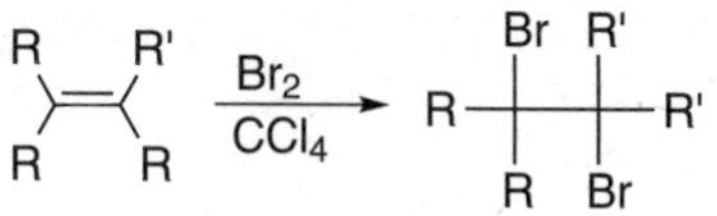

Figure 10-15:
The bromin-
ation of an
alkene.

The mechanism for bromination is somewhat unusual. Reaction of the alkene and bromine results in a cyclic three-membered intermediate called a bromonium ion (or chloronium ion for chlorine), plus the free halide, as shown in Figure 10-16. The free halide then attacks the backside of the bromonium ion to make the dibromide with anti-stereochemistry.

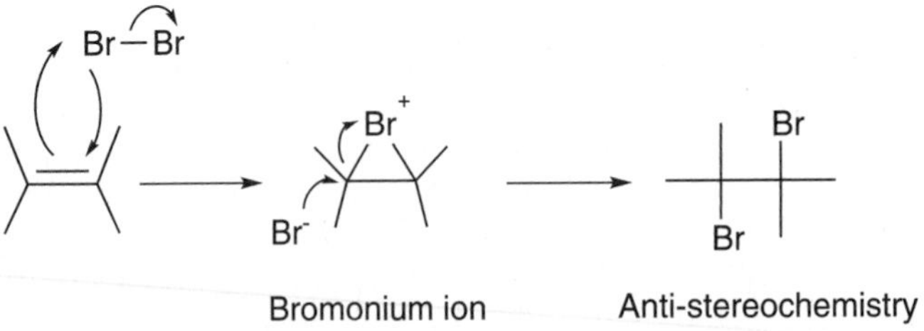

Figure 10-16: The mechanism for the bromination of an alkene.

Bromonium ion   Anti-stereochemistry

Chopping Up Double Bonds: Ozonolysis

Ozonolysis is a way of cleaving carbon-carbon double bonds into two fragments using ozone (O_3) as a reagent. The fragments formed are either aldehydes or ketones, depending on the nature of the R groups attached to the double bond (see Figure 10-17). If both R groups on one side of the double bond are alkyl groups, that side of the double bond will become a ketone fragment; if only one R group is an alkyl group and the other R is a hydrogen, that side of the double bond will become an aldehyde fragment.

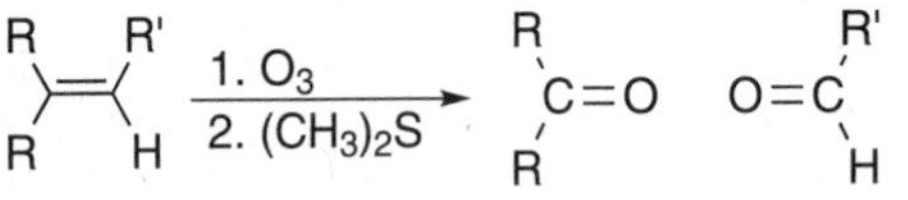

Figure 10-17: The ozonolysis of an alkene.

A quick way of determining the products of ozonolysis is to visually snip the double bond as shown in Figure 10-18, and then to cap both sides with oxygens to make the carbonyl compounds.

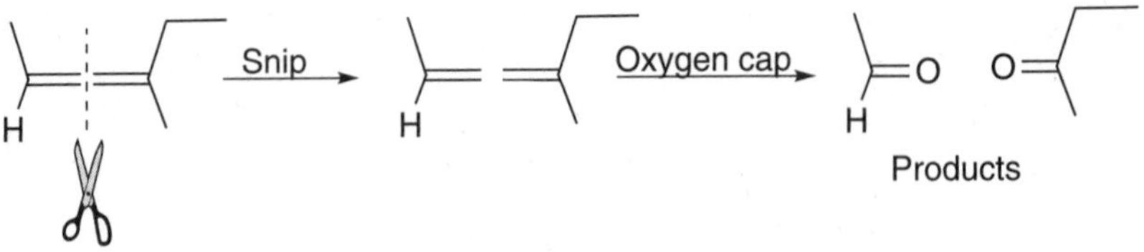

Figure 10-18: Determining products of ozonolysis.

Snip   Oxygen cap   Products

TIP

A common exam problem is one that asks you to determine the structure of the starting alkene given the products of ozonolysis. If you remember how to determine the products of ozonolysis and then work backward from that reaction (snip off the oxygens, and then smoosh the two pieces together), you get the starting alkene.

Double-Bond Cleavage: Permanganate Oxidation

Permanganate oxidation, shown in Figure 10-19, performs essentially the same reaction as ozonolysis, except that permanganate is a somewhat stronger oxidizing reagent than ozone. In permanganate oxidation, all products that would have become aldehydes using ozonolysis are further oxidized to carboxylic acids with potassium permanganate ($KMnO_4$); ketone fragments stay the same.

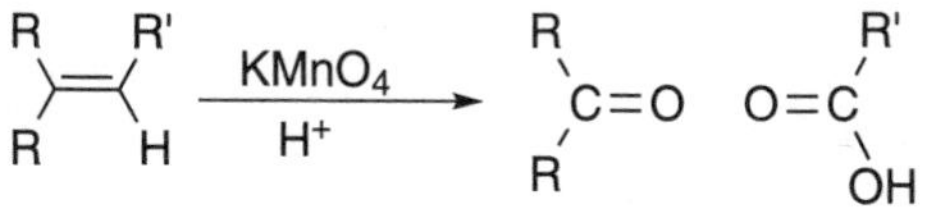

Figure 10-19:
The permanganate oxidation of an alkene.

$$R_2C{=}CHR' \xrightarrow[H^+]{KMnO_4} R_2C{=}O \quad O{=}CR'(OH)$$

Making Cyclopropanes with Carbenes

Alkenes can be converted into cyclopropane rings by reacting with an unusual species called a carbene, as shown in Figure 10-20.

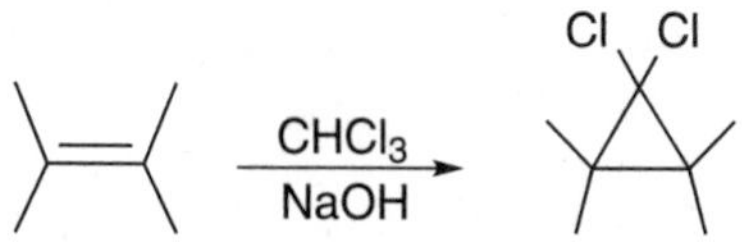

Figure 10-20:
Making a cyclopropane ring from an alkene.

$$\xrightarrow[NaOH]{CHCl_3}$$

A carbene is a neutral carbon atom with two substituents and one lone pair of electrons. Dichlorocarbene ($Cl_2C{:}$) is made by reacting chloroform ($CHCl_3$) with a base like sodium hydroxide. The base deprotonates chloroform to make the conjugate base anion, which eliminates a chloride ion to make dichlorocarbene, as shown in Figure 10-21.

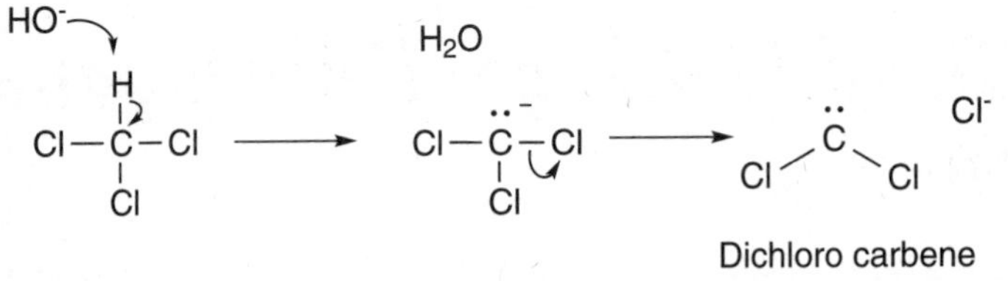

Figure 10-21:
Making dichloro-
carbene.

Dichloro carbene

Dichlorocarbene can then react with double bonds to make cyclopropane rings, by the mechanism shown in Figure 10-22.

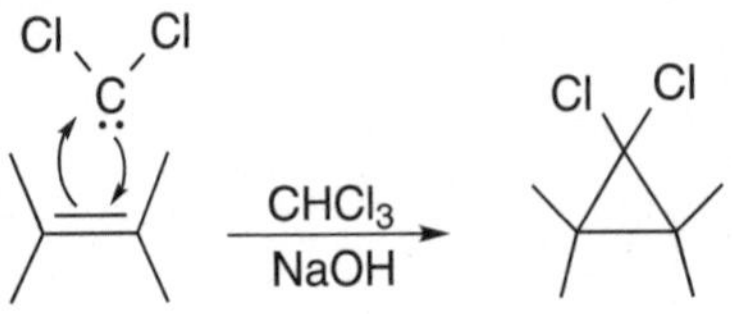

Figure 10-22:
The dichloro-
carbene addition to an alkene.

Making Cyclopropanes: The Simmons–Smith Reaction

A good way to make unsubstituted cyclopropanes from carbon-carbon double bonds is by using the Simmons–Smith reaction, shown in Figure 10-23. While the carbene methylene ($H_2C{:}$) is not actually made in this reaction, a zinc species (ICH_2ZnI) is used that reacts as if it were the carbene methylene. Because the reagent behaves like a carbene but really isn't one, this zinc species is said to be *carbenoid* (carbene-like).

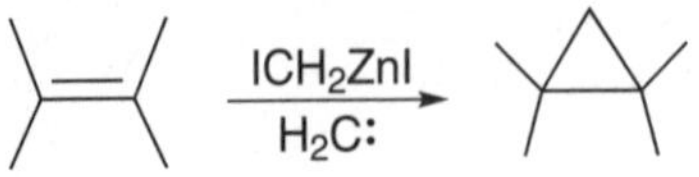

Figure 10-23:
The Simmons–Smith reaction.

Making Epoxides

Epoxides are ethers in three-membered rings. They're valuable synthetic reagents and are used in epoxy glues. Epoxides can be made by reacting an alkene with a peroxy acid (a carboxylic acid with an extra oxygen), as shown in Figure 10-24.

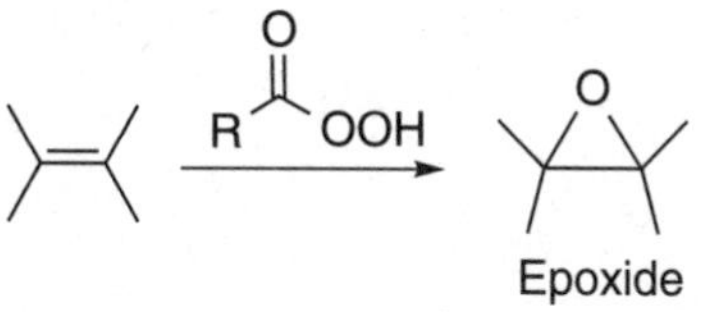

Figure 10-24:
The epoxidation of an alkene.

Epoxide

Adding Hydrogen: Hydrogenation

Alkenes can be converted into alkanes by hydrogenation. Passing hydrogen gas through a solution containing a catalyst — usually palladium on carbon (Pd/C) or platinum (Pt) — and the alkene causes hydrogen to be added across the double bond in a syn addition (see Figure 10-25).

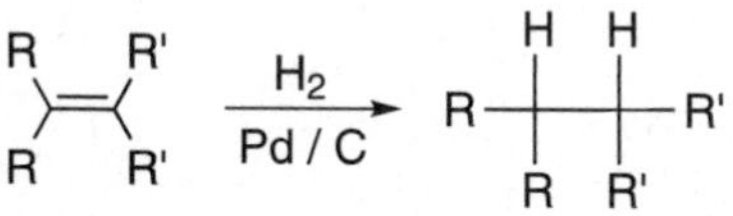

Figure 10-25:
The hydrogenation of an alkene.

Chapter 11

It Takes Alkynes: The Carbon-Carbon Triple Bond

*A*s a functional group, alkynes are somewhat less common than alkenes, but they still have some very interesting features and reactivities that are unique to them.

Alkynes are molecules that contain carbon-carbon triple bonds. As you may expect, the reactivities and properties of alkynes are very similar to those of alkenes, although a few interesting differences between the two types of molecules also exist. In this chapter, I discuss the properties of alkynes, show you how to name alkynes, and cover the fundamental reactions that form alkynes and convert them into other functional groups.

Naming Alkynes

Alkynes are named under the systematic nomenclature scheme in the same way that alkenes are (refer to Chapter 9), except that instead of the name ending with the suffix *–ene*, the names of alkynes end with the suffix *–yne*. As with alkenes, a number in the prefix is used to indicate the position of the alkyne in the molecule, as demonstrated in Figure 11-1.

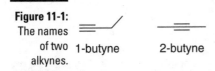

Figure 11-1:
The names of two alkynes.

1-butyne   2-butyne

Alkynes are often known by their common names; these common names are derivatives of the simplest alkyne, acetylene. Under the common name system, the two R groups are named as substituents of acetylene (see Figure 11-2 for some examples). An alkyne with two methyl groups would be dimethyl acetylene, an alkyne with two isopropyl groups, diisopropyl acetylene, and so on.

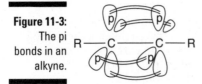

Figure 11-2:
The common names of some alkynes.

Alkyne	$HC\equiv CH$	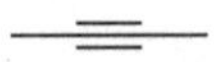	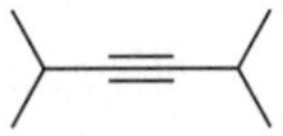
IUPAC Name	Ethyne	2-butyne	2,5-dimethyl-3-hexyne
Common Name	Acetylene	Dimethyl acetylene	Diisopropyl acetylene

Seeing Alkyne Orbitals

Alkyne carbons, by having only two substituents, are *sp* hybridized, and so have linear bonds that are oriented at 180 degrees to each other. (See Chapter 2 for more on hybridization.) The triple bond in an alkyne is made up of one sigma bond and two pi bonds. The two pi bonds result from the side-by-side overlap of the two *p* orbitals on each carbon atom (see Figure 11-3). As in the case with alkenes, the pi bonds are the reactive bonds in alkynes.

Figure 11-3:
The pi bonds in an alkyne.

The overlapping *p* orbitals that make up the two pi bonds keep the alkyne in a linear geometry. Because alkynes have linear geometries, they're drawn in a straight line, as shown in Figure 11-4.

Figure 11-4:
The correct
way to draw
an alkyne.

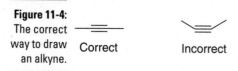

Figure 11-4:
The correct
way to draw
an alkyne.

Correct   Incorrect

Alkynes in Rings

Because alkynes prefer linear geometries, alkynes in small rings are highly unstable. Small- to medium-size cycloalkynes have bonds that must stretch out of their preferred geometries, and so inflict a good amount of ring strain on the atoms. Imagine grabbing your leg (which, like an alkyne, prefers to be linear) and touching your big toe to your nose, forming a ring with your body. The strain on your body is similar to the strain an alkyne feels. Just as you probably wouldn't feel particularly stable forming a ring by touching your toe to your nose, alkynes in rings smaller than seven carbons are too unstable to be made or isolated. Cycloheptyne, the smallest cycloalkyne isolated, has been characterized, but it's highly reactive (see Figure 11-5). Alkynes in rings containing more than seven carbons can be made and are fairly stable compounds.

Figure 11-5:
The relative
stabilities of
alkynes in
small rings.

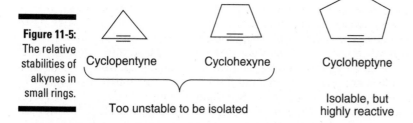

Cyclopentyne   Cyclohexyne   Cycloheptyne

Too unstable to be isolated

Isolable, but
highly reactive

Making Alkynes

To make alkynes, two reactions are readily available, one that you may expect based on reactions that form alkenes, and another one that you may not expect based on what you know about the chemistry of alkenes.

Losing two: Dehydrohalogenation

To form alkenes (see Chapter 8), you can react one equivalent (unit) of base with an alkyl halide to eliminate one equivalent of acid and form the alkene. You may expect that to make an alkyne, you may treat a dibromide with two equivalents of base, eliminating two equivalents of acid to make the alkyne. This is, indeed, the case, as shown in Figure 11-6. Usually, the base that's used is the strong base sodium amide ($NaNH_2$, sometimes called sodamide), although sodium hydroxide (NaOH) can also be used.

Figure 11-6:
The double dehydroha-logenation of an alkyl dihalide.

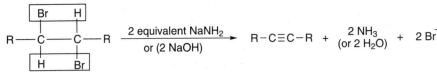

TIP

Recall that dibromides can be made by reacting an alkene with bromine (Br_2). This is a convenient way to convert alkenes into alkynes: Brominate the alkene using Br_2 to make the dibromide, and then treat the dibromide with two equivalents of base to make the alkyne.

Coupling alkynes: Acetylide chemistry

The major difference between the reactivities of alkenes and alkynes is the result of the acidity of the protons attached directly to terminal alkynes. (*Terminal alkynes* are alkynes in which the triple bond is at the end of a chain; *internal alkynes* are alkynes whose triple bond is in the middle of a chain.) The pKa of a terminal alkyne proton is roughly 25, which means that the molecule is not terribly acidic, but it's acidic enough for the hydrogen to be pulled off by a very strong base. (See Chapter 4 for a discussion of acids, bases, and pKa values.) The base that's usually used to deprotonate terminal alkynes is sodium amide ($NaNH_2$). Deprotonating the alkyne makes the acetylide anion, as shown in Figure 11-7.

Figure 11-7:
Making the acetylide anion.

$$R-C\equiv C-H \quad NH_2^- \longrightarrow R-C\equiv C:^- \;+\; NH_3$$
Acetylide ion

This acetylide ion is a very useful *nucleophile* (or "nucleus lover"), and will react with primary alkyl halides to form a new, internal alkyne, as shown in Figure 11-8. This reaction is highly useful because it forms a carbon-carbon bond where one did not exist before. Note, however, that only primary alkyl halides can be used in this reaction; that is, the halide must be attached to a methylene (CH_2) unit. Secondary and tertiary alkyl halides will not react in this way.

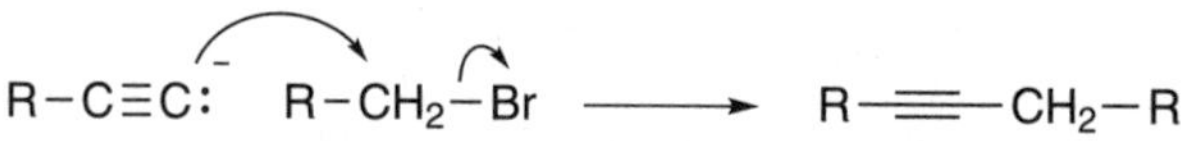

Figure 11-8:
Acetylide addition.

$$R-C\equiv C\colon^{-} \quad R-CH_2-Br \longrightarrow R-\!\!\equiv\!\!-CH_2-R$$

Alkyne reactions are similar to those of the alkenes, and these reactions use reagents similar to those used in alkene reactions. Alkynes can form a variety of functional groups, as shown in Figure 11-9.

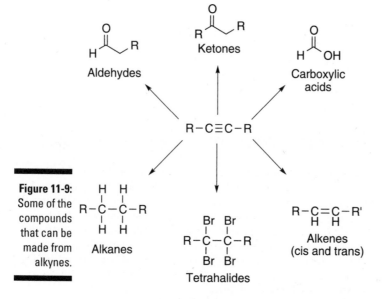

Figure 11-9:
Some of the compounds that can be made from alkynes.

Aldehydes

Ketones

Carboxylic acids

$R-C\equiv C-R$

Alkanes

Tetrahalides

Alkenes (cis and trans)

Brominating alkynes: Double the fun

Bromine reacts with the pi bond in alkynes to make the dibromide via the same mechanism as bromine addition to alkenes. (Do you recall the mechanism of the alkene reaction? Think "bromonium ion," and see Chapter 8 if you need to refresh your memory.) Because there are two pi bonds in an alkyne, two equivalents (units) of bromine can be added to make a tetrabromide, as shown in Figure 11-10.

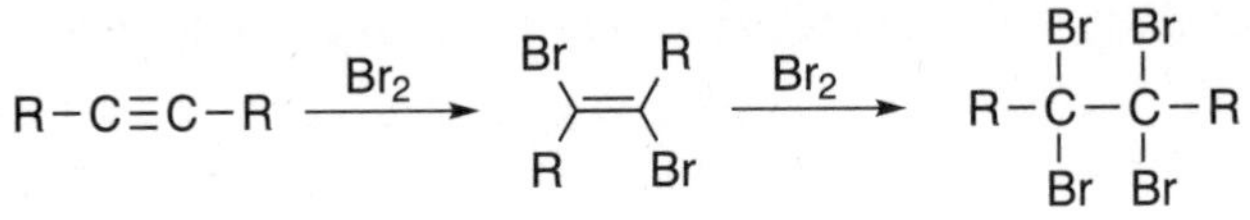

Figure 11-10: The bromination of an alkyne.

$$R-C\equiv C-R \xrightarrow{Br_2} \; \underset{R \quad Br}{\overset{Br \quad R}{>\!=\!<}} \; \xrightarrow{Br_2} \; R-\underset{\underset{Br}{|}}{\overset{\overset{Br}{|}}{C}}-\underset{\underset{Br}{|}}{\overset{\overset{Br}{|}}{C}}-R$$

Saturating alkynes with hydrogen

Alkynes can also be reduced to alkanes by bubbling two equivalents of hydrogen gas (H_2) over the alkyne in the presence of a metal catalyst as shown in Figure 11-11. This catalyst is usually palladium on carbon (Pd/C), but platinum (Pt) is also sometimes used.

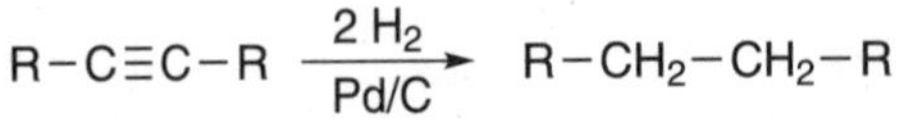

Figure 11-11: The saturation of an alkyne with hydrogen.

$$R-C\equiv C-R \xrightarrow[Pd/C]{2\,H_2} R-CH_2-CH_2-R$$

Adding one hydrogen molecule to alkynes

Stopping the reaction of alkynes with hydrogen at the alkene stage is possible because alkenes are somewhat less reactive than alkynes, but this reaction requires a special catalyst. To reduce an alkyne to the *cis* alkene (see Chapter 9 for more about *cis* and *trans* alkenes), you use Lindlar's catalyst, which is a cocktail of palladium (Pd) powder made less reactive with added lead (Pb) and quinoline (C_9H_7N). (In reaction diagrams, instead of writing out all the components of the catalyst, chemists often write "Lindlar's catalyst" over or under the arrow.) Lindlar's catalyst is not as reactive as palladium on carbon (Pd/C) and generates the *cis* alkene, as shown in Figure 11-12.

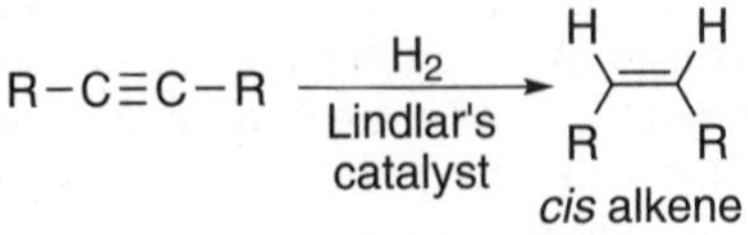

Figure 11-12: Using Lindlar's catalyst to make *cis* alkene.

$$R-C\equiv C-R \xrightarrow[\substack{Lindlar's \\ catalyst}]{H_2} \; \underset{R \quad R}{\overset{H \quad H}{>\!=\!<}}$$

cis alkene

To convert an alkyne to the *trans* alkene, you use sodium metal (Na) in liquid ammonia (NH_3), as shown in Figure 11-13.

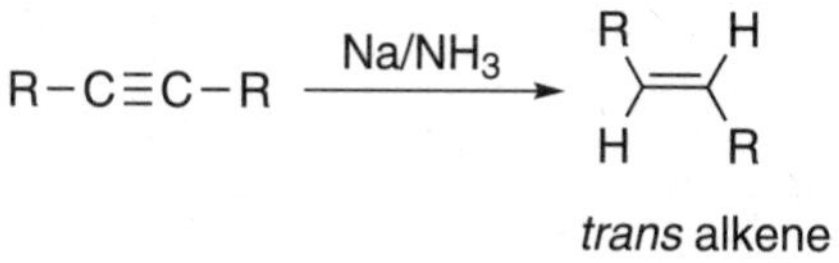

Figure 11-13:
Making a trans alkene from an alkyne.

$$R-C\equiv C-R \xrightarrow{\text{Na/NH}_3} \underset{\text{\textit{trans} alkene}}{\text{(trans alkene structure)}}$$

trans alkene

Oxymercuration of alkynes

Reacting alkynes with mercury, water, and acid, you may expect the reaction to make the alcohol on the double bond, just as mercuric acetate reacts with alkenes to make the Markovnikov alcohol (see Chapter 10 for a review of Markovnikov products and the oxymercuration of alkenes). Indeed, the Markovnikov *enol* (an alcohol on a double bond) is formed in which the alcohol group is placed on the more-substituted carbon. However, enols are unstable and rapidly convert into ketones, as shown in Figure 11-14.

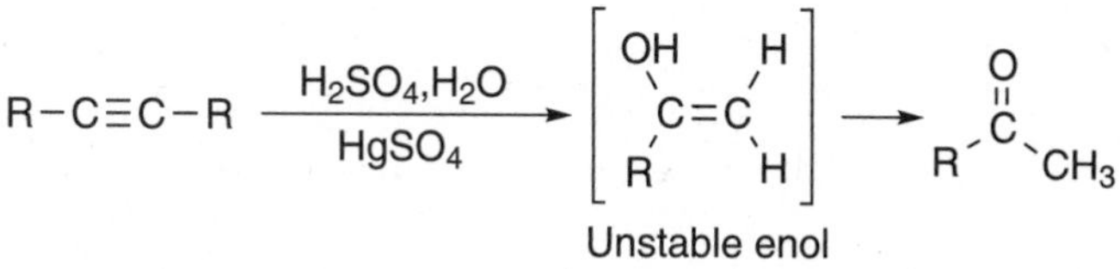

Figure 11-14:
The oxymercuration of an alkyne.

$$R-C\equiv C-R \xrightarrow[\text{HgSO}_4]{\text{H}_2\text{SO}_4,\text{H}_2\text{O}} \left[\text{Unstable enol} \right] \longrightarrow R\overset{\overset{\text{O}}{\|}}{C}\text{CH}_3$$

Unstable enol

The term *enol* is a combination of the suffixes *–en* (from alkene) and *–ol* (from alcohol).

The reaction that converts the enol into the ketone is called a *tautomerization reaction,* and both the enol and the ketone are considered *tautomers* of each other. Tautomers are molecules that differ only in the placement of a double bond and a hydrogen.

Hydroboration of alkynes

The hydroboration reaction of alkynes works in the same way as the hydroboration reaction of alkenes, and forms the anti-Markovnikov product. In the case of the hydroboration of alkynes, the product is an enol with the alcohol group on the least-substituted carbon, as shown in Figure 11-15. As with the oxymercuration reaction, this enol is unstable and tautomerizes to the aldehyde.

Figure 11-15:
The hydroboration of an alkyne.

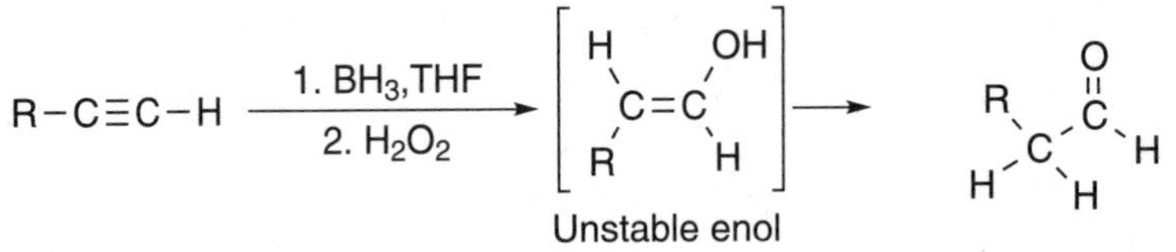

$$R-C\equiv C-H \xrightarrow[\text{2. H}_2\text{O}_2]{\text{1. BH}_3,\text{THF}} \left[\begin{array}{c} \underset{R}{\overset{H}{\diagdown}}C=C\underset{H}{\overset{OH}{\diagup}} \end{array}\right] \longrightarrow \underset{\underset{H}{\overset{R}{\diagdown}}C\underset{H}{\diagdown}}{\overset{O}{\parallel}}C\diagdown_H$$

Unstable enol

 Oxymercuration and hydroboration of carbon-carbon triple bonds are useful primarily with terminal alkynes because with terminal alkynes, you get only a single product (with hydroboration, you get the aldehyde; with oxymercuration, you get the ketone). With internal alkynes, both sides of the alkyne are equally substituted, so water can be added equally well to either side of the triple bond, so these reactions yield mixtures of two products.

Part III
Functional Groups

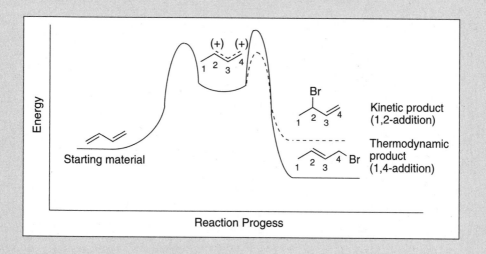

In this part . . .

- ✔ Recognize different functional groups.
- ✔ See reactions to make alkyl halides and alcohols.
- ✔ Find out what makes a ring system aromatic.
- ✔ Understand the reactions of alkyl halides, alcohols, and aromatics.

Chapter 12

Replacing and Removing: Substitution and Elimination Reactions

*A*lthough you see countless reactions in organic chemistry, you usually don't sweat the details too much, details like what kinds of solvents are ideal for that reaction, or typical side pathways that lead to undesirable byproducts. The substitution and elimination reactions are exceptions, because these are some of the most widely applicable and versatile reactions that you see in organic chemistry. As such, they deserve a closer look. In this chapter, I present the substitution and elimination reactions, and show you how to recognize these types of reactions. With just these two reaction types, you can synthesize more organic molecules than you can shake a stick at.

Group Swap: Substitution Reactions

Substitution reactions follow the form shown in Figure 12-1. The overall reaction is fairly simple — one group simply substitutes for another in the reaction.

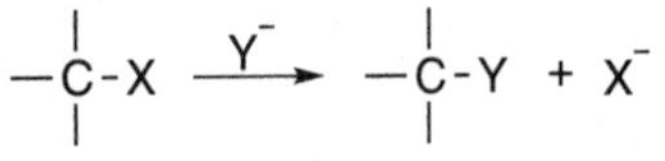

Figure 12-1:
A substitution reaction.

$$-\overset{|}{\underset{|}{C}}-X \xrightarrow{\ Y^{-}\ } -\overset{|}{\underset{|}{C}}-Y\ +\ X^{-}$$

Two mechanisms for substitution are possible, the S_N1 mechanism, and the S_N2 mechanism, and they're shown in Figure 12-2. Both involve substituting one group on a molecule for another. If you think of these substitution reactions in relationship terms, the S_N2 mechanism is analogous to dumping your current significant other and immediately starting a relationship with a new romantic partner. The S_N1 mechanism is analogous to breaking up with your current significant other, staying single for a while, and only after you've been single, becoming attached to a new romantic partner.

So how do you know which mechanism will occur for a given substitution reaction? The answer, unfortunately, is, "It depends." It depends on the solvent, the nature of the substrate, and the substituting group (called the nucleophile, or nucleus lover, sometimes abbreviated Nuc). To figure out which mechanism will occur, you need to see the details of each mechanism.

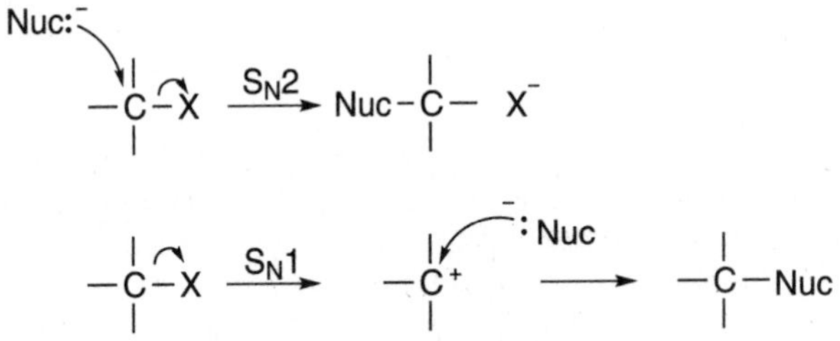

Figure 12-2:
Two different substitution mechanisms.

Seeing Second-Order Substitution: The S_N2 Mechanism

The mechanism for the S_N2 reaction is shown in Figure 12-3. This reaction is called the *substitution nucleophilic bimolecular reaction,* which, thankfully, is called S_N2 for short. The S_N2 reaction occurs in a single step: A nucleophile (Lewis base) attacks a carbon that's attached to an electronegative leaving group (labeled X), and gives the leaving group the boot, taking its place.

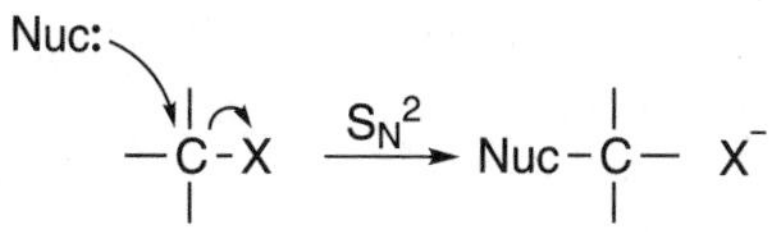

Figure 12-3:
The S_N2
mechanism.

$$\text{Nuc:} \quad -\overset{|}{\underset{|}{C}} - X \quad \xrightarrow{\text{S}_N2} \quad \text{Nuc} - \overset{|}{\underset{|}{C}} - \quad X^-$$

Why does the nucleophile attack the carbon? One way to think of this reaction is in terms of the attraction between opposite charges. The carbon-leaving group bond (C-X) is polarized; that is, the electronegative leaving group pulls electron density away from the carbon to which it's attached, leaving the carbon with a partially positive charge, as shown in Figure 12-4. (See Chapter 2 for more on electronegativity.) Therefore, the molecule that contains the leaving group (called the *substrate*) acts as an *electrophile* (an electron lover). *Nucleophiles* (electron-rich species that love nuclei) attack the partially positive carbon, in part due to the electrostatic attraction between the two nuclei.

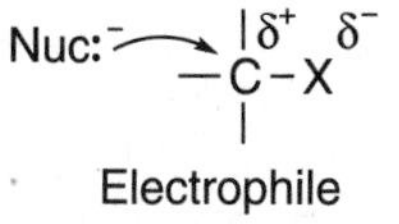

Figure 12-4:
Nucleophile-
electrophile
attraction.

$$\text{Nuc:}^- \quad -\overset{|}{\underset{|}{C}}\overset{\delta^+}{-} \overset{\delta^-}{X}$$

Electrophile

If you understand nucleophile-electrophile attraction in this way, you can understand many, many organic reactions. The basic templates for these reactions are the same: Some electron-rich atom (a nucleophile) attacks an electron-poor atom (an electrophile). The details change, but that's pretty much the gist of it.

How fast? The rate equation for the S_N2 reaction

The rate of the S_N2 reaction follows the rate equation: rate = k[substrate] [nucleophile]. Looking at this rate equation, you see that the concentration of both the substrate and the nucleophile determine the rate of the S_N2 reaction, making the S_N2 reaction second order (thus, the "2" in S_N2).

The reaction diagram for the S_N2 mechanism is shown in Figure 12-5. A reaction diagram displays the energy (labeled E) as a function of time. Every reaction has at least one energy "hill" that must be climbed to transform the starting materials into the products; this "hill" is a barrier called the activation energy, or Ea. The activation barrier is the minimum amount of energy that needs to be supplied to transform the starting material into product.

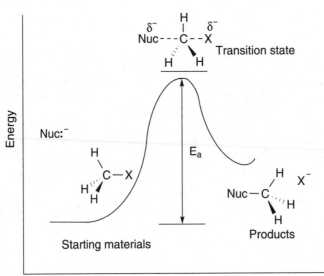

The size of the activation barrier determines the rate of the reaction. If the reaction only has to climb a very small energy hill, the reaction will proceed quickly. If the energy hill is high, the reaction will occur much more slowly, just as an athlete runs faster over a small hill than over a mountain.

The top of the energy hill is the transition state of the reaction. This is the point at which the starting materials are midway through the reaction, when the reactants are at their highest energy level, and when the bonds in the reactants are partially broken and bonds in the products partially formed. A transition state cannot be isolated; it's not a long-lived intermediate, but simply a transition between the starting materials and products. It's just a point on the road to transforming one material into the next. A reaction passes through the transition state faster than a chemistry professor can mark an X on an exam.

Effect of the substrate on the S$_N$2 reaction

A nucleophile approaching a substrate in order to undergo an S$_N$2 reaction is a lot like a fan approaching a celebrity to get an autograph. If a celebrity has no bodyguards, the fan can easily approach and ask for the autograph.

With one bodyguard (I'll call these bodyguards "R groups"), the fan has a little more difficulty slipping by the bodyguard, but getting the autograph is still doable. The fan may need to wait until the bodyguard lets his guard down and allows the fan to slip past. With two R group bodyguards, the fan would have an even harder time (or a longer wait) before he or she could slip through to the celebrity. But with three bodyguards blocking all lines of approach to the celebrity, the fan might as well forget about getting that precious autograph.

R groups on the carbon that contains the leaving group act as bodyguards, and hinder the approach of the nucleophile, as shown in Figure 12-6. Therefore, S_N2 reactions work best with methyl and primary substrates, when the carbon containing the leaving group has no R group bodyguards attached to it (in the case of a methyl substrate) or just one (in the case of a primary substrate). Reactions with substrates that contain two R group bodyguards (secondary substrates) do proceed, but these reactions take place at a slower rate than reactions with primary substrates. But with tertiary substrates (substrates in which the carbon that contains the leaving group has three R groups attached to it), the S_N2 reaction says, "No dice," and you get no reaction. Why? Everything comes back to the mechanism. In the S_N2 mechanism, the nucleophile attacks the backside of the substrate. With three big fat bulky R groups blocking the rear approach, the nucleophile can't get close enough to attack the carbon. Such blocking by bulky groups is called *steric hindrance,* using organic-speak. Because of such steric hindrance, you won't see an S_N2 reaction on a tertiary substrate.

Figure 12-6:
Steric
hindrance
can prevent
an S_N2
reaction.

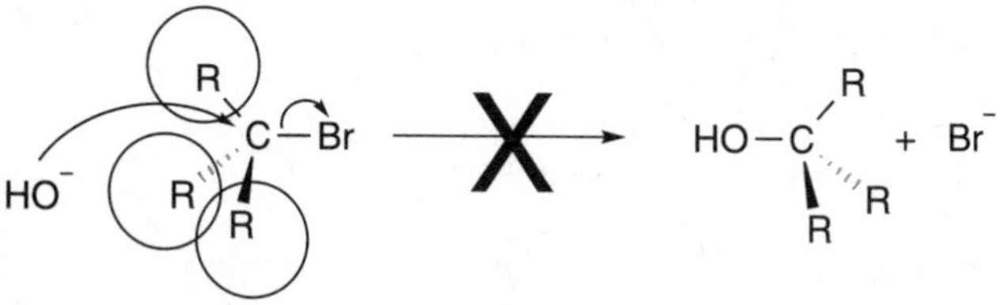

$$HO^- \quad \overset{R}{\underset{R}{\overset{|}{C}}}{-}Br \quad \longrightarrow \quad X \longrightarrow \quad HO-\overset{R}{\underset{R}{\overset{|}{C}}} + Br^-$$

S_N2 reactions prefer methyl (CH_3X) over primary substrates (R-CH_2-X), but primary over secondary substrates (R_2CHX). S_N2 reactions don't work with tertiary substrates (R_3CX).

Needs nucleus: The role of the nucleophile

Because the concentration of the nucleophile is included in the rate equation of the S_N2 reaction, a good nucleophile is required. That leads to the somewhat sticky question: What makes a good nucleophile? Unfortunately, that question can't be answered precisely, because the strength of a nucleophile can change by varying the solvent or reaction conditions. Still, some general rules about nucleophile strength can be applied.

Any molecule with a lone pair of electrons to donate can act as a nucleophile. The strength of a nucleophile (or the *nucleophilicity,* using organic-speak) generally goes hand-in-hand with basicity. A strong base is usually a strong nucleophile and vice versa. But basicity and nucleophilicity are not the same things. *Basicity* refers to the ability of a molecule to pluck off a proton, and is defined by the base's equilibrium constant; *nucleophilicity* refers to the ability of a lone pair to attack a carbon on an electrophile.

When wouldn't basicity and nucleophilicity coincide? Usually, these don't coincide when you have bases/nucleophiles substituted with bulky groups. For example, both methoxide (CH_3O^-) and *t*-butoxide ($(CH_3)_3CO^-$), shown in Figure 12-7, are strong bases, but methoxide is a good nucleophile while *t*-butoxide is not. The reason for this difference is that *t*-butoxide has three bulky methyl groups that hinder approach to the substrate (another example of steric hindrance). Therefore, even though *t*-butoxide is a strong base, it's a poor nucleophile.

Figure 12-7:
The nucleo-
philicity of
two strong
bases.

H_3C-O^-

Methoxide
Strong base
Strong nucleophile

$$H_3C-\overset{\overset{\displaystyle CH_3}{|}}{\underset{\underset{\displaystyle CH_3}{|}}{C}}-O^-$$

t-butoxide
Strong base
Poor nucleophile

Steric hindrance decreases nucleophilicity.

In addition to the basicity of a molecule, two other major factors can help you compare the nucleophilicities of molecules:

- ✔ Negatively charged nucleophiles are stronger nucleophiles than neutral nucleophiles (just as negatively charged atoms are more basic than neutral atoms). Thus, OH^- is a stronger nucleophile than H_2O, and HS^- is a stronger nucleophile than H_2S.

- ✔ Typically, nucleophilicity increases as you go down the periodic table. Therefore, H_2S is a better nucleophile than H_2O because sulfur is one row down on the periodic table from oxygen. Likewise, iodide (I^-) is a better nucleophile than bromide (Br^-) because iodine is one row down from bromine on the periodic table.

Seeing the S$_N$2 reaction in 3-D: Stereochemistry

In the S$_N$2 reaction, the nucleophile approaches from the back side of the substrate. Therefore, in the product, the three groups on the carbon have inverted, like an umbrella blown inside out by the wind. With the S$_N$2 reaction, then, you get inversion of stereochemistry (see Chapter 6 for more on stereochemistry). For example, with the reaction of 2-bromobutane and hydroxide (see Figure 12-8), the chiral center has an S configuration in the starting material but an R configuration in the product as a result of the inversion of configuration in the S$_N$2 reaction.

The S$_N$2 reaction leads to an inversion of the stereochemistry.

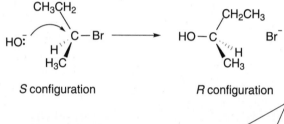

S configuration   R configuration

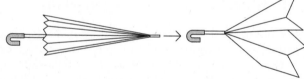

Figure 12-8:
The S$_N$2
reaction of
2-bromobu-
tane.

Seeing solvent effects

The choice of solvent also affects the S$_N$2 substitution reaction. Not all solvents are created equal; some solvents work better than others in a given reaction. In the S$_N$2 reaction, the preferred solvents are those that are both polar (see Chapter 2) and aprotic. *Protic* is organic-speak for solvents that contain O-H or N-H bonds; these solvents include alcohols, water, and amines. Aprotic solvents have no N-H or O-H bonds. Good polar aprotic solvents for the S$_N$2 reaction include DMSO (dimethyl sulfoxide), CH_2Cl_2 (dichloromethane), and ethers (R-O-R).

Why do you need aprotic and not protic solvents for the S_N2 reaction? Protic solvents tend to form a solvent "cage" around the nucleophile, as shown for the chloride ion caged by water in Figure 12-9. This cage makes the nucleophile less nucleophilic. A nucleophile in a protic solvent is like a college student surrounded by televisions — it doesn't feel particularly inclined to go out and do the work that needs to be done. Polar solvents that are aprotic still have the ability to dissolve the polar reactants (because "like dissolves like" as you discovered in your first year of chemistry), but these polar aprotic solvents don't cage the nucleophile, so they don't make the nucleophile less nucleophilic.

Protic solvents decrease nucleophilicity in S_N2 reactions.

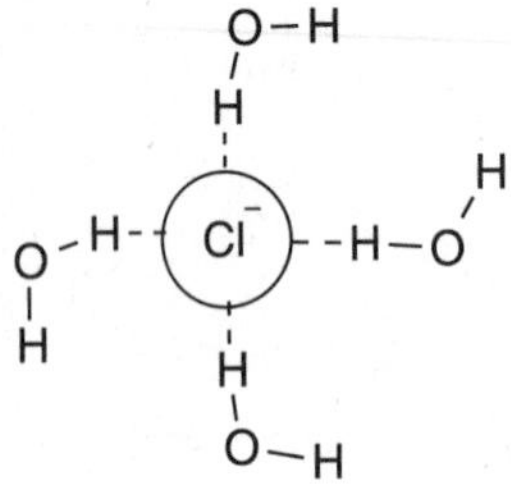

Figure 12-9:
The solvent cage formed by the protic solvent (water).

I'm outta here: The leaving group

Another requirement of the S_N2 reaction is that the substrate needs to have a good leaving group. The leaving group is the piece that's displaced by the nucleophile (often labeled X). Good leaving groups are typically weak bases, and weak bases are the conjugate bases of strong acids (see Chapter 4 for more on acids and bases). Therefore, if you can find a strong acid and deprotonate it, you've probably found a good leaving group. The hydrohalic acids (HBr, HCl, and so on) are strong acids, so the halides (like I^-, Br^-, Cl^-) are excellent leaving groups. The halides, in fact, are the most common leaving groups in S_N2 reactions. Another good leaving group is tosylate. Tosylate is the conjugate base of the strong acid p-toluenesulfonic acid, and is one of the best leaving groups available (better than all the halides). Figure 12-10 shows the most common leaving groups, and ranks them from best to worst.

Hydroxide ion (OH^-), alkoxides (RO^-), and amide ion, (NH_2^-) are bad leaving groups because they're strong bases. Show them as leaving groups at your own risk. Because these groups are bad leaving groups, you won't typically see an S_N2 reaction on an ether (which is why you can use ethers as solvents in S_N2 reactions), alcohol, alkyl fluoride, or an amine.

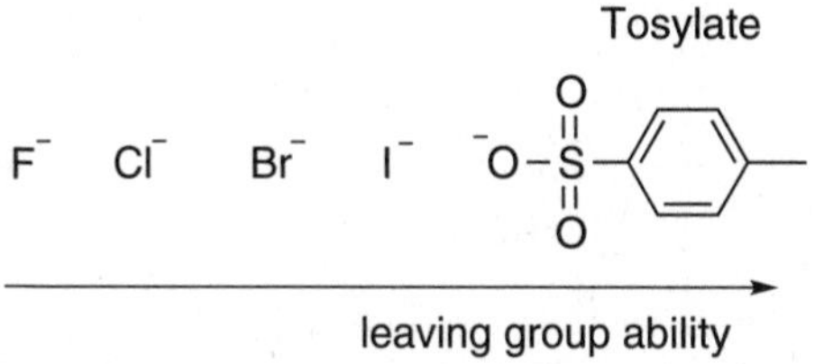

Figure 12-10:
Leaving
groups
for the S_N2
reaction.

Tosylate

F⁻ Cl⁻ Br⁻ I⁻

$$\longrightarrow$$

leaving group ability

First-Order Substitution: The S_N1 Reaction

The other substitution mechanism is the S_N1 mechanism, and is shown in Figure 12-11. The S_N1 reaction proceeds in two steps. In the first step, the leaving group decides to pack its bags and take off, leaving a carbocation intermediate. Then the nucleophile attacks the carbocation to make the substituted product.

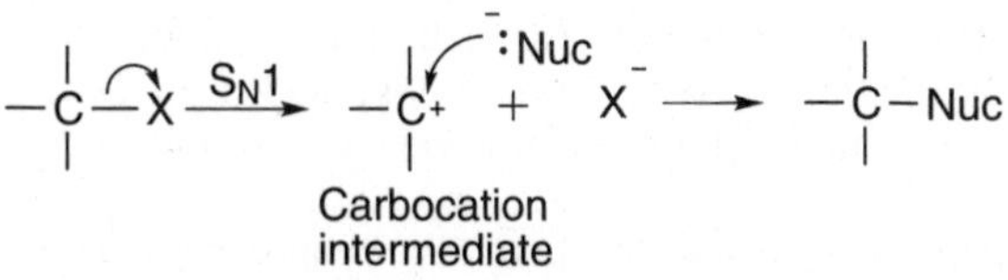

Figure 12-11:
The S_N1
mechanism.

$$-\overset{|}{\underset{|}{C}}-X \xrightarrow{S_N1} -\overset{|}{\underset{|}{C}}{}^{+} \; + \; X^{-} \longrightarrow -\overset{|}{\underset{|}{C}}-Nuc$$

:Nuc

Carbocation
intermediate

How fast? The rate equation for the S_N1 reaction

The rate for the S_N1 reaction follows the rate equation: rate = k[substrate].

You may be surprised to see that, unlike the case in the S_N2 reaction, the concentration of the nucleophile is not included in the rate equation for the S_N1 reaction. The rate equation is first order (thus, the "1" in S_N1) and depends only on the concentration of the substrate. Why is the nucleophile not included? You can see why by looking at the reaction diagram, shown in Figure 12-12.

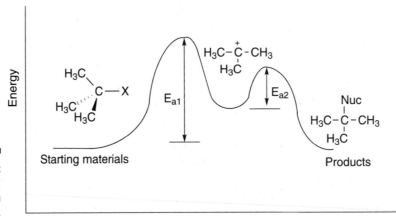

Figure 12-12:
The S_N1
reaction
diagram.

From the reaction diagram, you see that two hills need to be climbed to transform the starting materials into the product. The first hill is the taller one, and is the energy barrier for the step in which the leaving group makes like a tree and gets the heck out of there, forming the carbocation. The second step, the attack of the nucleophile on the carbocation, involves a smaller energy hill. Because the first step has the highest activation barrier, the first step is the slowest step, making it the *rate-determining step.*

You've probably heard the phrase "A chain is only as strong as its weakest link." Similarly, a reaction is only as fast as its slowest step, which is called the rate-determining step. Steps that follow the rate-determining step have no effect on the rate of a reaction. You have a similar situation when you have a tiny washing machine and a huge dryer, and want to know how quickly you can clean your clothes (see Figure 12-13). Is the rate at which you get clean clothes going to increase by increasing the size of your already too-big dryer? No, because the rate determining step is the washing step. Only increasing the size of the washer would increase the rate at which you get clean clothes out of this clothes-cleaning system.

That's why the nucleophile is not included in the rate equation for the S_N1 reaction, because the nucleophile only gets involved in the mechanism after the rate-determining step. Making the nucleophile stronger, or increasing the concentration of the nucleophile, is like making the dryer bigger — it has no effect on the rate of forming product.

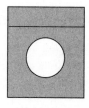

Figure 12-13:
With this tiny washer, washing is the rate-determining step.

Washer
Rate-determining step

Dryer

Seeing good S_N1 substrates

Good substrates for the S_N1 reaction are different from those that are good for the S_N2 reaction. Good substrates for the S_N1 reaction will be substrates that form a stable carbocation upon releasing the leaving group. If you lower the energy of the cation intermediate, you lower the activation energy and, thus, speed up the reaction. So, to find good substrates for the S_N1 reaction, you look for those that will lead to a stable intermediate carbocation.

Recall the features that contribute to carbocation stability (refer to Chapter 10 if you need a refresher on carbocations). Carbocations are stabilized by the presence of alkyl groups or by resonance. Thus, tertiary substrates form more stable carbocations than secondary or primary substrates. As a result, tertiary substrates are better S_N1 substrates than secondary substrates, which in turn are better S_N1 substrates than primary substrates (see Figure 12-14). Substrates with the ability to stabilize the cation through resonance are also good S_N1 substrates; these substrates include benzylic and allylic substrates.

Figure 12-14:
Substrates for the S_N1 reaction.

$$R-\underset{\underset{R}{|}}{\overset{\overset{R}{|}}{C}}-X \cong \text{(benzylic)}-X \; > \; R-\underset{\underset{R}{|}}{CH}-X \cong \text{(allylic)}-X \; > \; R-CH_2-X$$

3°   Benzylic   2°   Alylic   1°

Best substrate   **Worst substrate**

Good S_N1 substrates are those that can make stable carbocation intermediates.

Seeing solvent effects on the S_N1 reaction

S_N2 reactions prefer aprotic solvents, but S_N1 reactions prefer protic solvents. This preference comes from the ability of protic solvents to stabilize the intermediate carbocation (as shown in Figure 12-15), lowering the energy hill for the rate-determining step. Recall that in the S_N2 reaction you don't want protic solvents because they decrease the nucleophile strength. In the S_N1 reaction, the strength of the nucleophile has no effect on the rate, so protic solvents (like alcohols and water) are excellent for S_N1 reactions. You generally won't see S_N1 reactions in aprotic solvents.

Protic solvents stabilize carbocations in S_N1 reactions.

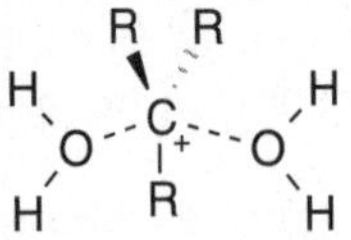

Figure 12-15:
The stabilizing interaction of a protic solvent (water) with a cation.

Stereochemistry of the S_N1 reaction

Stereochemistry in the S_N1 reaction is not as clear-cut as it is in the S_N2 reaction. The intermediate in the S_N1 reaction is a carbocation, which is sp^2 hybridized. As a result, these carbocations are planar, and have an empty p orbital. The empty p orbital could be attacked equally well by the nucleophile's lone pair from either the top or the bottom of the carbocation, as shown in Figure 12-16. Therefore, you get about a 50/50 mixture of enantiomers from the S_N1 reaction, which is called a *racemic mixture* (see Chapter 6).

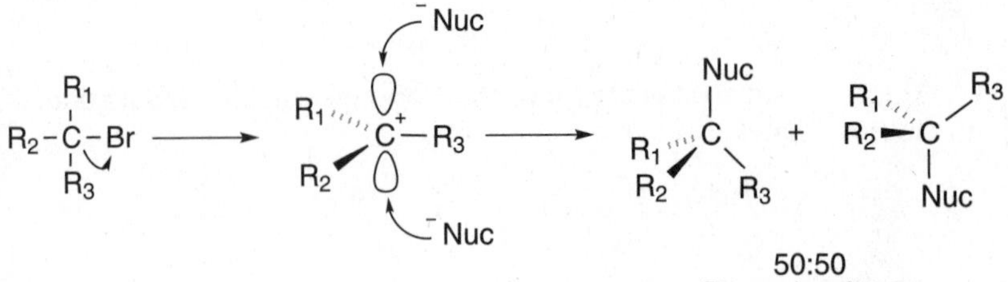

Figure 12-16:
The S_N1 reaction of a tertiary alkyl halide.

50:50
Racemic mixture

S_N1 reactions give racemic products.

Other fun facts about the S_N1 reaction

Because the strength of the nucleophile is unimportant, S_N1 reactions work best with weak nucleophiles. However, the S_N1 reaction requires a good leaving group (like a halide or tosylate), as the S_N2 reaction does. An important detail to remember about the S_N1 reaction is the consequence of going through a carbocation intermediate. Recall that carbocations are mischievous little devils, and will rearrange if doing so will lead to a stable cation (see Chapter 10 for more on carbocations). As a result, you occasionally see carbocation rearrangements in the S_N1 reaction.

Table 12-1 is a convenient tool for comparing the S_N1 and S_N2 reaction. If you want to determine whether a substitution reaction will go by the S_N1 or the S_N2 mechanism, look first at the substrate. If the substrate is primary or methyl, the reaction will most likely proceed by the S_N2 mechanism. If the substrate is tertiary, the reaction will proceed by the S_N1 mechanism. Secondary substrates are something of a gray area; both reactions will work with these substrates, so you need to look at other features of the reaction (like the solvent) to see which mechanism may be favored over the other. With secondary alkyl substrates, you may also get a mixture of the two mechanisms.

Table 12-1	Comparing S_N1 and S_N2 Reactions	
	S_N1	*S_N2*
Substrate	Prefers 3° over 2°	Prefers 1° over 2°
Rate equation	rate = k[substrate]	rate = k[substrate] [nucleophile]
Nucleophile	Nucleophile unimportant	Requires a good nucleophile
Reaction	Racemic products	Inversion of stereochemical configuration
Leaving groups	Good leaving group required	Good leaving group required
Solvent	Prefers polar protic solvents (alcohols, water, and so on)	Prefers polar aprotic solvents (ethers, halogenated solvents, and so on)
Rearrangements	Possible	Not possible

Seeing Elimination Reactions

Elimination reactions often compete with substitution reactions. The general form of an elimination reaction is shown in Figure 12-17. In this reaction, a substrate (typically, an alkyl halide) eliminates one equivalent (unit) of acid to make an alkene. As in substitution reactions, two possible mechanisms are available for this elimination reaction — the E1 and E2 mechanism — and both elimination reactions have similarities to their substitution counterparts.

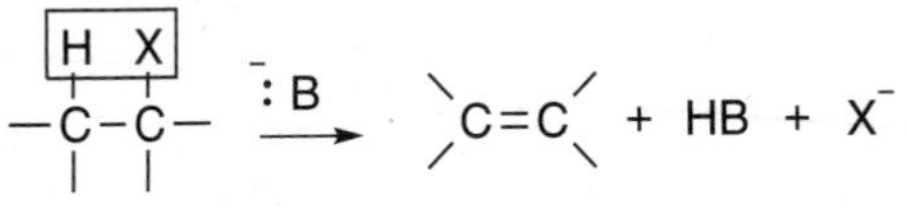

Figure 12-17:
The elimination reaction.

Seeing second-order eliminations: The E2 reaction

Second-order elimination is called the *E2 reaction.* Like the S_N2 mechanism, the E2 mechanism takes place in a single step, as shown in Figure 12-18. A base plucks off a proton on a carbon adjacent to the leaving group, forming the double bond and giving the leaving group the boot.

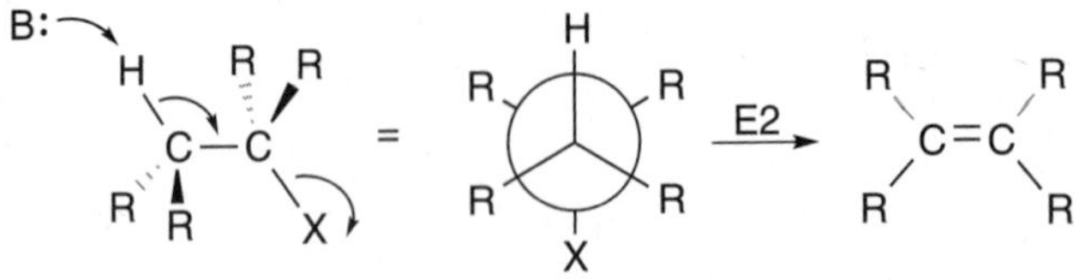

Figure 12-18:
The E2 elimination mechanism.

One requirement of the E2 reaction is that the hydrogen to be eliminated and the leaving group must typically be in anti-periplanar geometry. To be anti-periplanar means that the hydrogen and the leaving group (as well as the two carbons that will form the double bond) must be on the same plane and on opposite faces of the carbon-carbon bond, as shown in Figure 12-18.

The rate equation for an E2 reaction is as follows: rate = k[base][substrate]. Because the base is included in the rate equation, the base strength affects the rate of the reaction. The E2 reaction requires a strong base, and is the most common pathway for elimination reactions.

Seeing first-order elimination: The E1 reaction

First-order elimination, or the E1 reaction, is somewhat less common than second-order (E2) elimination. The mechanism for the E1 reaction, like the mechanism for the S_N1 reaction, has two steps and is shown in Figure 12-19. First, the leaving group pops off to make the carbocation; this is the same first step as in the S_N1 reaction. Then the base plucks off the hydrogen on an adjacent carbon to form the double bond.

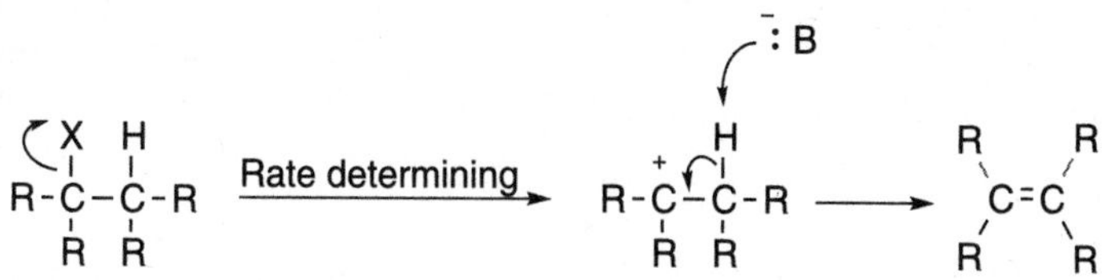

Figure 12-19:
The E1 elimination mechanism.

Because the first step of the mechanism — the formation of the carbocation — is the rate-determining step, the rate follows the equation: rate = k[substrate]. Because the base is not included in the rate equation, the strength of the base is unimportant to the rate of the reaction. E2 elimination is favored with strong bases, so you see E1 elimination typically only with weak bases.

Help! Distinguishing Substitution from Elimination

Unfortunately, chemists live in an imperfect world. Reactions that are exclusively substitution or elimination are rare in the laboratory. Instead, chemists often get a mixture of the reaction products of both substitution and elimination reactions, because good nucleophiles generally are also good bases and vice versa. Here are a few generalizations, however, that will help you distinguish which of these reactions will predominate:

✔ Strong bases/nucleophiles force the reaction into second-order reactions. Thus, with strong bases and nucleophiles (such as OH⁻), you get S_N2 or E2 reactions, or both. With weak bases/nucleophiles, you more often get first-order products (those produced by either S_N1 or E1 reactions).

✔ Reactions of primary substrates generally proceed via S_N2 reactions (methyl substrates always proceed by S_N2). When very strong bases/ nucleophiles are used with primary substrates, you get a mixture of both S_N2 and E2 reactions.

✔ Reactions of tertiary substrates produce E1 and S_N1 reactions with weak bases/nucleophiles plus a protic solvent; with strong bases, reactions of tertiary substrates produce E2 reactions.

✔ The reactions of secondary substrates are the hardest to predict. Under the right conditions, secondary substrates can undergo reactions by all four mechanisms. Weak bases/nucleophiles plus a protic solvent will typically give you a mixture of E1 and S_N1 products; strong bases/strong nucleophiles will typically give you a mixture of E2 and S_N2 products.

✔ Spotting nucleophiles that are not basic will help you distinguish substitution from elimination reactions. For example, the halides (I^-, Br^-, Cl^-) and thiols (R-SH) are nucleophilic but not terribly basic. The reactions of these molecules typically proceed exclusively by substitution. *t*-Butoxide ($(CH_3)_3CO^-$), on the other hand, is a poor nucleophile but a powerful base, and almost exclusively forces the reaction to go via an E2 elimination.

Chapter 13

Getting Drunk on Organic Molecules: The Alcohols

*W*hen most people think of alcohol, they think of ethyl alcohol, the alcohol found in beer and wine. Many people, though, don't realize that there are many thousands of different alcohols. Ethylene glycol is an alcohol found in antifreeze; methanol is a wood alcohol that can cause blindness if ingested; isopropyl alcohol is an alcohol used for sterilizing cuts. Alcohols, like the alkenes, are an extremely important functional group, not just because they're found in so many valuable products (and not just beers and wines), but also because they're so versatile: They're easily formed and converted into many other functional groups.

In this chapter, I introduce you to the alcohol functional group, show you how to name and classify alcohols, and show you reactions that both make alcohols and convert alcohols into other functional groups.

Classifying Alcohols

Alcohols are molecules that contain a hydroxyl (OH) group, and they're typically classified by the carbon to which the hydroxyl group is attached. If the carbon bonded to the OH is attached to one other alkyl group, the alcohol is classified as primary (1°); if the carbon is attached to two other alkyl groups, the alcohol is classified as secondary (2°); if the carbon is attached to three alkyl groups, the alcohol is classified as tertiary (3°), as shown in Figure 13-1.

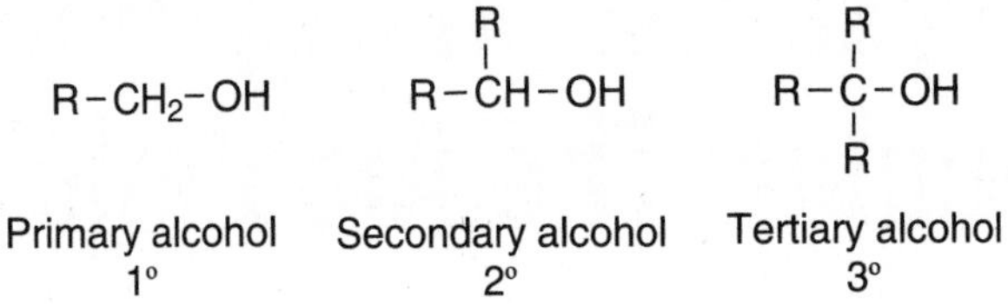

Figure 13-1:
Alcohol
classification.

R–CH$_2$–OH

Primary alcohol
1°

R–CH–OH
|
R

Secondary alcohol
2°

R
|
R–C–OH
|
R

Tertiary alcohol
3°

An Alcohol by Any Other Name: Naming Alcohols

You can name alcohols just by extending the nomenclature rules used for alkanes (refer to Chapter 7 for an alkane refresher). To name an alcohol you follow these five steps:

1. **Determine the parent name of the alcohol by looking for the longest chain that includes the alcohol.**

   Snip the *e* off the suffix for the alkane and replace it with the suffix *–ol,* which stands for alcohol. For example, a two-carbon alcohol would not be ethane but ethanol.

2. **Number the parent chain. Start numbering from the side closer to the hydroxyl group.**

3. **Identify all the substituents of the parent chain and name them.**

4. **Order the substituents alphabetically in front of the parent name.**

5. **Identify the location of the hydroxyl group by placing a number in front of the parent name.**

Now try naming the alcohol shown in Figure 13-2.

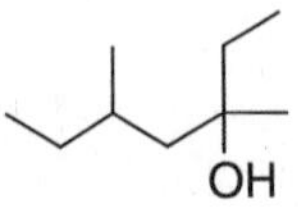

Figure 13-2:
An alcohol.

OH

First, find the parent chain, as shown in Figure 13-3. The parent chain is the longest chain of carbons that contains the hydroxyl group. In this case, the parent chain is seven carbons long, so this is a heptanol.

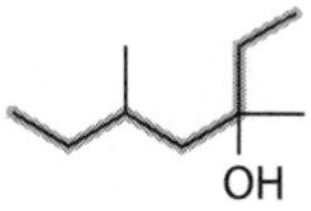

Figure 13-3:
The parent group.

Then number the parent chain, starting from the end that reaches the hydroxyl group sooner. In this case, that's from right to left (see Figure 13-4).

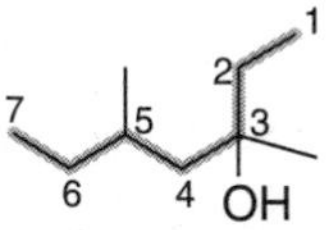

Figure 13-4:
Numbering the chain.

Find and name the substituents. The molecule shown in Figure 13-5 has two methyl group substituents — one at the number-three carbon and one at the number-five carbon.

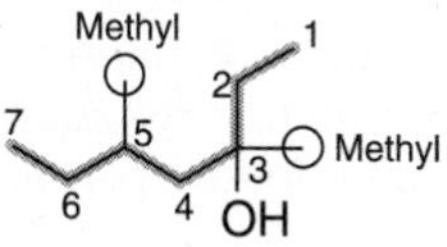

Figure 13-5:
Identifying substituents.

Then place the substituents (in alphabetical order) in front of the parent group. Indicate the position of the hydroxyl group by placing a number in front of the parent name. The two methyl groups combine to make dimethyl, and the name for this alcohol is, therefore 3,5-dimethyl-3-heptanol.

Alcohol-Making Reactions

So, now, take a look at how you can make alcohols.

Adding water across double bonds

Chapter 10 shows you two reactions that produce alcohols from alkenes. The oxymercuration of alkenes makes the Markovnikov alcohol (the alcohol with the OH on the most highly substituted carbon), while hydroboration makes the anti-Markovnikov alcohol (the alcohol with the OH on the least-substituted carbon). This conversion is outlined in Figure 13-6.

Figure 13-6:
Adding water to an alkene.

$$\underset{/}{\overset{\backslash}{C}}=\underset{\backslash}{\overset{/}{C}} \quad \xrightarrow[\text{or:}\left(\begin{array}{l}\text{1. } BH_3,\ THF\\ \text{2. } H_2O_2\end{array}\right)]{\begin{array}{l}\text{1. } Hg(OAc)_2\\ \text{2. } NaBH_4\end{array}} \quad \overset{H}{\underset{|}{\overset{|}{-C}}}-\overset{OH}{\underset{|}{\overset{|}{C}}}-$$

Reduction of carbonyl compounds

Alcohols can also be formed by the reduction of carbonyl (C=O) compounds. Two reagents are widely used to reduce carbonyl compounds, sodium borohydride ($NaBH_4$), a weaker reducing agent, and lithium aluminum hydride ($LiAlH_4$), a stronger reducing reagent. In metaphorical terms, sodium borohydride acts as a reductive popgun, while lithium aluminum hydride acts as a reductive cannon, as illustrated in Figure 13-7.

Figure 13-7:
The relative strengths of reducing agents.

pop

$NaBH_4$
Mild reducing agent

BOOM

$LiAlH_4$
Strong reducing agent

What do I mean by sodium borohydride being a weaker reducing agent than lithium aluminum hydride? Well, sodium borohydride can reduce carbonyl compounds that are easy to reduce to alcohols, like aldehydes and ketones. But this reagent is not strong enough to reduce the more stubborn carbonyl compounds like esters and carboxylic acids. For that job, you bring out the cannon reagent (lithium aluminum hydride), which is strong enough for the task.

Reducing aldehydes with sodium borohydride (or lithium aluminum hydride) is a very convenient way of making primary alcohols, as shown in Figure 13-8. Similarly, the reduction of ketones with sodium borohydride is a convenient way of preparing secondary alcohols. Tertiary alcohols cannot be prepared by reduction.

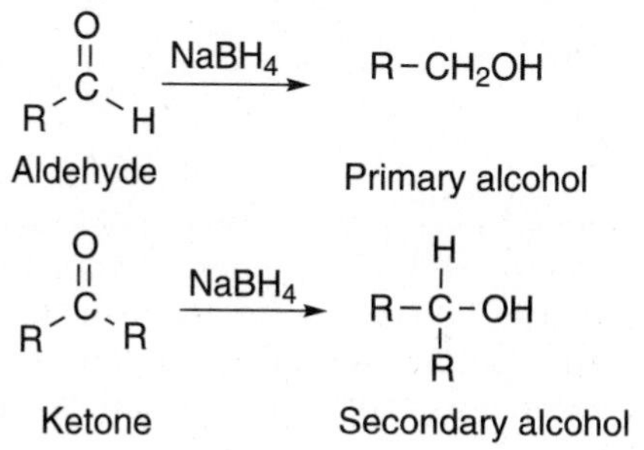

Figure 13-8: The formation of an alcohol by the reduction of an aldehyde or ketone.

Lithium aluminum hydride is strong enough to reduce both carboxylic acids and esters, as shown in Figure 13-9. Both carboxylic acids and esters are reduced to primary alcohols with this reagent.

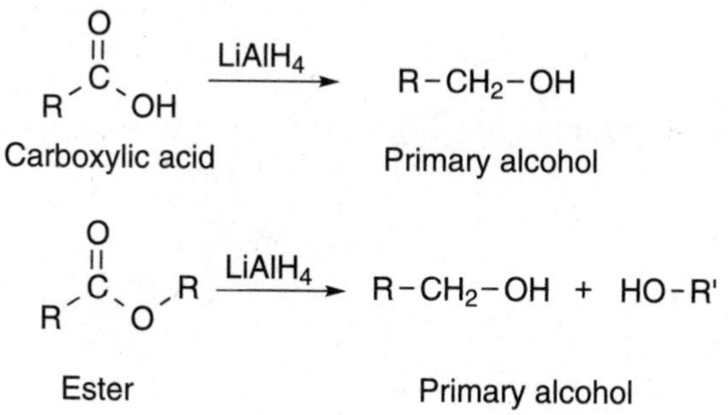

Figure 13-9: The formation of an alcohol via a lithium aluminum hydride reduction.

The Grignard reaction

Another extremely useful reaction for making alcohols is the Grignard reaction (pronounced *grin*-yard). To make alcohols using the Grignard reaction, you react a "Grignard reagent" with a carbonyl compound. Making a Grignard reagent is fairly simple: You simply add magnesium to an alkyl halide, as shown in Figure 13-10, which inserts the magnesium into the C-X bond to make the Grignard reagent.

Figure 13-10:
Making a
Grignard
reagent.

$$R-X \xrightarrow[\text{Ether}]{\text{Mg}} R-MgX$$

(X=Cl, Br, I) A Grignard reagent

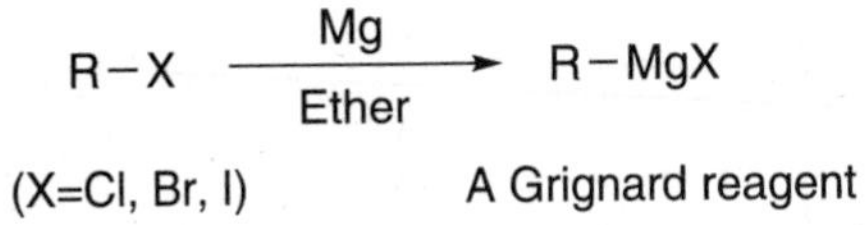

A Grignard reagent is an extremely powerful nucleophile (nucleus lover), and can react with electrophiles like carbonyl compounds. To determine the products made in a Grignard reaction, you can ignore the magnesium halide portion of the reagent (because this portion doesn't get involved in the reaction) and think of the Grignard reagent as acting as a carbanion in disguise. Figure 13-11 illustrates this idea.

Figure 13-11:
A Grignard
reagent.

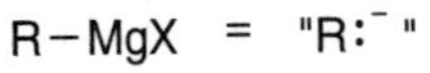

$$R-MgX = \text{"R:}^-\text{"}$$

The mechanism for the addition of a Grignard reagent to a carbonyl is shown in Figure 13-12.

Figure 13-12:
The
Grignard
reaction.

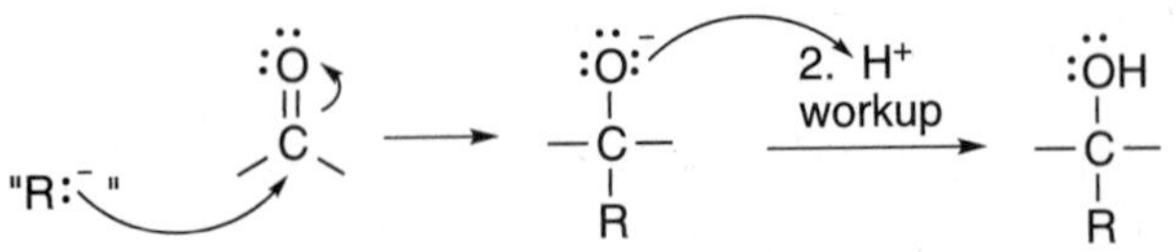

Although you can make only primary and secondary alcohols by reduction, you can make all kinds of alcohols using the Grignard reaction. If you react a Grignard reagent with formaldehyde, as shown in Figure 13-13, you can make primary alcohols. If you react it with an aldehyde, you get secondary alcohols. If you react it with ketones, you get tertiary alcohols.

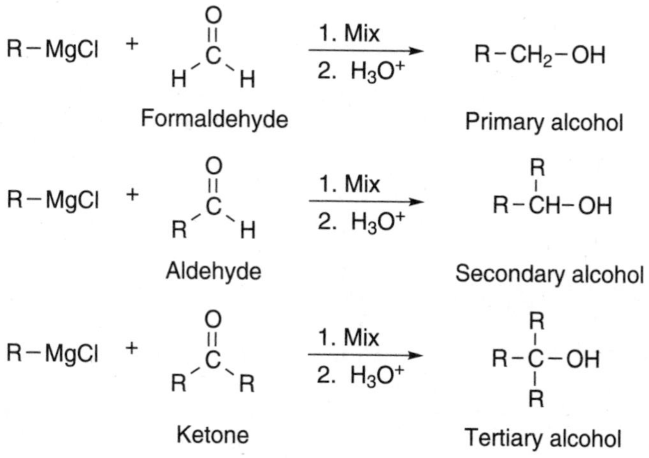

Figure 13-13: The formation of alcohols via the addition of Grignard reagents to carbonyl compounds.

Reactions of Alcohols

So, what can you do with alcohols? Now that you've seen reactions that make alcohols, I show you reactions that convert alcohols into other functional groups.

Losing water: Dehydration

One of the best reagents for converting alcohols into alkenes is phosphorous oxychloride ($POCl_3$). This reagent causes the alcohol to lose water to make the alkene, as shown in Figure 13-14.

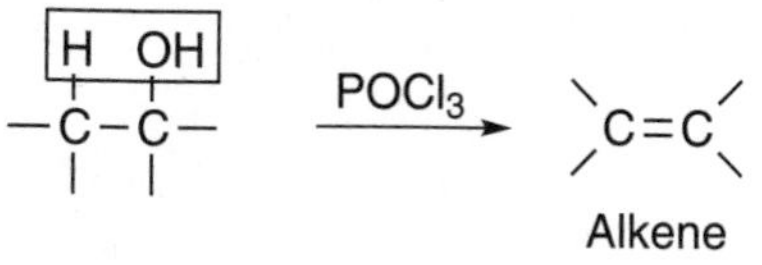

Figure 13-14: The dehydration of an alcohol.

Making ethers: Williamson ether synthesis

Ethers can also be made from alcohols. Reacting an alcohol with sodium metal forms an alkoxide, the alcohol's conjugate base, as shown in Figure 13-15. Alkoxides are strong nucleophiles that can react with primary halides to make ethers in an S_N2 reaction (refer to Chapter 12).

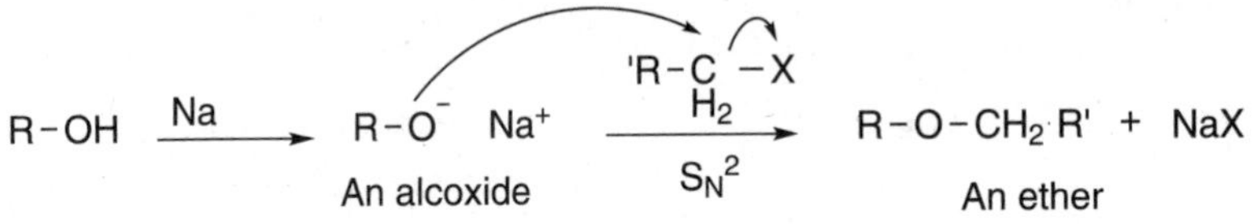

Figure 13-15:
The Williamson ether synthesis.

$$R-OH \xrightarrow{\text{Na}} R-O^- \ Na^+ \xrightarrow[S_N2]{'R-\underset{H_2}{C}-X} R-O-CH_2 R' + NaX$$

An alcoxide   An ether

Oxidation of alcohols

In the previous section, I showed how alcohols can be made by the reduction of carbonyl compounds. Running that reaction in reverse is also possible, and oxidizing alcohols into carbonyl compounds (ketones and aldehydes) can be accomplished with either of two different reagents — PCC and the Jones reagent. PCC (which stands for pyridinium chlorochromate) is a weaker oxidizing agent than the Jones reagent (a mix of chromium trioxide [CrO_3] and acid). For example, PCC oxidizes primary alcohols into aldehydes. The Jones reagent, on the other hand, oxidizes the primary alcohol even further into the carboxylic acid (see Figure 13-16). Both reagents oxidize secondary alcohols into ketones.

You cannot oxidize a tertiary alcohol.

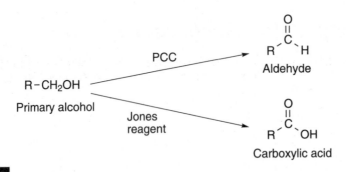

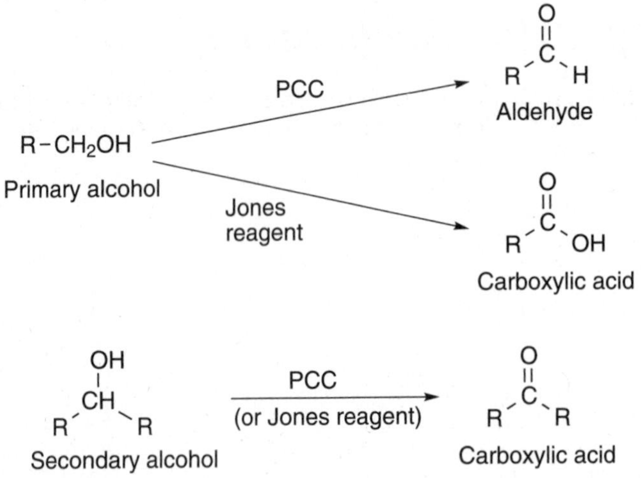

Figure 13-16:
The oxidation of alcohols.

Chapter 14

Side-by-Side: Conjugated Alkenes and the Diels–Alder Reaction

In This Chapter

▶ Seeing the difference between conjugated and isolated alkenes

▶ Examining 1,2- and 1,4-additions to alkenes

▶ Distinguishing kinetics from thermodynamics

▶ Seeing the Diels–Alder reaction of conjugated dienes

Some of the first reactions that you saw in organic chemistry were likely the reactions of alkenes (discussed in Chapter 10), but those reactions don't close the book on this important functional group. Multiple double bonds that alternate in a carbon chain, called *conjugated alkenes,* have different properties and reactivities than double bonds that exist all by their lonesome. In this chapter, I explore some of these differences, and use the reaction of conjugated double bonds with acids to explain the difference between kinetics and thermodynamics. I also show you perhaps the most interesting reaction in organic chemistry, the Diels–Alder reaction, which lets you easily form rings and bizarre bicyclic structures.

Seeing Conjugated Double Bonds

Conjugated double bonds are molecules in which two or more double bonds are separated by one carbon-carbon single bond, as shown in Figure 14-1. The double bonds are alternating, in other words. Isolated double bonds are those that are separated by more than one carbon-carbon single bond.

Figure 14-1:
Conjugated
versus
isolated
dienes.

Conjugated double bonds   Isolated double bonds

One feature of conjugated alkenes is that they're more stable than isolated alkenes. You can justify this added stability by writing out the resonance structures for a conjugated alkene (see Chapter 3 for more on resonance structures) and recalling that resonance is a stabilizing feature of molecules. Figure 14-2 shows the resonance structures for the conjugated diene butadiene. In contrast, isolated double bonds have no resonance structures and so are less stable.

Figure 14-2:
One
resonance
structure for
butadiene.

Conjugated diene

Addition of Hydrohalic Acids to Conjugated Alkenes

In addition to their greater stability over isolated double bonds, conjugated double bonds react differently than isolated double bonds. The addition of hydrohalic acids (like HBr, HCl, and so forth) to double bonds is shown in detail in Chapter 10. But hydrohalic acids react somewhat differently with conjugated double bonds than they do with isolated double bonds.

You can see how this difference in reactivity arises by looking at the mechanism for this reaction, shown in Figure 14-3. The first step in the mechanism is the protonation of one of the double bonds to form the carbocation. (Note that in general the cation prefers to be secondary rather than primary; see Chapter 10 for more on carbocations.) This cation has two resonance structures, one in which the positive charge resides on the number-two carbon and one in which the positive charge resides on the number-four carbon. Thus, the true structure of the cation lies somewhere in between these two resonance structures, with a structure that has some positive charge located on the number-two carbon and some positive charge on the number-four carbon. The halide can attack either of these two positively charged spots. Attacking the number-two carbon leads to the 1,2-addition product (because the hydrogen has added to position one and the halide to position two), while attacking at the number-four carbon leads to the 1,4-addition product (because the hydrogen has added to position one and the halide to position four).

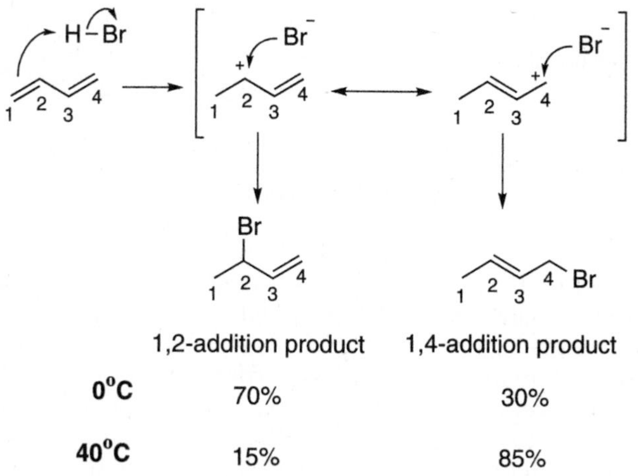

Figure 14-3: Mechanism of 1,2-addition and 1,4-addition.

	1,2-addition product	1,4-addition product
0°C	70%	30%
40°C	15%	85%

Seeing the reaction diagram of conjugate addition

So, which of the two conjugate addition products will be favored? Interestingly, the relative amount of each material produced depends on the temperature of the reaction. While you get mostly 1,2-addition at lower temperatures (at 0°C), you get mostly 1,4-addition at higher temperatures (at 40°C). To see why this is the case, you need to look at the reaction diagram for the two different pathways, shown in Figure 14-4. Note that a reaction diagram plots the progress of the reaction versus the energy of the reaction (refer to Chapter 12 for more on reaction diagrams).

Both pathways are initially identical because both go through the same carbocation. Then the pathways diverge. You can see from the reaction diagram that the 1,2-addition product has a lower energy of activation for the second step (a smaller energy hill to climb), but gives a higher-energy product, while the 1,4-addition reaction pathway has a higher energy of activation for the second step, but leads to a more stable, lower-energy product.

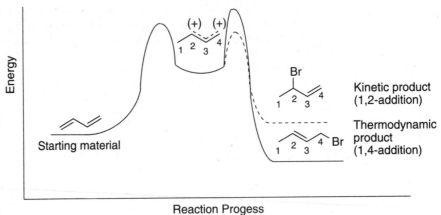

Figure 14-4:
Reaction
diagram for
conjugate
addition.

At low temperatures, the reaction supplies enough energy for the reaction to
go over the smaller barrier to reach the 1,2-addition product, but not enough
energy to go over the higher-energy barrier to reach the 1,4-addition product.
Because at low temperatures the reaction is irreversible — that is, after the
reaction has gone over the energy hill to make product, the reaction doesn't
have enough energy to climb the higher hill back to the cation intermediate —
after the starting material is converted to product, it stays as product and
doesn't go back to the intermediate carbocation. In this way, at low temperatures you favor the 1,2-addition product that has the lower activation barrier,
even though it results in a higher-energy product. The product formed with the
lower activation barrier is called the *kinetic product* using organic-speak.

At higher temperatures (around 40°C), the reaction is reversible (has enough
energy to return to the cation after it's made the kinetic product) and can
reach equilibrium. Any system at equilibrium will favor the lower-energy product, called the *thermodynamic product*. The reaction at high temperature has
enough energy to go over the higher-energy activation barrier to reach the
more stable product. Additionally, because enough heat is present to make
the 1,2-addition reaction reversible, eventually most of the 1,2-product will go
back up the hill to the carbocation state. Once there, the reaction will eventually lead to 1,4-addition, because this reaction gives the lower energy product.

Comparing kinetics and thermodynamics of conjugate addition

Kinetics and thermodynamics are easy to confuse, but here's how you can
remember the difference between the two. *Kinetics* refers to the rates of
reactions. Reaction rates depend on the heights of the energy barriers in a

reaction (the activation energies). *Thermodynamics,* on the other hand, refers to the relative energies of the starting material and the product (the change in energy). If you want a reaction to be under kinetic control, you run the reaction at low temperature so that only low-energy hills will be climbed by the reaction, and the kinetic product will be favored. At higher temperatures, when a reaction is at equilibrium, the reaction is under thermodynamic control, and the lowest-energy product is favored, regardless of the height of the activation barrier.

In general, 1,2-addition typically leads to the kinetic product, while 1,4-addition leads to the thermodynamic product (refer to Figure 14-4 for the specific case of the HBr addition to butadiene). So to get the 1,2-addition product on a conjugated alkene you typically run the reaction at lower temperature, and to get the 1,4-addition product you run the reaction at higher temperatures.

The Diels–Alder Reaction

One of the most interesting differences between isolated alkenes and conjugated alkenes is that conjugated alkenes can undergo the Diels–Alder reaction. The Diels–Alder reaction is probably the funkiest reaction you encounter in organic chemistry. It's also one of the most valuable reactions in organic chemistry, because in a single step this reaction produces two valuable carbon-carbon bonds (so valuable to chemists that the discoverers, Otto Diels and Kurt Alder, shared the Nobel Prize in chemistry for discovering this reaction). The reaction is extremely useful in the construction of six-membered rings and bicyclic molecules.

The general form of the Diels–Alder reaction is shown in Figure 14-5. In this reaction, a conjugated diene reacts with an alkene (called the *dienophile,* or diene lover) to form a six-membered ring. The reaction occurs in a single step; it produces no intermediates. As shown in the mechanism drawing, the three double bonds in the two starting materials are used to make two new carbon-carbon single bonds and one new carbon-carbon double bond in the product.

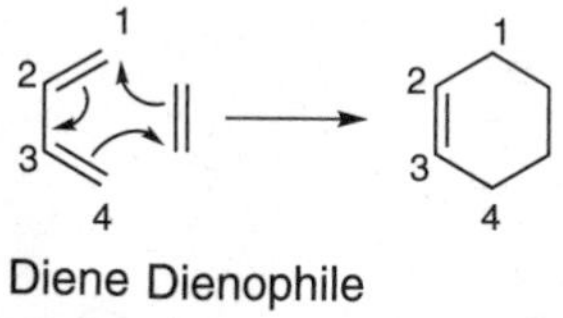

Figure 14-5:
The mechanism of the Diels–Alder reaction.

Diene   Dienophile

Seeing the diene and the dienophile

The Diels–Alder reaction prefers certain dienes and dienophiles. Dienes that are substituted with electron-donating groups react faster than those that are unsubstituted. Conversely, dienophiles react the fastest when they're substituted with electron-withdrawing groups.

Good electron-donating groups (abbreviated EDG) to put on dienes for the Diels–Alder reaction include ethers (OR), alcohols (OH), and amines (NR_2). Good electron-withdrawing groups (abbreviated EWG) for the dienophile include cyano groups (CN), nitro groups (NO_2), and all the carbonyl compounds (including esters, aldehydes, acids, ketones, and so on). These groups are shown in Figure 14-6.

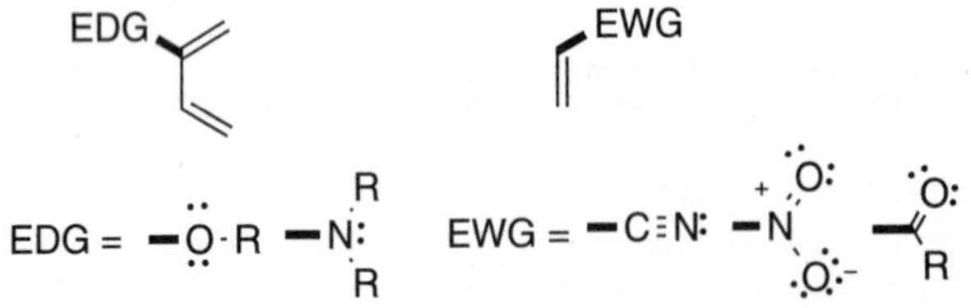

Figure 14-6:
Some diene and dienophile preferences.

Another requirement of the diene is that it adopt the *s-cis* conformation so that the two double bonds are in the proper orientation to undergo the reaction. The *s-cis conformation* is the one in which both of the double bonds are on the same side of the carbon-carbon single bond that separates the two double bonds, as shown in Figure 14-7. Dienes in rings react very quickly in the Diels–Alder reaction because they're locked in the *s-cis* conformation.

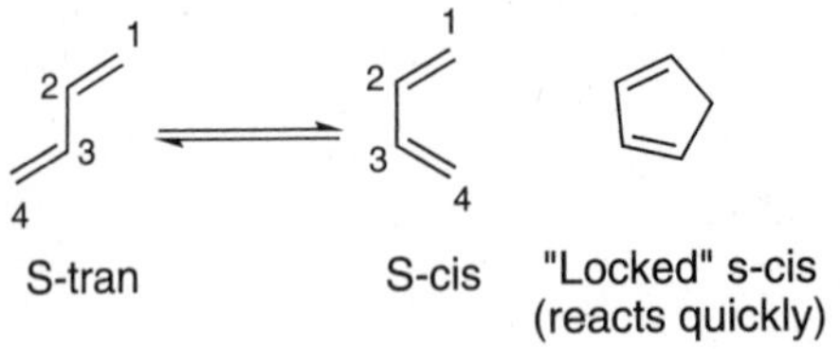

Figure 14-7:
The *s-cis*
and *s-trans*
conforma-
tions.

S-tran   S-cis   "Locked" s-cis
                          (reacts quickly)

The stereochemistry of addition

One of the most valuable features of the Diels–Alder reaction is that the reaction is *stereoselective* — that is, it preferentially forms one stereoisomer over another. (For a discussion of stereochemistry, check out Chapter 6.) With a disubstituted dienophile, if the two substituents on the dienophile start *cis,* they end up *cis* in the product; if they start *trans,* they end up *trans* in the product, as shown in Figure 14-8.

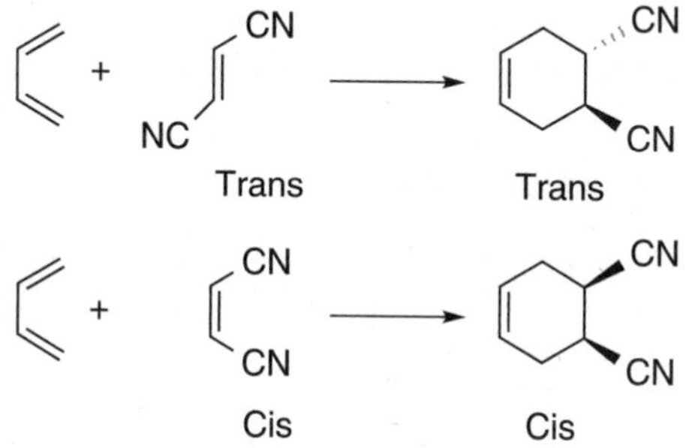

Figure 14-8:
Stereochemistry of addition in a Diels–Alder reaction.

Seeing bicyclic products

If the diene is in a ring, you get bicyclic products from the Diels–Alder reaction, as shown in Figure 14-9. And if the two substituents on the dienophile are *cis,* the two stereoisomers could end up going down relative to the carbon bridge in the product, called *endo addition,* or sticking up toward the carbon bridge, called *exo addition.* The Diels–Alder reaction preferentially forms the endo product.

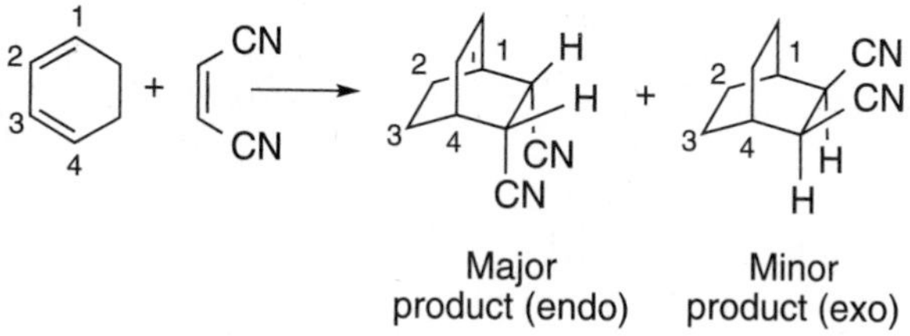

Figure 14-9:
The endo and exo products of a Diels–Alder reaction.

Problem Solving: Determining Products of Diels–Alder Reactions

Diels–Alder reactions can look confusing, but following these four simple steps will help you to determine the products of these reactions:

1. **Orient the diene and the dienophile correctly.**

   In this step, you make sure that the double bonds are oriented correctly (the diene double bonds are pointing in the direction of the dienophile), and that the diene is in the *s-cis* conformation (if it isn't, you need to rotate it so that it is).

2. **Number the diene carbons (1 through 4).**

   You can start the numbering on either end of the diene. The numbering is just a way for you to keep track of where bonds should be formed to make the product.

3. **Work the reaction.**

   Make a bond from the number-one carbon on the diene to one side of the dienophile. Then make a bond from the number-four carbon to the other side of the dienophile. Get rid of the two double bonds in the starting material between carbons 1 and 2 and 3 and 4, and put a double bond between carbon 2 and 3 in the product.

4. **Make sure you have the correct stereochemistry.**

   If you start with a *cis* disubstituted dienophile the product should be substituted *cis*; if the dienophile starts *trans*, it should end up *trans*. With bicyclo products, make sure that you show endo stereochemistry in the product.

Now try taking the problem shown in Figure 14-10 through the four steps.

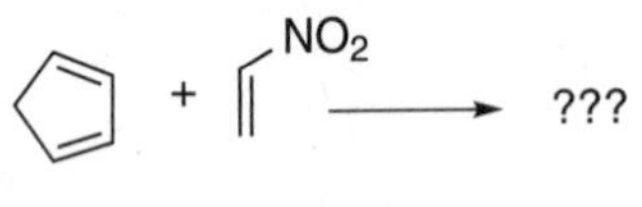

Figure 14-10:
A Diels–
Alder
reaction.

Step 1: Orienting the molecule

Start by orienting the diene so that the two double bonds in the diene point in the direction of the dienophile (with the single bond that connects the double bonds oriented away from the dienophile). In this case, you need to rotate the diene 180 degrees, as shown in Figure 14-11. Because this diene is in a ring, you don't have to worry about twisting the diene to the *s-cis* conformation because dienes in rings are locked in the *s-cis* conformation.

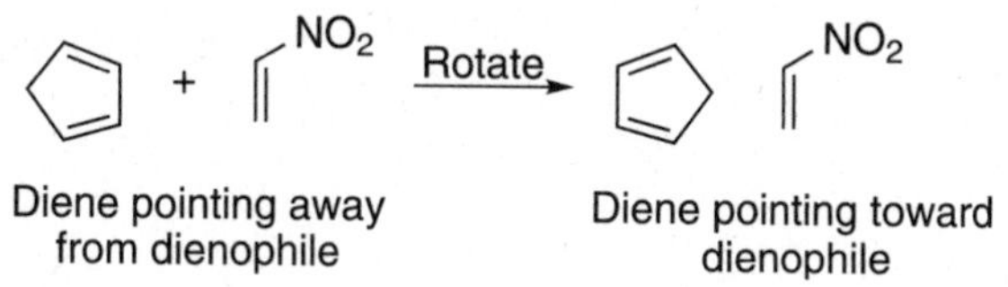

Figure 14-11:
Orienting
the diene.

Diene pointing away
from dienophile

Rotate

Diene pointing toward
dienophile

Step 2: Numbering the diene

Next, number the carbons on the diene (1 through 4). Start the numbering on any side of the diene. In Figure 14-12, the numbering begins on the top.

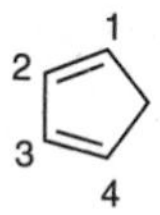

Figure 14-12:
Diene
numbering.

Step 3: Making the bonds

Now that everything's lined up, take a deep breath and work the reaction. You make bonds between the number-one carbon on the diene and the dienophile, and between the number-four carbon and the dienophile. You end with a double bond between carbons two and three. Notice that in the reaction shown in Figure 14-13, the fifth carbon in the diene ring is flipped up, forming a pointed "bridge" between carbons one and four.

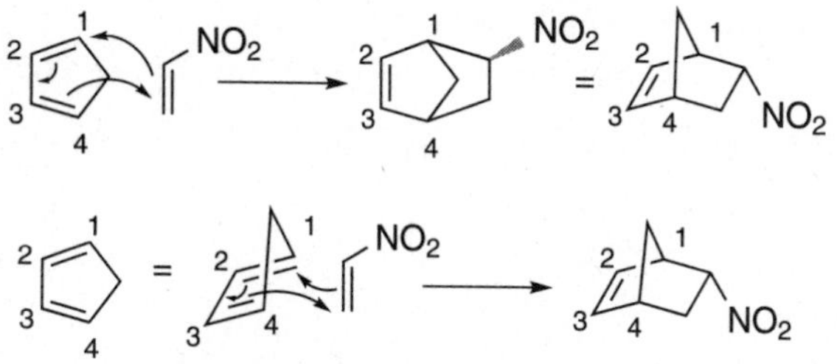

Figure 14-13:
Two differ-
ent ways
of visual-
izing this
Diels–Alder
reaction.

Step 4: Checking the stereochemistry

Make sure you have the correct stereochemistry. Because you have a bicyclic product, you need to make sure that the nitro substituent is *endo* (points down from the carbon bridge at the top of the molecule). And that's it!

Chapter 15

Lord of the Rings: Aromatic Compounds

According to organic legend, the structure of the parent aromatic system, benzene, came to the chemist August Kekulé in a dream. For a long time the structure of benzene — a liquid isolated from a tarry residue from burning petroleum gas — had eluded chemists, and many structures for the compound had been proposed by big names in the field. Kekulé claimed that in a dream he saw a snake devouring its own tail, forming a circle. Benzene must be a ring, he thought when he awoke, and jotted down the correct structure of the compound (or so he claimed, somewhat dubiously, many years later). Kekulé's intuition about the cyclic structure of benzene was correct, and soon led to the discovery of aromatic rings other than benzene.

Benzene and other *aromatics* are a highly interesting class of ringed compounds of exceptional stability. In this chapter, I show you how to name aromatics, and discuss aromatic compounds other than benzene that have properties and reactivities similar to those of benzene. I also show you how to determine whether a ring system is aromatic and show you what makes aromatic compounds so stable.

Defining Aromatic Compounds

Aromatics are a class of ring compounds containing double bonds. The name *aromatic* comes from the fact that many of the simple aromatic compounds that were first isolated were highly fragrant; the lovely odors of such substances as vanilla, almond, and wintergreen are due to the presence of aromatic compounds in these products. But many aromatic compounds are unpleasant smelling, or are odorless. Most simple aromatics, in fact, are obtained commercially from coal tar. Aromatic compounds contain double bonds, but they don't react like alkenes, so they're classified as a separate functional group. Benzene, for example, doesn't behave as if it were 1,3,5-cyclohexatriene. Although bromine reacts rapidly with alkenes to make dibromides (see Figure 15-1), bromine is completely unreactive with benzene. This lack of reactivity results from aromatics being much more stable than alkenes.

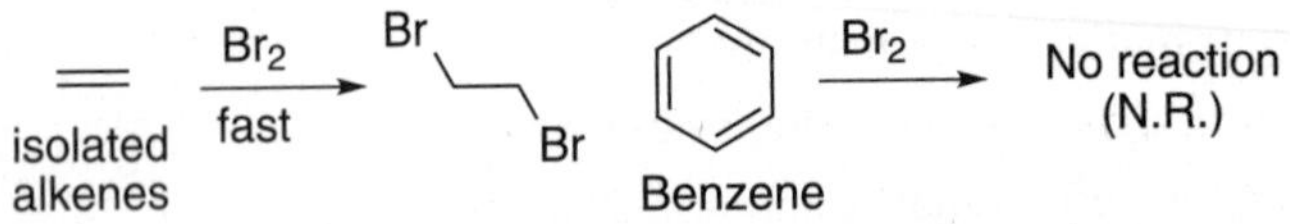

Figure 15-1:
The relative stability of an alkene and benzene in the presence of bromine.

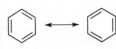

Because of this greater stability, aromatic compounds require significantly more vigorous reaction conditions to make them react compared to the conditions required to react simple alkenes. Aromatic rings scoff when exposed to the relatively mild reagents that react with alkenes (reagents such as those used in hydroboration, oxymercuration, and HBr addition, which are shown in Chapter 10). Trying to react aromatics with these reagents is like shooting popguns at a well-defended castle. These reagents simply glance off the rings and remain impotent in the reaction flask. You need to unleash the howitzer reagents to get aromatic rings to react, and I show you these in Chapter 16.

The structure of benzene

The actual structure of benzene is the hybrid of its two resonance structures, shown in Figure 15-2. (For a review of resonance structures, see Chapter 3.) Every bond in benzene is exactly the same length — neither as long as a single bond nor as short as a double bond — and its bond character can best be described as being 1.5, the bond length between that of a single bond and a double bond.

Figure 15-2:
Benzene resonance structures.

Because benzene is a hybrid of two resonance structures, to more accurately show the location of the pi electrons with a single structure, all the double-bond electrons are often represented with a circle (as shown in Figure 15-3) rather than with individual double bonds. This circle drawing of benzene more accurately depicts the electron distribution of benzene because it shows all atoms and bonds as equal. But because the number of pi electrons is unclear in these drawings (and because reactions are better shown from the double bond–containing structures), I avoid using this circle representation for benzene in this book. Just keep in mind that when the double bond–containing structures for benzene are drawn, these hybrid structures are assumed.

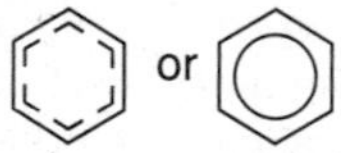

Figure 15-3:
Benzene.

Diversity of aromatic compounds

Although benzene is the parent aromatic compound, many other aromatic compounds that have the same exceptional stability as benzene are known. Some of these aromatics consist of benzene rings smooshed together to make fused rings, such as benzopyrene (shown in Figure 15-4). These fused aromatic compounds, which are found in coal, have been found to be toxic. Chimney sweeps and other workers exposed to these compounds for long periods of time run the risk of developing cancers because of exposure to the benzopyrene and other aromatic compounds found in chimney soot.

Aromatic compounds come in rings of all sizes, and many include *heteroatoms* (non-carbon atoms) like oxygen, nitrogen, and sulfur; aromatics that contain such heteroatoms include furan and pyridine, shown in Figure 15-4. All the DNA bases, including adenine, which is also shown in Figure 15-4, have aromatic character. Because they're so stable, aromatic compounds are ubiquitous in nature.

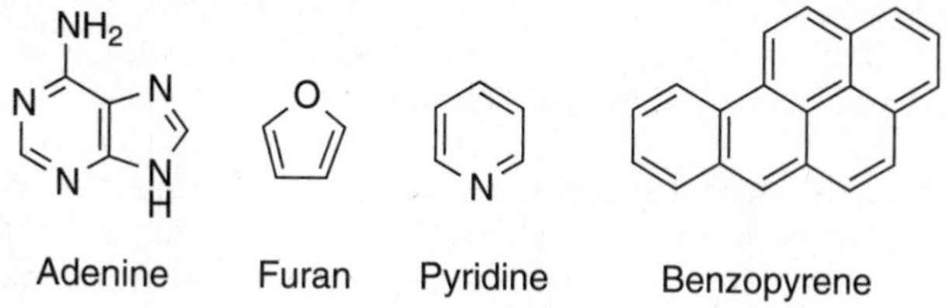

Figure 15-4:
Some
aromatic
rings found
in nature.

Adenine   Furan   Pyridine   Benzopyrene

So, what exactly makes a molecule aromatic?

At first, many chemists thought that all rings that contained alternating double bonds all the way around a ring would have the same aromatic character as benzene — and the same exceptional stability that comes along with this structure. This assumption is not correct, however. Cyclobutadiene, for example, shown in Figure 15-5, is an extremely unstable compound, and rapidly *dimerizes* (reacts with another molecule of cyclobutadiene) in solution via the Diels–Alder reaction from Chapter 14 (you get brownie points if you can draw the three-ringed product!). Benzene, as a result of its aromaticity, is more stable than its open chain analog, 1,3,5-hexatriene, but cyclobutadiene is actually less stable than its open-chain counterpart, 1,3-butadiene. Therefore, some rings are stabilized by these double bond–containing rings and become aromatic, while others (like cyclobutadiene) are destabilized and become what are called *anti-aromatic* rings.

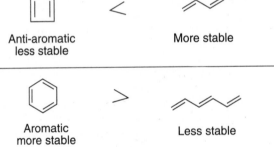

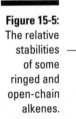

Anti-aromatic
less stable

More stable

Figure 15-5:
The relative stabilities of some ringed and open-chain alkenes.

Aromatic
more stable

Less stable

Hückel's 4n + 2 rule

A German chemist named Erich Hückel explained this somewhat confusing result. He observed that planar ring systems that contained $4n + 2$ pi electrons (where n is an integer), were stabilized and aromatic, while ring systems with $4n$ pi electrons were destabilized and anti-aromatic. Thus, benzene, with its 6 pi electrons (2 pi electrons from each double bond) is aromatic because it follows the $4n + 2$ rule, where $n = 1$. Cyclobutadiene, on the other hand, is anti-aromatic because of its 4 pi electrons; it follows the $4n$ equation, where $n = 1$.

Hückel's equation, useful as it is, doesn't explain why the rings that contain a *Hückel number of pi electrons* ($4n + 2$) have this stabilizing feature of aromaticity, while those with a *non-Hückel number* ($4n$) are destabilized. Although you know that benzene has resonance structures, and that resonance structures generally contribute to a molecule's stability, the best explanation for the stability of aromatic compounds comes from molecular orbital theory, which is covered in the next section.

Explaining Aromaticity: Molecular Orbital Theory

Molecular orbital (MO) theory offers a more sophisticated model of bonding than the valence bond model (which is represented with Lewis structures). Of course, this greater sophistication comes at a cost: MO theory is not as user-friendly as the valence bond model. It's been whispered (and not only among students) that while valence bond theory is too good to be true, MO theory is too true to be good. That description is somewhat unfair to both models, but it's true that organic chemists have staved off using MO theory even when an MO model gives the correct description of a molecule and the valence bond model is wrong. For example, organic chemists prefer to use resonance structures to correct flaws in the valence model's depiction of pi and nonbonding electrons rather than switch to the more precise (but more unwieldy) MO theory.

What the heck is molecular orbital theory?

In valence bond theory, electrons are located in bonds between atoms or as nonbonding electrons on an atom. In molecular orbital theory, electrons are not restricted to being localized between two atoms or on an atom itself. Instead, the electrons are allowed to spread across the entire molecule. If you think of orbitals as being like apartments for electrons, the electron has just been allowed to move from the small, cloistered hovel of valence bond theory into the much more spacious suite of molecular orbital theory. Because electrons can delocalize across the entire molecule in MO theory, the orbitals that these electrons reside in are called *molecular orbitals*, abbreviated Ψ (psi, pronounced "sigh," as in, "Sigh! Why do I have to deal with molecular orbital theory?").

Valence bond theory and molecular orbital theory are just models for depicting the actual location of electrons in a molecule. Although valence bond theory is the more user friendly of the two, it's not as accurate as MO theory because it doesn't do as good a job at modeling how electrons can spread out over an entire molecule in certain instances (like in aromatic rings).

Making molecular orbital diagrams

A very informative orbital schematic is called a *molecular orbital diagram*. A molecular orbital diagram shows the number and relative energies of the different molecular orbitals for a molecule, and the electron occupancy in each of these orbitals. These diagrams are just like the atomic electron configurations you drew in introductory chemistry (reviewed in Chapter 2), except that these diagrams are for molecules instead of atoms, and in these diagrams you fill molecular orbitals rather than atomic orbitals.

To make an MO diagram, you need to follow three guidelines:

- ✔ **Molecular orbitals are made by combining atomic orbitals (like *s* and *p* orbitals).** The number of atomic orbitals combined must equal the number of molecular orbitals in the molecule. If you put six atomic orbitals in, for example, you must get six molecular orbitals out.

- ✔ **The MOs that go down in energy are called *bonding molecular orbitals*; the MOs that go up in energy are called *antibonding molecular orbitals*.**

- ✔ **Electrons fill molecular orbitals in the same way that electrons fill atomic orbitals in an atom.** In other words, the lowest-energy orbitals fill first with a maximum of two electrons per orbital, following Hund's rule. (See Chapter 2 for a review of atomic orbitals and Hund's rule.)

Every carbon atom in benzene is *sp²* hybridized. Therefore, every carbon atom in benzene has one *p* orbital orthogonal (at a right angle) to the plane of the benzene ring, as shown in Figure 15-6. The pi orbitals in benzene are made from the overlap of these *p* orbitals. When making these MO diagrams for benzene and other aromatics, you usually only care about the pi orbitals (because only the pi electrons contribute to aromaticity). So, on these MO diagrams you leave out all the sigma molecular orbitals, and combine just the *p* orbitals to make your pi molecular orbitals (recall that only *p* orbitals make pi bonds).

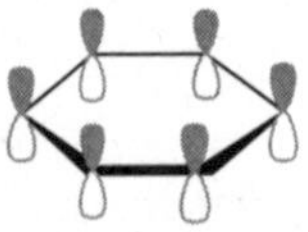

Figure 15-6:
Benzene *p*
orbitals.

Two rings diverged in a wood: Frost circles

Because benzene has six *p* orbitals (one from each carbon), exactly six pi molecular orbitals must be formed from these six orbitals. A very handy guide to making the molecular orbital diagrams for these ring systems is a Frost circle. A *Frost circle* enables you to quickly place the six orbitals on the molecular orbital diagram and gives you the relative energies of the orbitals.

Making a Frost circle is easy: You simply draw a circle, and within the circle draw the carbon ring with one of its corners pointing down. Each point where the ring touches the circle is the location of one of the molecular orbitals. After you have all the orbitals situated at the correct energy level, you fill the orbitals with the number of pi electrons in the ring, and — *voilà!* — you have the molecular orbital diagram.

After drawing a couple of these molecular orbital diagrams for ring systems, you notice that in many diagrams two orbitals are at the same energy level. Such orbitals are called *degenerate orbitals,* using organic-speak.

Making the molecular orbital diagram of benzene

The construction of the molecular orbital diagram of benzene is shown in Figure 15-7. First, a circle is drawn, and then a six-membered ring is plopped inside the circle with the point down.

One of the most common mistakes in constructing MO diagrams using a Frost circle is to forget to put the point of the ring *down* in the circle.

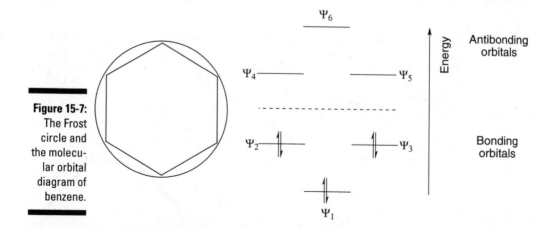

Figure 15-7:
The Frost circle and the molecular orbital diagram of benzene.

Every point where the six-membered ring touches the circle represents the location of one of the molecular orbitals (which I have transposed to the right of the circle in the figure for clarity). The orbitals lower than the center of the circle are called *bonding orbitals,* while those higher than the center of the circle are called *antibonding orbitals.* Bonding orbitals are lower in energy than antibonding orbitals. Because benzene has six pi electrons, you fill the lowest energy orbital (Ψ_1) plus the two degenerate orbitals above it (Ψ_2 and Ψ_3). Note how lovely the electronic structure of benzene is — all three of these low-energy bonding orbitals are completely filled with electrons, while each of the high-energy antibonding orbitals remains empty. Because each of the filled orbitals in benzene is low in energy, benzene is exceptionally stable.

Seeing the molecular orbitals of benzene

Of course, these MO diagrams don't tell you what the individual molecular orbitals look like. Each molecular orbital in benzene is made from a combination of the six atomic *p* orbitals. Recall that *p* orbitals have a *node* — a region of zero electron density at the center of the orbital where the wave function that describes the orbital changes sign. (This change in sign is made apparent by shading one half of the *p* orbital.) For a bonding interaction to occur between two *p* orbitals, the orbitals must line up so that each lobe that overlaps with another *p* orbital has the same sign. When orbitals line up with an opposite sign, it creates an antibonding (high-energy) interaction and a node between the orbitals. An example for the overlap of *p* orbitals is shown in Figure 15-8.

Node

Figure 15-8:
Bonding and
antibonding
p orbitals.

Bonding   Antibonding

The MOs of benzene are shown in Figure 15-9. When all the *p* orbitals circle the benzene such that all the signs match up, every single orbital overlaps with the other in a bonding interaction and makes the molecular orbital (Ψ_1) with the lowest possible energy. When one side of the ring has orbitals that line up one way, and the other side has orbitals that line up the opposite way, a molecular orbital is created that has one node across the benzene ring. The molecular orbitals Ψ_2 and Ψ_3 have one node each. A general feature of pi molecular orbitals is that the lowest energy level orbitals have zero nodes, the second energy level has one node, the third has two nodes, and so on. The more nodes a molecular orbital contains, the higher the energy of the orbital. In the highest energy molecular orbital of benzene (Ψ_6), every p orbital is antibonding with its neighboring orbital.

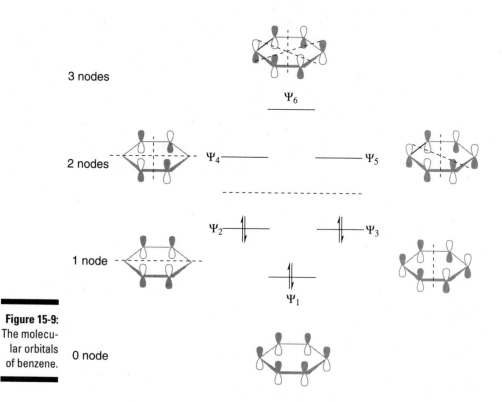

3 nodes

Ψ_6

2 nodes Ψ_4 Ψ_5

Ψ_2 Ψ_3

1 node

Ψ_1

0 node

You might wonder how an electron is able to hop across nodes, those regions of zero electron density where electrons are not allowed. Continuing the apartment analogy, these nodes are like having rooms in your apartment, but no doorways. How do you get from one room into the next? The confusion comes from thinking of electrons only as particles, like little planets orbiting the nucleus sun, and ignoring the wave properties of electrons. Here a wave analogy is useful. When you thumb a guitar string on the fret of the guitar and then pluck the string, you see the vibration propagate past your thumb, a place where the vibration is zero. In the same way, an electron, because it can behave as a wave, can hop past the nodes in a molecular orbital. When thinking of node hopping, think of the electron as a wave rather than as a particle.

Making the molecular orbital diagram of cyclobutadiene

Now try making the MO diagram for cyclobutadiene, which is unstable and anti-aromatic (see Figure 15-10). You make the Frost circle the same as you did with benzene, except that you plunk a four-membered ring point down in the circle to obtain the MO levels. (Note that you get four molecular orbitals, because you start with four p atomic orbitals.) Because cyclobutadiene has four pi electrons (again, two from each double bond), you completely fill the first molecular orbital (Ψ_1). The next two electrons are placed in the Ψ_2 and Ψ_3 orbitals, which are degenerate. According to Hund's rule (refer to Chapter 2), you must place one electron into each degenerate orbital instead of pairing them in the same orbital. Thus, unlike benzene, in which all bonding orbitals are filled with electrons, two orbitals in cyclobutadiene are only partially filled. Cyclobutadiene, therefore, is essentially a *diradical* (a species containing two unpaired electrons) and diradicals are highly unstable! Certainly, you wouldn't have guessed that from looking at the misleading Lewis structure.

Figure 15-10:
The Frost circle and the molecular orbital diagram of cyclobutadiene.

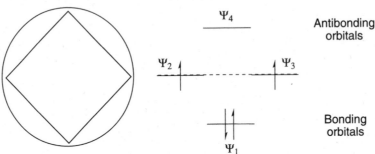

In fact, you see that this trend is followed for all aromatic and anti-aromatic rings. Aromatic rings have only filled bonding orbitals, while anti-aromatic rings have two unfilled orbitals, making them extremely grouchy and unstable.

Problem Solving: Determining Aromaticity

Now that I've shown you why aromatic rings are so stable, and why anti-aromatic rings are so unstable, I show how you can determine whether a ring system is aromatic, anti-aromatic, or non-aromatic.

For a molecule to be aromatic, it must meet the following four conditions:

- ✔ It must be a ring.

- ✔ It must be flat (planar).

- ✔ It must have in each atom of the ring a *p* orbital that's orthogonal to the plane of the ring. In other words, no atom in the ring can be sp^3 hybridized.

- ✔ It must have a Hückel number of pi electrons, following the $4n + 2$ rule.

If the molecule meets the first three conditions, but only has $4n$ pi electrons, the molecule is *anti-aromatic*. If the molecule fails any or all of the first three conditions, then the molecule is *non-aromatic*. Following are some details about each of these points.

The first condition is that only rings can be aromatic. Acyclic systems cannot be aromatic.

The second condition involves the shape of the ring. Ringed systems can be flat or three-dimensional. Most conjugated ring systems are flat in order to maximize the overlap between the *p* orbitals. But some exceptions exist. Cyclodecapentaene, for example, shown in Figure 15-11, is puckered because two hydrogens in the ring are pushed into each other's space, and the ring is forced out of planarity so that the hydrogens can occupy separate spaces. Because of its nonplanar structure, cyclodecapentaene is non-aromatic. Removing the two hydrogens and replacing them with a carbon bond to make naphthalene erases this problem; naphthalene is planar (and aromatic).

Figure 15-11:
The non-planar and planar rings of cyclo-decapen-taene and naphthalene.

Cyclodecapentaene   Naphthalene

Not planar   planar
(non-aromatic)   (aromatic)

Cyclooctatetraene (Figure 15-12) is another example of a molecule that has alternating double bonds around the ring but isn't planar. Instead, this ring system adopts a puckered tub-shaped conformation. Why? Because if it remained flat, it would become anti-aromatic as a result of having a non-Hückel number of pi electrons (8)! Because anti-aromatic systems are so unstable, this compound puckers and sacrifices the *p* orbital overlap so that it won't have to endure the instability associated with being anti-aromatic.

The third condition involves the orbitals. Aromatic systems must have an unbroken ring of p orbitals, so any ring that contains an sp^3 hybridized carbon will not be aromatic. Cycloheptatriene, shown in Figure 15-13, is non-aromatic because one of the ring carbons is sp^3 hybridized. However, carbocations (positively charged carbons) are sp^2 hybridized (and have an empty p orbital), so the cycloheptatriene cation has an unbroken ring of p orbitals and is an aromatic compound.

Figure 15-13:
The nonaro-
matic cyclo-
heptatriene
molecule
and the
aromatic
cyclohepta-
trienecation.

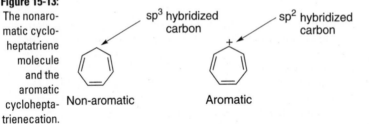

sp^3 hybridized carbon

sp^2 hybridized carbon

Non-aromatic   Aromatic

Finally, to be aromatic, a ring must have a Hückel number of electrons ($4n + 2$). Table 15-1 gives solutions to these equations for low values of n.

Table 15-1	Pi Electron Counts	
Integer (n)	*Aromatic Numbers (4n + 2)*	*Anti-aromatic Numbers (4n)*
0	2	—
1	6	4
2	10	8
3	14	12
4	18	16

One tricky aspect of counting the number of electrons in the pi system is when the ring contains heteroatoms (like O, S, N). How do you know which lone pairs to count as part of the pi system and which to ignore (see Figure 15-14)?

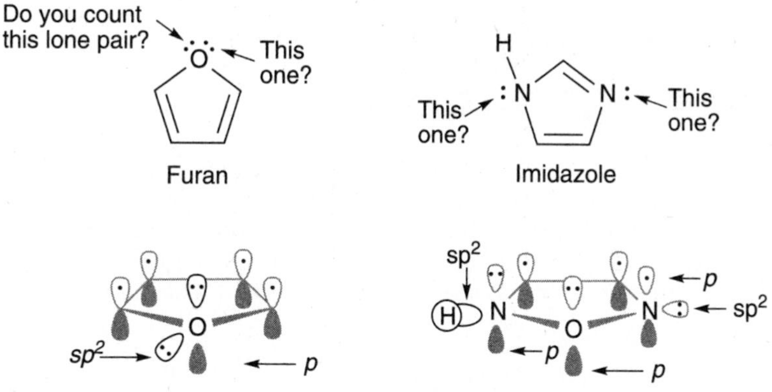

Do you count
this lone pair?   ←   This
one?

Furan

Imidazole

This
one?   This
one?

Figure 15-14:
Counting pi
electrons on
heteroatoms.

sp^2

p

sp^2

p

p

p

sp^2

The long answer is that you have to draw out the orbitals to see which lone pair electrons would go into the pi system and which ones wouldn't. For furan, one lone pair from the oxygen would go into the p orbital (pi system), and one would go into an sp^2 hybridized orbital orthogonal to the ring. So, the oxygen in furan would contribute two electrons to the pi system.

But wait a minute! Mustn't the oxygen in furan be sp^3 hybridized because the oxygen has four substituents (and lone pairs count as a substituent; see Chapter 2)? Not necessarily. Although the general rule of counting substituents to determine hybridization holds true in most cases, it fails when an atom with a lone pair is adjacent to a double bond (that is, when it's conjugated). In those cases, the atom with the lone pair will *rehybridize* from sp^3 to sp^2 so that it can conjugate the lone pair of electrons with the double bond.

Both furan and imidazole have six pi electrons and are aromatic. (Can you look at the orbital diagrams in Figure 15-14 and see that that's the case?)

Here's a timesaving tip so you don't have to draw out the orbitals of heteroatom-containing rings every time. Heteroatoms (O, S, N) double bonded to other atoms in a ring cannot contribute their lone pairs to the pi system, because their p orbital is already being used to make the double bond. (Refer to Figure 15-14, and notice the orbitals of the double-bonded nitrogen in imidazole.) Heteroatoms only singly bonded to other atoms in a ring can contribute one lone pair to the pi system, but not two because only one lone pair can be in the pi system; the other lone pair will be in an sp^2 hybridized orbital orthogonal to the pi system. (Refer to Figure 15-14, and notice the orbitals of the singly bonded oxygen in furan.)

Problem Solving: Predicting Acidities and Basicities

Two of the most common problem types, when dealing with aromatic compounds, are predicting the acidities and the basicities of double bond–containing rings, including aromatic rings. These problems are designed to see if you can apply aromaticity on your own and understand that aromaticity is a stabilizing feature of molecules and ions.

Comparing acidities

One common question is to ask which one of two double bond–containing rings is more acidic, such as the molecules in Figure 15-15. As with any acid-base reaction in which you need to compare acidity, you should look to see which acid has the more stable conjugate base.

Figure 15-15:
Cyclopenta-
diene and
cyclohepta-
triene.

Cyclopentadiene Cycloheptatriene

Stronger acids have more stable conjugate bases

Looking at the conjugate bases of both molecules (see Figure 15-16), you see that both compounds have rings that are entirely conjugated, but one has six pi electrons (cyclopentadienyl anion), while the other has eight pi electrons (cycloheptatrienyl anion). Thus, the conjugate base with six pi electrons is aromatic and should be more stable than the ring with eight pi electrons, which cannot be aromatic. Based on this analysis, cyclopentadiene must be more acidic than cycloheptatriene.

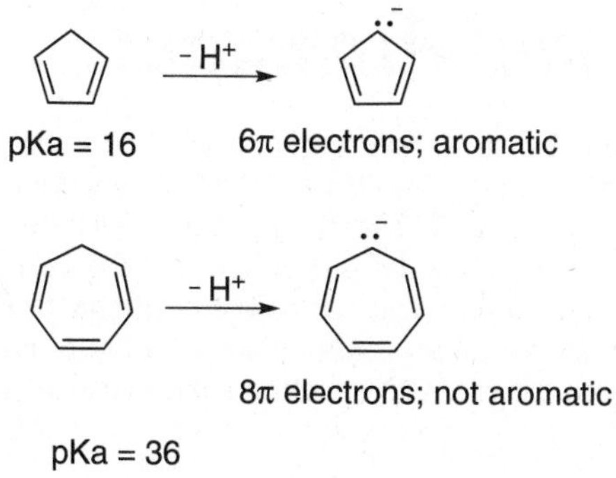

- H+
pKa = 16
6π electrons; aromatic
- H+
8π electrons; not aromatic
pKa = 36

Figure 15-16: Comparing acidities of rings.

Comparing basicities

Another common question type is to compare or predict the basicity of nitrogen atoms in aromatic compounds. For example, can you predict which nitrogen is more basic in the aromatic molecule imidazole, shown in Figure 15-17?

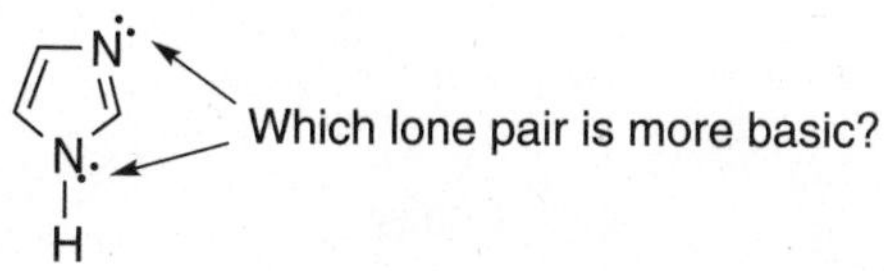

N
N
H
Which lone pair is more basic?

Figure 15-17: Comparing the basicities of the nitrogens in imidazole.

To determine the strength of a base, you look at the stability of the conjugate acid. Protonating the bottom nitrogen in the imidazole in Figure 15-17 disrupts the aromaticity of the ring, because that nitrogen lone pair is part of the pi system. (Upon protonation, the lone pair is tied up in a bond to hydrogen and can no longer contribute to the aromaticity.) Protonating the top nitrogen doesn't disrupt the pi system (or the aromaticity) of imidazole because the lone pair is situated in an sp^2 hybridized orbital and is not part of the pi system. (Refer to Figure 15-14 for the orbital drawings of imidazole.) Therefore, the top nitrogen is much more basic than the bottom nitrogen, because the top nitrogen is much happier with an added hydrogen ion than the bottom nitrogen is. The general feature of these problems is that nitrogens whose lone pairs are not involved in the pi system are more basic than nitrogens with lone pairs in the pi system.

Naming Benzenes and Aromatics

Substituted benzenes are named with the parent name of — *surprise!* — benzene. Thus, benzene with an ethyl group is named ethylbenzene, as shown in Figure 15-18. When you have an alkyl chain that has more carbons than the aromatic ring, the alkyl portion becomes the parent group. When benzene is named as a substituent, it's called a *phenyl group* (often abbreviated Ph), as in 3-phenylheptane. A general term for an aromatic ring as a substituent is the word *aryl*. A benzene with a methylene (CH_2) portion is known as a *benzyl group*, as shown in Figure 15-19.

REMEMBER

Don't confuse a phenyl ring with a benzyl group.

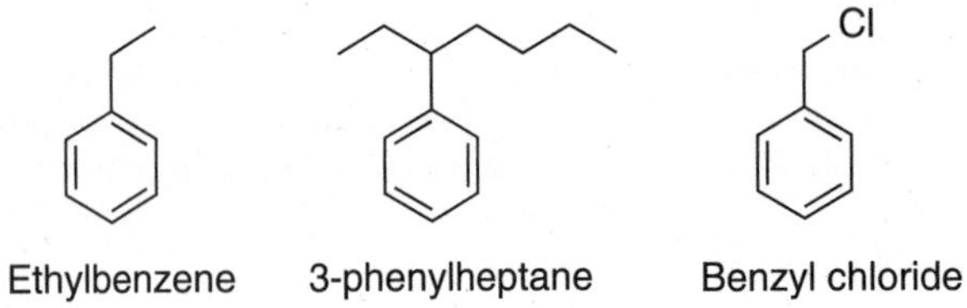

Figure 15-18:
The names
of some
substituted
benzenes.

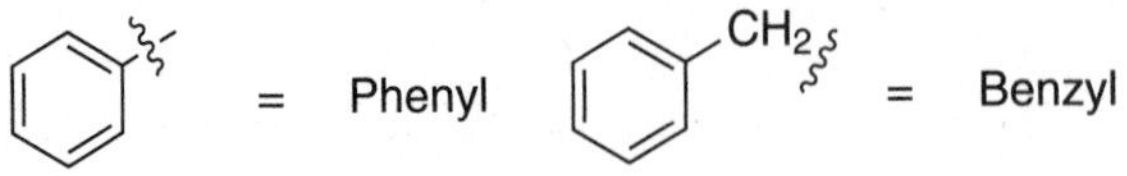

Ethylbenzene   3-phenylheptane   Benzyl chloride

Figure 15-19:
The phenyl
ring and
the benzyl
group.

= Phenyl   = Benzyl

Common names of substituted benzenes

One thing that you might find frustrating about aromatic compounds is that substituted benzenes are known by many common names. Methylbenzene, for example, is almost always called *toluene*. These names simply must be learned. Some of the most common substituted benzenes are shown in Figure 15-20.

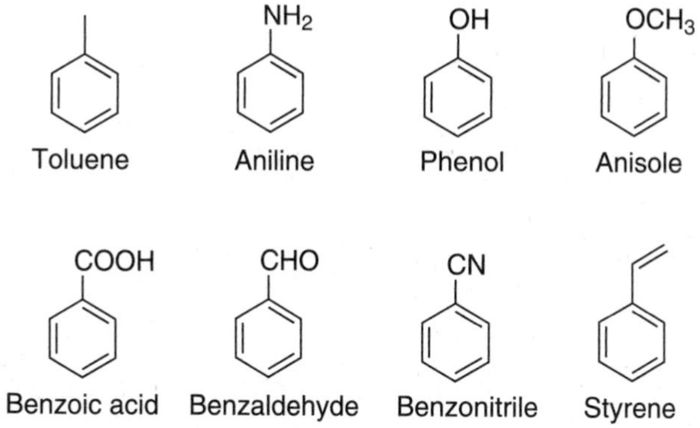
NH2
OH
OCH3
Toluene
Aniline
Phenol
Anisole
COOH
CHO
CN
Benzoic acid
Benzaldehyde
Benzonitrile
Styrene

Figure 15-20: The common names of some substituted benzenes.

TIP

Many professors like to test whether you can remember the common names of compounds by giving synthesis problems on exams using the common names of the starting materials. For example, you might get a problem that says, "Starting with toluene, propose a synthesis to make compound X." If you don't know the structure of toluene, you're up a creek on that problem without a you-know-what, even if you could have done the synthesis had you known the structure.

Names of common aromatics

The names of aromatics other than benzene are shown in Figure 15-21. For practice, you might want to see if you can determine how many pi electrons each of these rings contains. (They all have six pi electrons, but from where do these electrons come?)

O
S
H
N
N
Furan
Thiophene
Pyrrole
Pyridine

Figure 15-21: The names of some common aromatics.

Chapter 16

Bringing Out the Howitzers: Reactions of Aromatic Compounds

*I*n Chapter 15, I discuss how you can determine whether a ring system is aromatic and what makes such aromatic compounds so stable. In this chapter, I show you how you can get such aromatic compounds to react by introducing a much hotter set of reagents than those used to react normal alkenes. These aromatic reactions are used to introduce multistep synthesis problems, or problems that require you to synthesize a complex structure by stringing together a sequence of known reactions.

Electrophilic Aromatic Substitution of Benzene

Unlike the double bonds of alkenes, the double bonds in benzene are only weakly *nucleophilic* (nucleus loving), so you need powerful *electrophiles* (electron lovers) to make benzene react. You usually have to throw a full positive charge on the electrophile to make it strong enough to react with benzene. For example, Br_2 reacts with alkenes but does not react with benzene. However, positively charged bromine, Br^+, does react with benzene, because positively charged species are much more electrophilic than neutral species are. Unlike the reaction of bromine with alkenes (in which you get an addition reaction across the double bond), when benzene is reacted with the bromine ion (Br⁺), you get substitution reactions (in which the electrophile substitutes for hydrogen).

Reacting benzene with electrophiles is a reaction called *electrophilic aromatic substitution,* and the general mechanism for this reaction is shown in Figure 16-1. In the first step of the mechanism, one of the double bonds in benzene attacks the positively charged electrophile (E⁺) to make a cation. This intermediate cation is stabilized by resonance (can you draw out the other two resonance structures?), but it's non-aromatic (because the ring has an sp^3-hybridized carbon). In the second step, when a base (abbreviated B: ⁻) plucks off a proton adjacent to the carbocation, the aromatic ring is re-formed to make the substituted benzene.

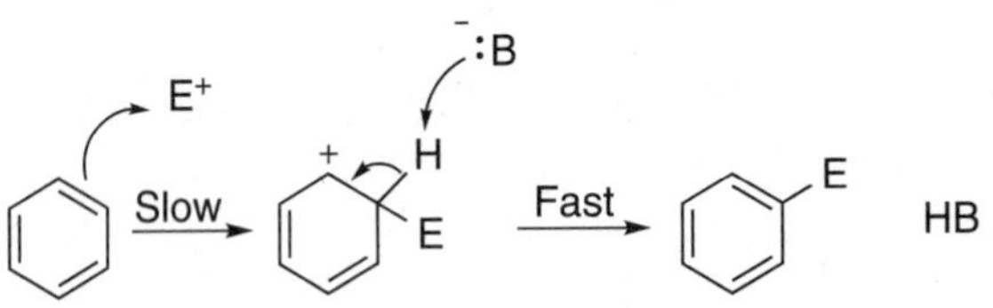

Figure 16-1: The mechanism of electrophilic aromatic substitution.

Figure 16-2 shows the reagents that you use to make different types of positively charged electrophiles. These reagents can be used to nitrate, brominate, chlorinate, or sulfonate benzene rings. These electrophiles all add to benzene in the fashion shown in Figure 16-1. The alkylation and acylation reactions are discussed in more detail in the following sections.

	Reagents	Electrophile (E⁺)	Byproducts
Nitration	$HNO_3 + H_2SO_4$	$O=\overset{+}{N}=O$	H_2O / HSO_4^-
Bromination	$Br_2 + FeBr_3$	Br^+	$FeBr_4^-$
Chlorination	$Cl_2 + FeCl_3$	Cl^+	$FeCl_4^-$
Sulfonation	$SO_3 + H_2SO_4$	$O=\overset{+}{\underset{O-H}{S}}=O$	HSO_4^-
Alkylation	$R-Cl + AlCl_3$	R^+	$AlCl_4^-$
Acylation	⠀ + $AlCl_3$	⠀	$AlCl_4^-$

Figure 16-2: Generating electrophiles for electrophilic aromatic substitution reactions.

Adding alkyl substituents: Friedel–Crafts alkylation

One way to alkylate benzene rings was proposed by Charles Friedel and James Crafts. Reacting alkyl chlorides with the Lewis acid aluminum trichloride makes carbocations, they discovered, as shown in Figure 16-3. Carbocations are electrophilic enough to react with benzene to form alkyl benzenes by the same mechanism outlined in Figure 16-1.

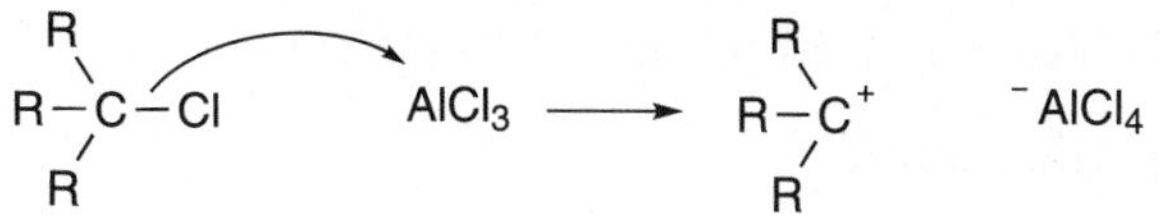

Figure 16-3:
Making carbocations.

$$R-\underset{\underset{R}{|}}{\overset{\overset{R}{|}}{C}}-Cl \quad AlCl_3 \longrightarrow R-\underset{\underset{R}{|}}{\overset{\overset{R}{|}}{C}}{}^+ \quad {}^-AlCl_4$$

Carbocations, though, are like naughty little children. As illustrated in Chapter 10 with the addition of HCl to alkenes, carbocations can — and will — rearrange if doing so makes a more stable carbocation.

Tertiary cations are more stable than secondary cations, which are more stable than primary cations.

For example, reacting benzene with propyl chloride and aluminum trichloride makes isopropyl benzene as the major product, with the expected product, propyl benzene, as the minor product. See Figure 16-4 for the products of this reaction.

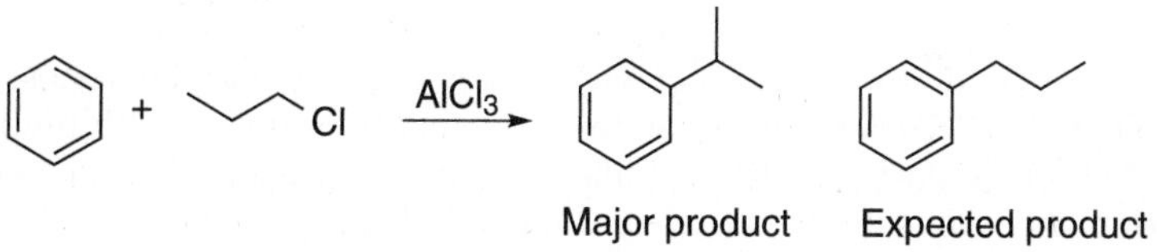

Figure 16-4:
The Friedel–Crafts alkylation.

Major product   Expected product

The reason for this reaction not giving the expected product is that the primary cation can rearrange to the more stable secondary carbocation by a hydride shift, as shown in Figure 16-5. The addition of this secondary cation to benzene yields isopropyl benzene as the major product. Because of these rearrangements, adding straight-chain alkyl groups to benzene is difficult using the Friedel–Crafts alkylation reaction.

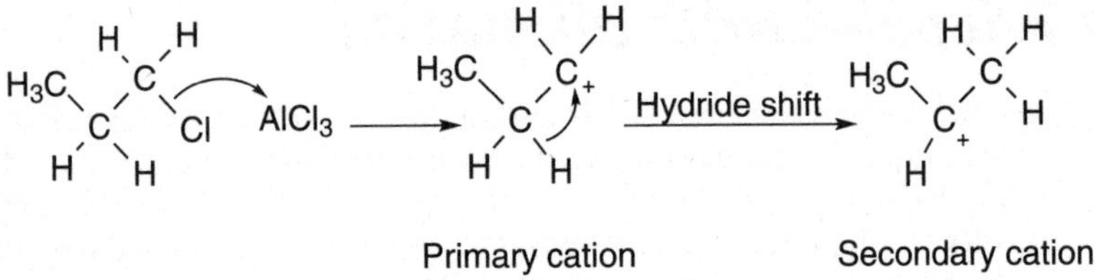

Figure 16-5:
The carbocation rearrangement in the Friedel–Crafts reaction.

Primary cation   Secondary cation

Overcoming adversity: Friedel–Crafts acylation

To stop these pesky rearrangements that lead to undesired products, Friedel and Crafts came up with an acylation reaction (pronounced ay-sill-*ay*-shun). Taking an acid chloride (RCOCl) and adding aluminum trichloride makes an acylium cation, as shown in Figure 16-6. Acylium ions are stabilized by resonance and don't rearrange.

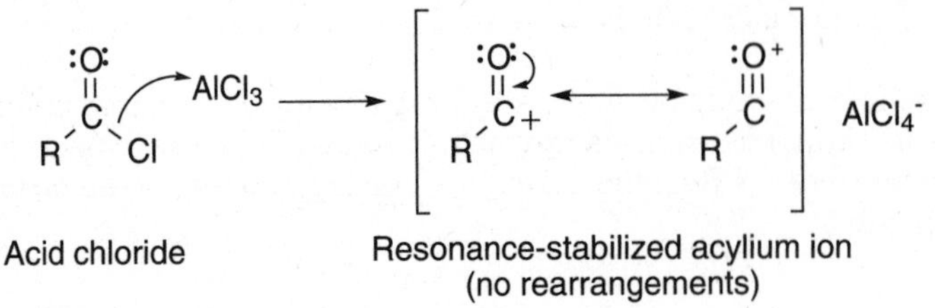

Figure 16-6:
Making acylium cations.

Acid chloride   Resonance-stabilized acylium ion
(no rearrangements)

These acylium ions then react with benzene to make aryl ketones, as shown in Figure 16-7. Aryl ketones can be conveniently reduced with hydrogen and palladium on carbon (Pd/C) to alkyl aromatics. (Regular ketones not next to benzene rings, however, are untouched by these reagents.) Although it requires an additional step, this reaction sequence — acylation followed by reduction — is a handy way of making alkyl benzenes without worrying about carbocation rearrangements giving you undesired products.

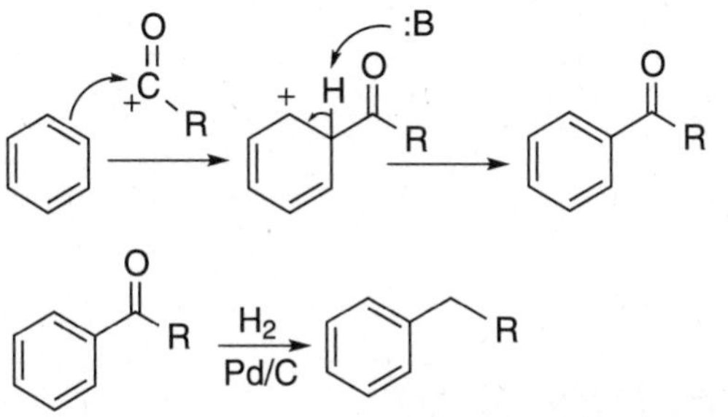

Figure 16-7:
The Friedel–
Crafts
acylation
mecha-
nism and a
convenient
follow-up
reaction
to reduce
the aryl
ketone to
the alkane.

Reducing nitro groups

A convenient way of making aryl amines is by reducing nitro groups. These groups can easily be reduced by the addition of tin chloride ($SnCl_2$) in a little acid, as shown in Figure 16-8.

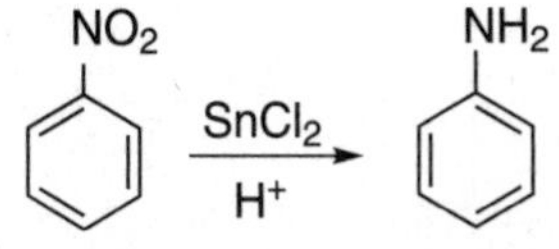

Figure 16-8:
Nitro reduc-
tion and the
formation
of an aryl
amine.

Nitrobenzene   Aryl amine (aniline)

Oxidation of alkylated benzenes

Potassium permanganate is a powerful oxidizing reagent. In the presence of acid, this reagent takes alkyl benzenes, chews them up, and spits out aryl carboxylic acids (benzoic acids; refer to Figure 15-20), as shown in Figure 16-9. Any alkyl side chain that has a hydrogen adjacent to the phenyl ring will be oxidized to a carboxylic acid (a COOH group). If the alkyl side chain has no hydrogen adjacent to the ring, it will be left untouched.

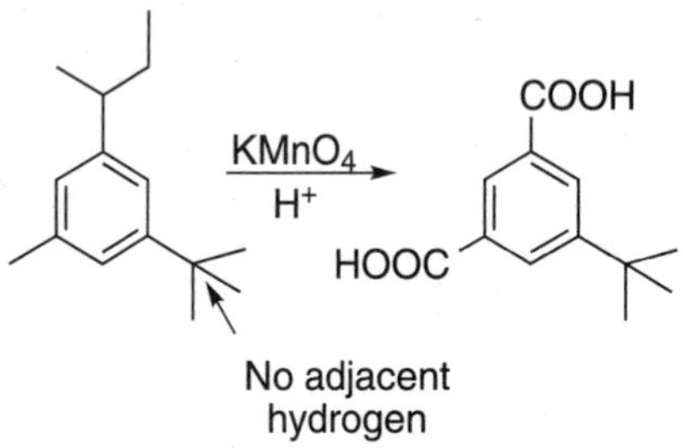

Figure 16-9:
Permanga-
nate oxida-
tion of alkyl
benzenes.

Adding Two: Synthesis of Disubstituted Benzenes

Now you know the reagents that add different groups to benzene. But what if you want to synthesize disubstituted benzenes? After adding the first group, three locations are possible for the next group to add to, as shown in Figure 16-10. One possibility is for the next group to add *ortho* — organic-speak for adding to the carbon adjacent to the substituent. The group could also add *meta* (two carbons away from the first group) or it could add *para* (on the opposite side of the ring from the original group).

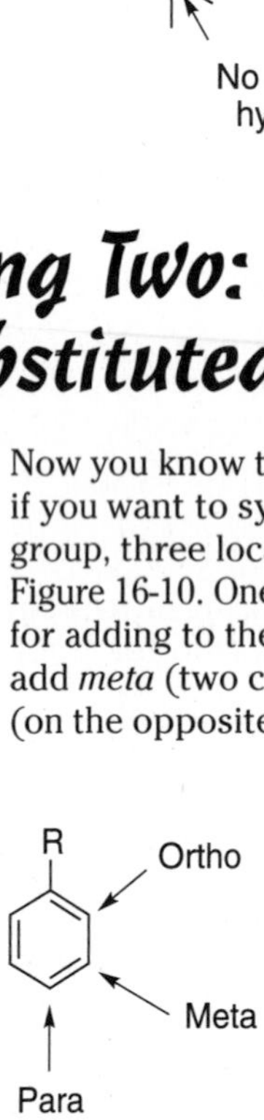

Figure 16-10:
The ortho,
meta,
and para
positions.

Where the second substituent adds — ortho, meta, or para to the first substituent — depends on the nature of the group already attached to the benzene ring. Think of the substituents already on the ring as traffic controllers at an airport, directing the electrophile to land at certain runways (ortho, meta, or para, in this case). Electron-donating substituents activate the ring (that is, they make the ring react faster than an unsubstituted benzene reacts), and they tell the incoming electrophile to land either at the ortho or the para position. With electron-donating substituents, you usually get a mixture of the ortho and para products. Electron-withdrawing groups on the ring, on the other hand, deactivate the benzene ring toward electrophilic attack (making the reaction slower than is the case with an unsubstituted benzene) and these electron-withdrawing groups tell the incoming electrophile to land at the meta position.

Why do you get different substitution products depending on whether the ring is substituted with electron donors or electron acceptors? The answer is that electrophiles add to the benzene ring on the atom that generates the most stable intermediate cation. With electron donors on the ring, the intermediate carbocation resulting from ortho or para addition is more stable than the cation deriving from meta addition. With an electron-withdrawing substituent on the ring, the intermediate carbocation from meta addition is more stable than the carbocation resulting from ortho or para addition.

In the following sections, I show you why the cation resulting from ortho-para addition with electron donors is more stable than the meta-derived cation. I also discuss why the cation resulting from meta addition with electron acceptors is more stable than the cations resulting from ortho or para addition.

Electron donors: Ortho-para activators

If you brominate anisole, for example, as shown in Figure 16-11, you get substitution of the bromine at the ortho and para positions, but not to the meta position. This is because methoxy groups (OCH_3) are pi electron donors, so they direct all incoming electrophile traffic into the ortho and para positions.

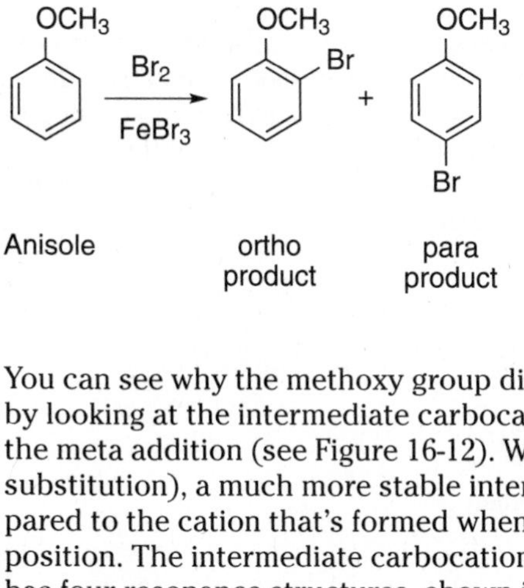

Figure 16-11: Bromine addition to anisole. Anisole — ortho product — para product

You can see why the methoxy group directs to the ortho and para positions by looking at the intermediate carbocation for both the para substitution and the meta addition (see Figure 16-12). With para substitution (and with ortho substitution), a much more stable intermediate carbocation is formed compared to the cation that's formed when the substituent adds in the meta position. The intermediate carbocation that results from para substitution has four resonance structures, shown in Figure 16-12, with one of these resonance structures being particularly good because all valence octets on all atoms are filled. (This good resonance structure is boxed in the figure.)

The carbocation resulting from meta substitution, on the other hand, has only three resonance structures, none of which have all atoms with filled valence octets. Recall the general rule that stability increases as the number of resonance structures increases. Thus, with electron donors on the aromatic ring, ortho-para products are selectively formed.

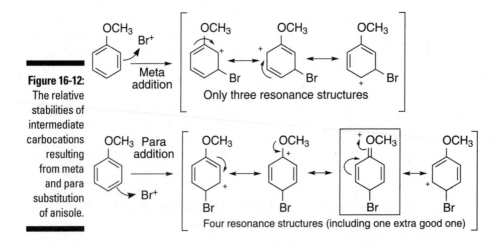

Figure 16-12:
The relative stabilities of intermediate carbocations resulting from meta and para substitution of anisole.

Electron-withdrawing groups: Meta directors

Of course, the situation is reversed when the aromatic ring is substituted with electron-withdrawing substituents rather than electron-donating substituents. Electron-withdrawing substituents usually direct incoming electrophiles to the meta position. Take the bromination reaction of nitrobenzene, for example, shown in Figure 16-13. Nitro groups are electron-withdrawing groups, so bromine adds to the meta position.

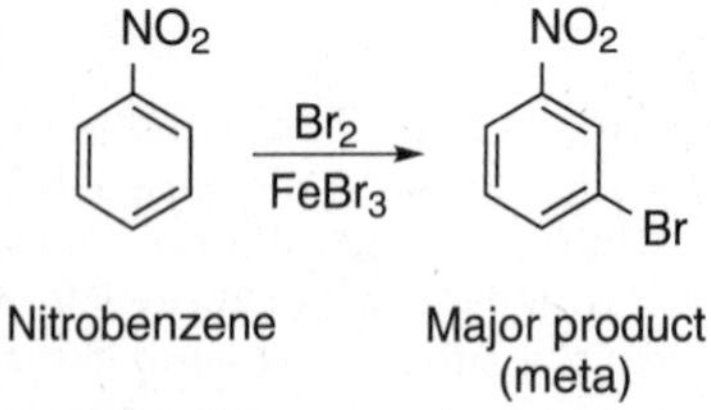

Figure 16-13:
The bromine substitution of nitrobenzene.

Nitrobenzene → (Br₂, FeBr₃) → Major product (meta)

To see why the meta product is formed instead of the ortho-para products, compare the intermediate cation formed as a result of para bromine addition to the cation generated from meta addition (see Figure 16-14). The intermediate carbocation resulting from para substitution (or ortho substitution) has three resonance structures, but one of them is a particularly bad resonance structure because the structure has two adjacent positive charges (and like charges repel). Thus, this bad resonance structure doesn't contribute much to the overall resonance hybrid. The cation resulting from meta substitution also has three resonance structures, but none of them is bad. Thus, for benzenes substituted with electron-withdrawing groups, the cation resulting from meta substitution is more stable than the cation resulting from either ortho or para substitution.

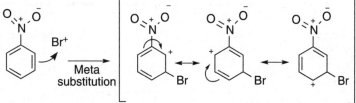

Figure 16-14:
The relative stabilities of intermediate carbocations resulting from para and meta substitution to nitrobenzene.

The main point to remember here is that electron-donating groups direct substitution to the ortho and para position, while pi electron-withdrawing groups direct substitution to the meta position.

So far, I've talked about electron-donating and electron-withdrawing substituents, without really saying what I mean. Any substituent whose first atom (the one that's attached to the benzene ring) has a lone pair will be a pi electron donor to the phenyl ring, as shown in the resonance structure in Figure 16-15. Note that the resonance structures show that substituents that are pi donors add electron density to the ortho and para positions of the ring. Thus, pi donors activate the benzene ring toward *electrophilic attack* (attack by incoming electrophiles) at the ortho and para positions. Adding electron density to the benzene ring makes the ring more *nucleophilic* (that is, more nucleus loving) and activates the ring. Therefore, pi donors are considered ring activators.

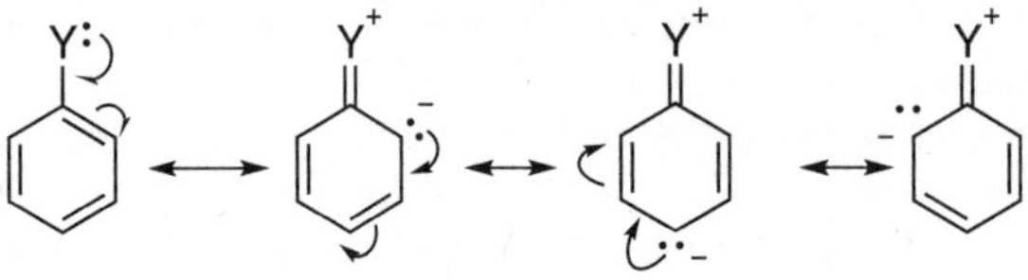

Figure 16-15:
Pi electron
donors to the
phenyl ring.

The only exceptions are the halogens, which are not terribly good pi donors. They deactivate the ring as a result of being highly electronegative groups, pulling electrons away from the benzene ring toward themselves, making the ring less nucleophilic. But even though halogens are ring deactivators, they're still ortho-para directors.

Pi-withdrawing groups (such as NO_2 groups, carbonyl groups, CN, and so on) pull electrons away from the ring and deactivate it, making the ring less nucleophilic. Pi electron-withdrawers are thus ring deactivators. A deactivator means that the reaction of benzenes substituted with these substituents will be slower than the reaction of unsubstituted benzene. Pi electron-withdrawing substituents are meta directors. Table 16-1 outlines the nature of different substituents.

Table 16-1	The Nature of Aromatic Substituents		
Ortho-Para Directing			*Meta Directing*
Strongly activating	*Weakly activating*	*Deactivating*	*Deactivating*
OH	Alkyl	Halogens (F, Cl, Br, I)	NO_2
OCH_3	Phenyl		—COR (COOH, COOR, CHO, and so on)
NH_2			CN
NR_2			SO_3H

Here's a tip for remembering Table 16-1: Any substituent whose first atom (the one attached to the benzene) contains a lone pair of electrons will be an ortho-para director (although not necessarily a ring activator). Those substituents without a lone pair on the first atom will likely be meta directors (with the exception of alkyl groups and aromatic rings, which are ortho-para directors).

For performing multistep synthesis, the reduction of a nitro group to make an aryl amine is a way of converting a meta director into an ortho-para activator. The oxidation of an alkylated benzene (to make an aryl carboxylic acid) is a way of converting an ortho-para activator into a meta director. Additionally, the reduction of an aryl ketone (the product of a Friedel–Crafts acylation) to an alkane is a way of converting a meta director into an ortho-para activator.

Problem Solving: Synthesis of Substituted Benzenes

When thinking about the synthesis of *polysubstituted benzenes* (benzenes substituted more than once), remember that the order of substituent addition is crucial. You have to put on the substituents in the right order so that they direct the next substituent to the right place.

For example, how would you synthesize 3-bromo-1-ethylbenzene, shown in Figure 16-16? First, you should notice that the substituents are meta to each other. That means that the first substituent that you add should be a meta director in order to get the next substituent in the meta position. But the substituents — the ethyl group and the bromine — are both ortho-para directors! How are you going to get them meta to each other? At some point, one of them must have been a meta director that then got converted into the ortho-para substituent.

Figure 16-16:
3-bromo-
1-ethylben-
zene.

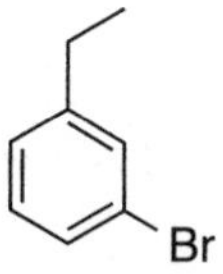

In the case of 3-bromo-1-ethylbenzene, the ethyl group could be added by a Friedel–Crafts acylation, as shown in Figure 16-17. Acyl groups are meta directors. The bromine could then be added and would go meta to the acyl group. The acyl group could then be reduced into the alkyl portion, as shown in the last step of Figure 16-17. The key to this synthesis is the order — switching any of the steps would lead to the wrong product.

Figure 16-17:
The syn-
thesis of a
disubstituted
benzene.

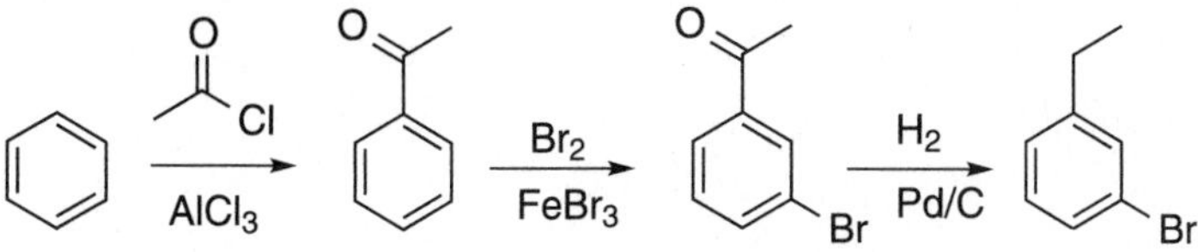

The key to mastering aromatic synthesis problems is to add substituents in the correct order.

Nucleophiles Attack! Nucleophilic Aromatic Substitution

I've shown you how benzene can act as a nucleophile (nucleus lover), reacting with powerful electrophiles in the electrophilic aromatic substitution reaction. Benzene can also act as an electrophile when the ring is sufficiently activated toward nucleophilic attack. When benzene is activated toward nucleophilic attack, strong nucleophiles can displace leaving groups substituted on the ring (usually a halide).

Benzene rings activated toward nucleophilic attack include those that are substituted with strong electron-withdrawing groups in the ortho and para positions (groups like NO_2, CN, COR, and so on). Substituted benzenes that are not activated with powerful electron-withdrawing groups will not undergo nucleophilic aromatic substitution. For example, 1-bromo-2,4-dinitrobenzene, shown in Figure 16-18, is activated toward nucleophilic attack because it has two powerful electron-withdrawing groups (NO_2) ortho and para to the leaving group (Br). The addition of hydroxide ion (HO^-) displaces the bromide.

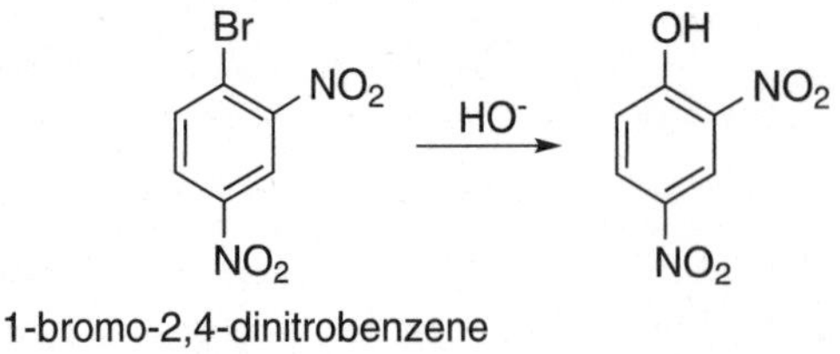

Figure 16-18:
Nucleophilic
aromatic
substitution.

1-bromo-2,4-dinitrobenzene

The mechanism of the addition is shown in Figure 16-19. First, the nucleophile (HO^-) attacks the benzene ring at the carbon holding the leaving group, generating a *carbanion* (a negatively charged carbon). This carbanion is stabilized by the electron-withdrawing groups (NO_2) on the ring (and by additional resonance structures that are not shown). The anion can then displace the bromide ion leaving group to regenerate the substituted aromatic ring.

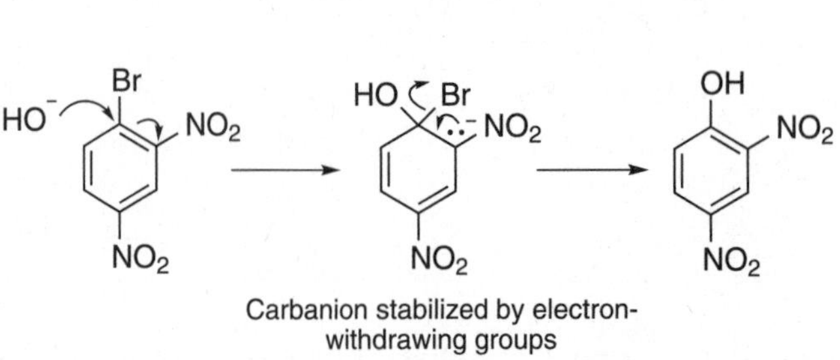

Figure 16-19:
The mecha-
nism of
nucleophilic
aromatic
substitution.

Carbanion stabilized by electron-
withdrawing groups

REMEMBER

Nucleophilic aromatic substitution only occurs with benzene rings activated by strong electron-withdrawing groups.

A somewhat less useful but still highly interesting reaction of benzene is the reaction of a halobenzene (like bromobenzene or chlorobenzene) with a strong base (like hydroxide) at high temperature and pressure. Because the aromatic ring is not activated for nucleophilic attack by electron-withdrawing substituents, you can't get nucleophilic aromatic substitution. Instead, you get elimination, as shown in Figure 16-20, to make a weird intermediate called a benzyne. *Benzyne* is a highly unstable intermediate, consisting of a benzene ring with a triple bond. Nucleophiles (abbreviated ⁻:Nuc in Figure 16-20) quickly add to benzyne to make substituted benzenes.

Figure 16-20:
Benzyne
reactions.

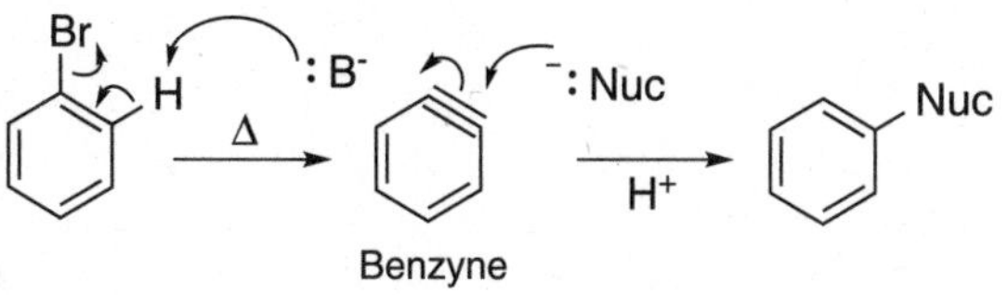

Benzyne

An example of benzyne addition is shown in Figure 16-21. In this case, the base that's used to make the benzyne and the nucleophile that reacts with the benzyne is hydroxide (⁻OH). Note that because the nucleophile can add to either side of the benzyne, you get a mixture of products. Because you often get product mixtures with substituted benzenes, and because this reaction requires high temperature and pressure, this reaction is not quite as generally useful as electrophilic or nucleophilic aromatic substitution.

Figure 16-21:
An example
of benzyne
substitution.

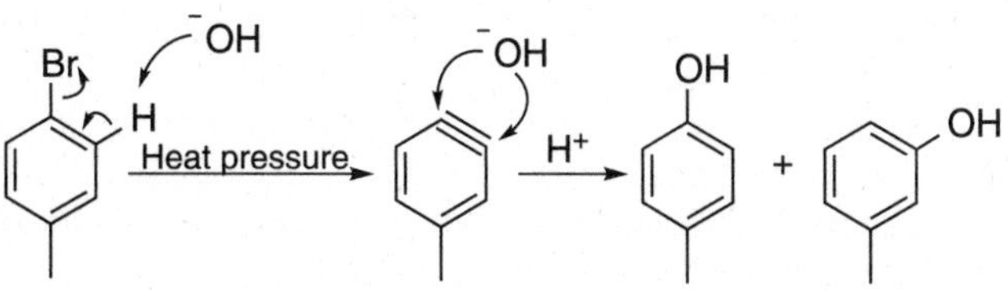

Heat pressure

Part IV
Spectroscopy and Structure Determination

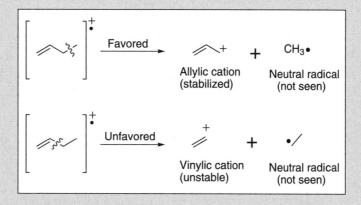

In this part . . .

- ✔ See how mass spec, NMR, and IR spectroscopy work.

- ✔ Use mass spec to look at fragmentation patterns of molecules.

- ✔ Look at how IR spectroscopy can allow you to determine the functional groups in a molecule.

- ✔ Do detective work to figure out the structure of unknown compounds using spectroscopy.

Chapter 17

A Smashing Time: Mass Spectrometry

*I*magine trying to determine how a watch is put together by whacking it with a mallet and examining all the little pieces. KAPLOW! A spring here, a cog there. This process may not be the most elegant examination — and it probably won't please your friend whose watch you "borrowed" — but it will give you some information about the different parts that made up that watch. Such an examination probably wouldn't give you enough information to determine exactly how the watch fit together, but you would certainly be more informed about what was inside the watch after your experiment than you were before.

Furthermore, certain brands of watches would likely break apart in different ways. If you smashed a particular brand of watch, say Watch Brand A, you may find that in most cases you break off the clasp when you hit it with a mallet; with Watch Brand B, you may find that you break off the casing more of the time but leave the clasp intact.

That's sort of how mass spectrometry works — with a molecule, of course, instead of a watch. A mass spectrometer works by smashing a molecule into bits. Instead of viewing all the shattered pieces, though, a mass spectrometer weighs them. The weights of these fragments give clues about the structure of the molecule itself. And, like different kinds of watches, different kinds of molecules break apart in distinct ways in a mass spectrometer.

In this chapter, I talk about how mass spectrometry works, how to interpret a mass spectrum, and how to figure out what fragments will most likely be formed from a given structure.

Defining Mass Spectrometry

Mass spectrometry is really quite different from spectroscopy. You often hear or read the words *mass spectroscopy*. But by definition, spectroscopy involves light, and mass spectrometry doesn't involve light at all. Still, like spectroscopy (see Chapters 18 and 19), mass spectrometry (or "mass spec") provides valuable information about the structure of a compound. In fact, modern organic chemists use mass spectrometry nearly as much as NMR (see Chapter 19) in order to determine the structures of compounds. Perhaps this is because, in addition to giving clues about the structures of unknown molecules, mass spectrometry appeals to chemists' childlike urges to smash things — albeit on a molecular scale.

Taking Apart a Mass Spectrometer

Here's a very general rundown of how a mass spectrometer works, free for the moment of technical names and jargon (see Figure 17-1). The sample containing the unknown compound is first injected into the spectrometer through an inlet, and is then vaporized by heating it under a vacuum. The vaporized sample is then pushed by an inert gas into the "smasher" where the sample is clobbered and broken into little bits. Some particles come out of the "smasher" with an electrical charge and some come out uncharged. All the charged bits (called *ions*) move along into the sorter, which separates the bits according to how much they weigh. All the uncharged bits are discarded. The remaining charged particles then hit a detector, which determines how many bits of each weight there are, and plots the data on the mass spectrum. This spectrum shows the molecular weight of each fragment versus the number of fragments with that weight that hit the detector.

Now I talk about the specifics of each of these parts.

Figure 17-1:
The basic parts of a mass spectrometer.

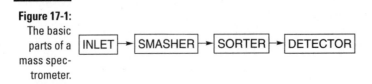

The inlet

Mass spectrometry is performed in the gas phase. Therefore, the first thing that happens after the sample is injected into the spectrometer is that the sample is vaporized by heating the sample under low pressure. After the sample is in the gas phase, it's pushed into the smasher by the flow of an inert gas (usually helium).

Electron ionization: The smasher

A mass spectrometer can use many different "mallets" to *ionize* molecules. *Ionize* is organic-speak for smashing the neutral molecules into charged fragments. The most common mallet is one that incorporates high-energy electrons (there are other kinds, but this is a popular one). Mass spectrometry using this type of smasher is termed *electron-ionization* mass spectrometry (EIMS). (On an exam, be sure to use this term, not "smasher.") In an electron-ionization mass spectrometer, the molecules are pushed through a stream of very high-energy electrons to ionize them. When one of these high-energy electrons crashes into the molecule, an electron is ejected from the molecule, as shown in Figure 17-2.

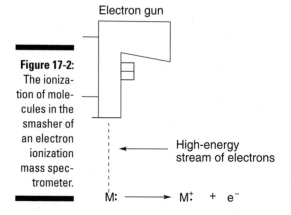

Figure 17-2: The ionization of molecules in the smasher of an electron ionization mass spectrometer.

Both the smashing electron from the electron source (diagrammed in the figure as an electron gun) and the electron that's kicked out of the molecule don't play any further roles in the process and are simply discarded. But each molecule that has had an electron knocked out becomes a species called a *radical cation* (abbreviated +·). The "cation" portion of the name results from the fact that the molecule has lost an electron and is now positively charged, and the "radical" portion of the name comes about because one of its electrons is unpaired. For more on radicals, see Chapter 8. These radical cations can either remain in one big piece or spontaneously fragment further into several smaller pieces. (I talk more about this later.) In either case, all the fragments move on toward the sorter and weigher.

The sorter and weigher

What happens after the molecules are smashed and become radical cations? Some of them stay as they are and move through the spectrometer to the weigher. Those pieces that stay whole give a peak in the mass spectrum called the *molecular ion peak,* or M+ peak. This molecular ion peak tells you the molecular weight of the molecule, because losing an electron doesn't really change the weight of the molecule (it's like a tractor trailer losing a lug nut). The molecular ion peak is the most important piece that's weighed by the mass spectrometer, because knowing the molecular weight of the unknown molecule is a very valuable piece of information when you're trying to determine its structure.

Some of the radical cations stay intact and lead to the molecular ion peak, but others break down spontaneously into smaller bits. Most commonly, the radical cations break into two pieces, one piece that's a neutral radical, and one piece that's a positively charged cation (see Figure 17-3). For reasons I talk about later, only the charged cationic species is "seen" and weighed by the mass spectrometer. The neutral radicals are discarded and go undetected.

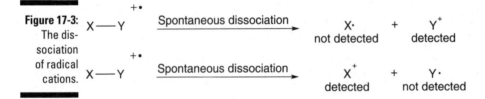

Figure 17-3:
The dissociation of radical cations.

After the pieces have become charged by the smasher, they're sent through the weigher. This weigher, though, isn't like the scale that collects dust in your bathroom. Instead, because the fragments are charged, they can be weighed by accelerating them through the poles of a magnet. When charged particles move through a magnetic field, they're *deflected* (pulled off course) by the magnetic field; all uncharged fragments are not deflected by the magnet and simply crash and burn into the walls of the spectrometer, never to be seen again. Therefore, only charged particles can hit the detector.

The weight of a fragment determines how much it will be deflected by the magnet. Light fragments are deflected a lot by the magnet, while heavier fragments are deflected less. With a low magnetic field strength, small particles will be bent the right amount by the spectrometer to hit the detector, while all the other fragments will crash and burn into the walls of the spectrometer. With a larger magnetic field strength, larger particles will curve the right amount to hit the detector. By varying the magnetic field strength (which is proportional to the weight of the fragment), the weights of the fragments can be determined.

Detector and spectrum

The number of ions that bend the right amount at a particular magnetic field strength to hit the detector is measured. On a mass spectrum, the molecular weight of the fragment is shown on the *x*-axis versus the relative number of fragments that hit the detector on the *y*-axis (the intensity). The most intense peak on the mass spectrum is arbitrarily assigned an intensity value of 100. The full mass spectrometer is shown in Figure 17-4.

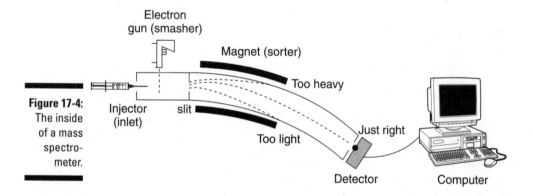

Figure 17-4: The inside of a mass spectrometer.

It's actually not the weight of the fragment that's measured on the *x*-axis of the mass spectrum, but the mass (*m*) to charge (*z*) ratio (called the *m/z value*). A fragment that has a +2 charge will require a smaller magnetic field to deflect it toward the detector than a fragment that has a +1 charge, and will therefore show up on the mass spectrum at half the *m/z* value. However, since most fragments simply have charges of +1 (and since the vast majority of undergraduate classes will not discuss fragments with more than one charge), the *m/z* axis is often simply said to represent the molecular weight of the fragment.

The Mass Spectrum

Figure 17-5 shows the mass spectrum for the molecule pentane (C_5H_{12}). This spectrum contains several peaks that are worth pointing out. The molecular ion peak, or M+ peak, is the peak at m/z 72, which represents the molecular weight of the molecule (72 amu). The tallest peak in the spectrum, which is always given a value on the y-axis of 100, is called the *base peak*. The base peak represents the most abundant fragment that hits the detector. In some spectra, the base peak is also the M+ peak, although in this case it's not.

The mass spectrum for pentane is fairly simple, but it still contains a lot of peaks, each of which represents a different fragment that was formed by electron ionization. Later in the chapter, I show you how to determine the structure of some of the most important fragments in the mass spectrum. Usually, though, you can't expect to be able to identify all the peaks in a mass spectrum (or even most of them).

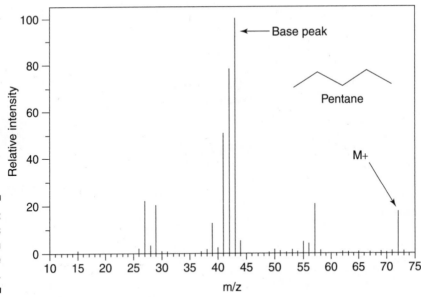

Figure 17-5:
The mass spectrum for pentane (C_5H_{12}).

Kind and Caring: Sensitivity of Mass Spec

One of the most valuable features of mass spectrometry is its sensitivity. NMR and IR spectroscopy usually require milligrams of material to do an analysis, but mass spec requires only nanograms (10^{-9} grams) of material. This sensitivity is particularly useful for characterizing compounds isolated from biological systems in which many substances are produced and isolated in miniscule amounts. (The structure of the ketone chiloglottone, discussed in Chapter 5, was determined solely using mass spectrometry because it was isolated in such small amounts.) Unlike NMR and IR spectroscopy, however, the sample used to take a mass spectrum is nonrecoverable, so the tiny amounts that are required in order to obtain a spectrum are destroyed and cannot be re-collected for later study.

Resolving the Problem: Resolution

In addition to its ability to detect tiny amounts of material, mass spectrometry can also weigh things with incredible accuracy. In other words, high-resolution mass spectrometers (HRMS) can determine the exact weight of a compound, enough to definitively identify a compound's molecular formula, which is a very valuable piece of information. For example, the compounds with molecular formulas C_3H_6 and C_2H_2O both have molecular weights of about 42 amu. So, how can a mass spectrometer distinguish between them?

The answer lies in the fact that the mass spectrometer determines the weight of the molecular ion (M+) fragment with accuracy of several digits after the decimal point. Both C_3H_6 and C_2H_2O have molecular weights that are approximately — but not exactly — 42 amu. The molecular mass system is set up so that ^{12}C has a mass of exactly 12.0000 amu, and the weights of all other atoms are calibrated against this number. Therefore, elements other than carbon don't have masses that are exact whole numbers. By this system, the exact mass of C_3H_6 is 42.0469 amu and the exact mass of C_2H_2O is 42.0105 amu.

In this example, the masses of these two compounds differ by about 0.036 amu; this is a very small difference, really, but a difference that's large enough to be distinguished by a good mass spectrometer. A mass spectrometer with a high enough resolution can detect these differences, and the correct molecular formula can therefore be deduced (usually, using a computer program). In this case, a mass spectrometer that's accurate to the second decimal place would be able to distinguish between the two compounds. Most high-resolution mass spectrometers, however, are much more accurate than that, and many are accurate to the fourth decimal place.

Changing the Weight: Isotopes

Because mass spectrometry determines the weights of fragments, atoms that naturally have heavy isotopes become important.

Isotopes are atoms that have the same number of protons and electrons, but different numbers of neutrons.

Isotope effects are most readily observed in compounds containing the halogens chlorine (Cl) and bromine (Br). In nature, chlorine consists of about 75 percent of the ^{35}Cl isotope and about 25 percent of the heavier ^{37}Cl isotope, in which the chlorine has two additional neutrons. This means that if 100 chlorine-containing molecules were chosen at random, on average 75 of them would contain the ^{35}Cl isotope, while 25 of them would weigh two mass units more because they contain the heavier ^{37}Cl isotope. Because mass spectrometry weighs molecules, these heavier-mass isotopes will be observed in the mass spectrum.

In fact, both of these isotopes will be seen in the mass spectrum. But because a sample of a chlorine-containing molecule has three times as many of the ^{35}Cl isotope as it has of the heavier ^{37}Cl isotope, the molecular ion peak (which corresponds to molecules containing the ^{35}Cl isotope) will be three times as intense as the peak two mass units heavier that corresponds to molecules containing the heavier isotope. Because this heavy isotope peak is two mass units heavier than the peak of the molecular ion containing the ^{35}Cl isotope, it's called the M+2 ion. (If the heavy isotope is heavier by one neutron, it's called the M+1 isotope peak; by two neutrons, the M+2 peak; by 3 neutrons, the M+3, and so on.)

These isotope effects allow you to spot molecules that contain chlorine, because the spectra of a chlorine-containing molecule will contain a peak that's two mass units heavier than the molecular ion (M+) and that's one-third as intense as the M+ peak (see Figure 17-6).

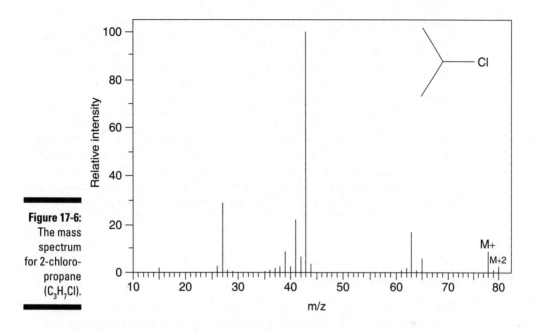

Figure 17-6:
The mass spectrum for 2-chloropropane (C$_3$H$_7$Cl).

Naturally occurring bromine consists of a mixture of isotopes that's approximately 50 percent ^{79}Br and 50 percent ^{81}Br. Therefore, bromine is easily spotted in the mass spectrum by observing a peak of approximately equal intensity two mass units higher than the molecular ion peak (see Figure 17-7).

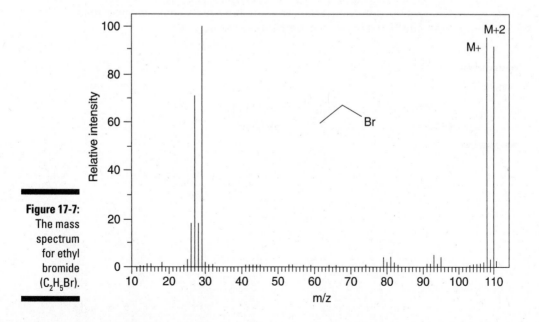

Figure 17-7:
The mass spectrum for ethyl bromide (C$_2$H$_5$Br).

Naturally occurring iodine is virtually 100 percent ^{129}I, so unlike the spectra of chlorine- and bromine-containing molecules, iodine-containing molecules don't show isotope peaks. However, iodo compounds can often be spotted by observing the iodonium peak (I^+) in the mass spectrum at m/z 129.

Other atoms have isotopes that are less abundant. Naturally occurring carbon contains about 1 percent of the ^{13}C isotope. Because a molecule that consists of many carbons will have a greater chance of having a ^{13}C atom, molecules with many carbon atoms will have a larger ^{13}C isotope peak in the mass spectrum than molecules that contain fewer carbon atoms. (Because ^{13}C is one mass unit heavier than ^{12}C, the isotope peak is called the M+1 peak.) This, in fact, is a crude way to determine the number of carbons in the molecule, because the size of the M+1 isotope peak relative to the molecular ion (M+) peak depends on the number of carbons in the molecule.

The Nitrogen Rule

Any molecule that contains only carbon, hydrogen, oxygen, or the halogens (F, Cl, Br, I), will have an even-numbered molecular weight. However, molecules that contain an odd number of nitrogens will have odd molecular weights. Knowing this, you can spot certain molecules that contain nitrogen, because if your molecular ion (M+) peak occurs at an odd-numbered m/z value, the molecule likely contains an odd number of nitrogens. This is because nitrogen has three bonds and a lone pair of electrons, rather than four bonds like carbon (thus, for every nitrogen you substitute for carbon, you lose one hydrogen). An even number of nitrogens will lead to an even-numbered molecular weight (see Figure 17-8).

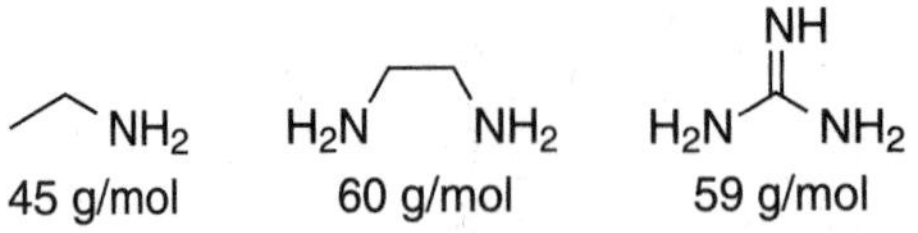

Figure 17-8:
Seeing the nitrogen rule.

45 g/mol   60 g/mol   59 g/mol

Identifying Common Fragmentation Patterns

Often, predicting what peaks will be observed in the mass spectrum is possible simply by looking at a structure and seeing which pieces would be easy to break off to make stable cations. In this section, I go through some common structural features that make stable cations.

The ionizer makes the molecules into radical cations, which then generally break apart into a cationic piece (which is seen by the detector) plus a neutral radical piece (not seen by the detector).

Smashing alkanes

Alkanes generally break apart to make the most highly substituted cation. *Tertiary cations* (cations substituted with three carbons) are more stable than *secondary cations* (cations substituted with two carbons), which in turn are more stable than *primary cations* (cations substituted with only one carbon). Therefore, breaks in a molecule that make tertiary cations are likely to give fragments that correspond to large peaks in the mass spectrum, because the most stable fragments produce the largest peaks (see Figure 17-9).

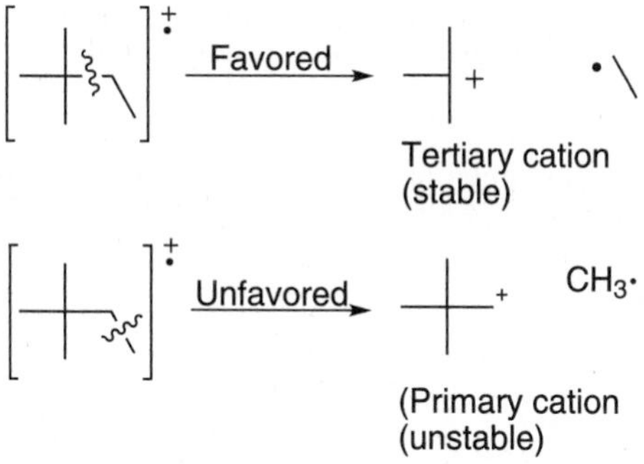

Figure 17-9:
Favored and unfavored bond cleavage.

Tertiary cation (stable)

(Primary cation (unstable)

Breaking next to a heteroatom: Alpha cleavage

When a molecule contains heteroatoms (elements such as oxygen, sulfur, and nitrogen), breaking next to these atoms makes cations that are resonance stabilized. For example, breaking the C-C bond next to an alcohol group creates a resonance-stabilized carbocation. This type of break is called *alpha cleavage* and is commonly seen in alcohols (as shown in Figure 17-10).

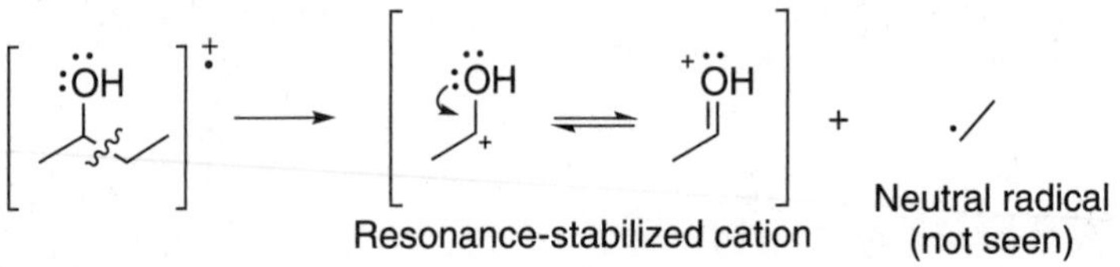

Figure 17-10: Alcohol alpha cleavage.

Resonance-stabilized cation

Neutral radical (not seen)

This same pattern of alpha cleavage is observed with amines, as shown in Figure 17-11.

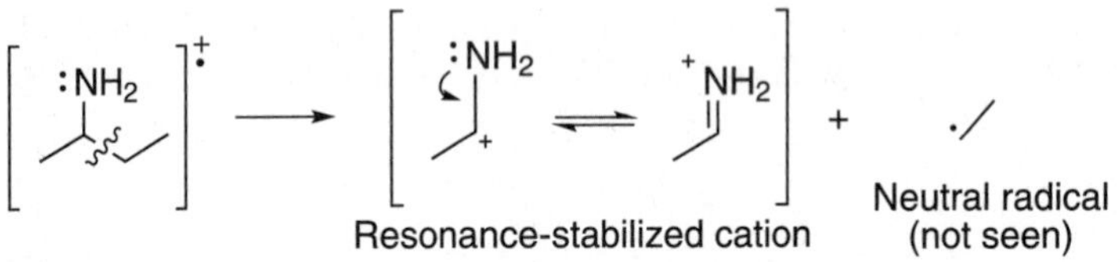

Figure 17-11: Amine alpha cleavage.

Resonance-stabilized cation

Neutral radical (not seen)

Alpha cleavage is also commonly seen in ethers, where the bond breaks adjacent to the oxygen (see Figure 17-12).

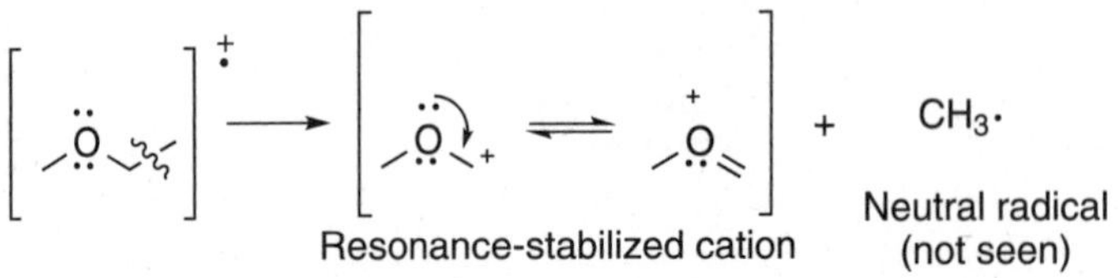

Figure 17-12: Ether alpha cleavage.

Resonance-stabilized cation

Neutral radical (not seen)

Breaks are often seen next to carbonyl groups (C=O groups), because this creates the resonance-stabilized cation (see Figure 17-13). (I hope this is getting repetitious.)

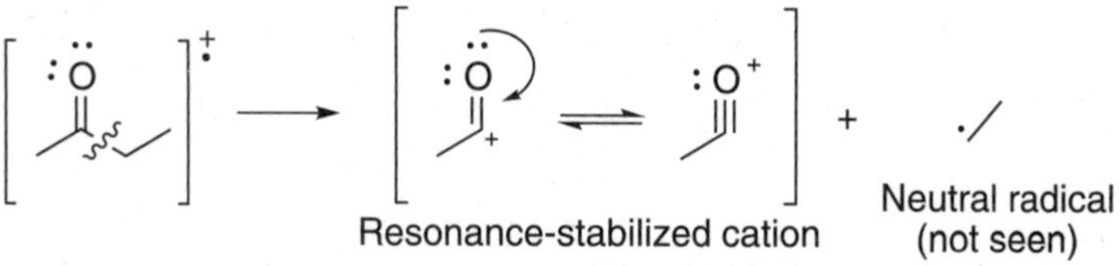

Figure 17-13:
Carbonyl
alpha
cleavage.

Resonance-stabilized cation

Neutral radical
(not seen)

Loss of water: Alcohols

In addition to alpha cleavage, alcohols readily lose a water molecule to form an *alkene* (a carbon-carbon double bond). This is why the mass spectrum of an alcohol often has a peak corresponding to the loss of 18 mass units (the weight of water). See Figure 17-14.

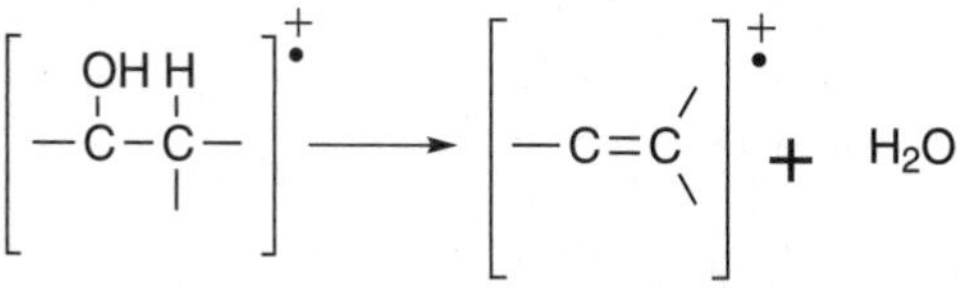

Figure 17-14:
Alcohol
dehydration.

Rearranging carbonyls: The McLafferty rearrangement

Some radical cations can also rearrange. The most famous rearrangement is called the *McLafferty rearrangement.* It can occur on carbonyl compounds (such as ketones and aldehydes; see Chapter 5 for more on carbonyl functional groups) that have a hydrogen on a carbon that's three carbons away from the carbonyl group. This third carbon position is called the gamma, or γ, position. The rearrangement involves a six-membered ring transition state in which the carbonyl group pulls off the γ proton, breaking the molecule into two pieces. These pieces consist of an enol radical cation (recall that an enol is a combination of an alk*ene* and an alcoh*ol*) and a neutral alkene fragment (a fragment with a carbon double bond). The enol radical cation is observed in the mass spectrum, while the neutral alkene fragment is not observed. Any carbonyl compound that has a hydrogen in the γ position is likely to have a peak in the mass spectrum corresponding to the enol radical cation that's formed by the McLafferty rearrangement of that carbonyl compound. An example of the McLafferty rearrangement is shown in Figure 17-15.

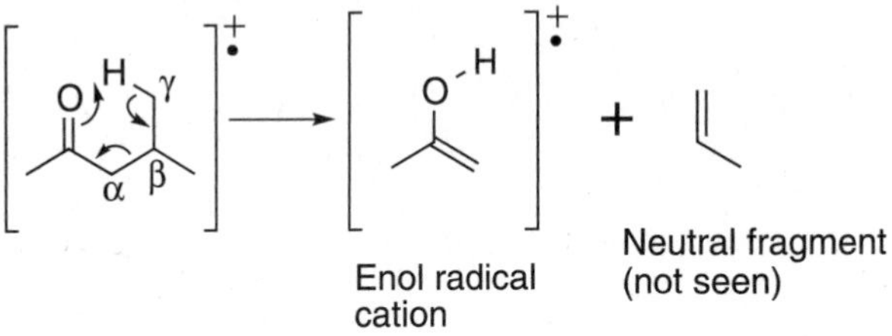

Figure 17-15:
The McLafferty rearrangement.

Enol radical cation

Neutral fragment (not seen)

Breaking benzenes and double bonds

Breaking one carbon away from an aromatic ring leads to the stable benzylic cation as shown in Figure 17-16. This benzene ring stabilizes the cation through resonance (see Chapter 3).

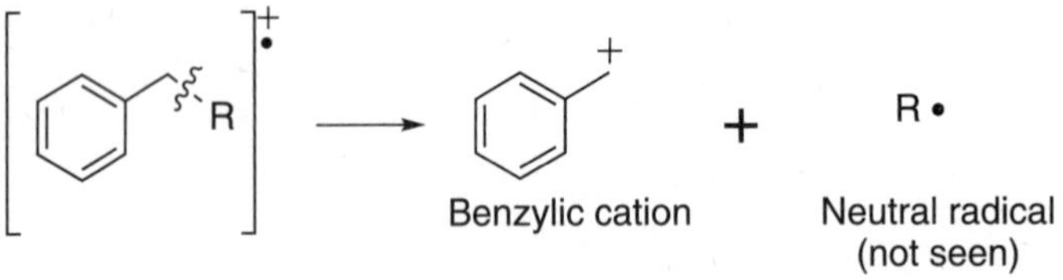

Figure 17-16:
Benzylic cleavage.

Benzylic cation

R•

Neutral radical (not seen)

Likewise, breaking the bond one carbon away from a double bond leads to an allylic cation that's stabilized by resonance. Note that breaking adjacent to the double bond leads to the unstable vinylic cation, as shown in Figure 17-17.

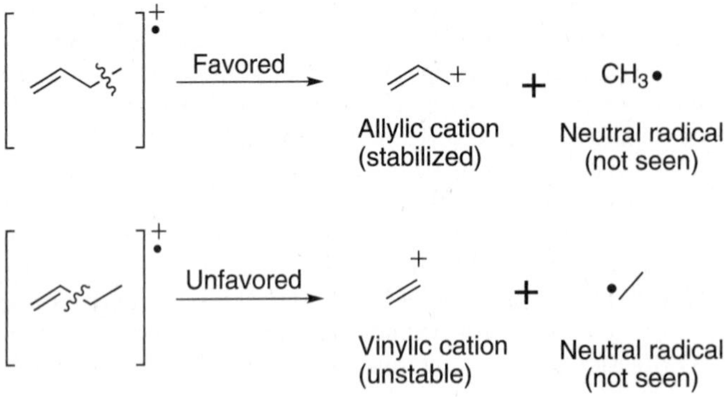

Favored

Allylic cation (stabilized)

CH_3•

Neutral radical (not seen)

Figure 17-17:
Fragmentation next to alkenes.

Unfavored

Vinylic cation (unstable)

•/

Neutral radical (not seen)

The loss of 15 mass units from the molecular ion generally indicates a loss of a methyl (CH_3) group. The loss of 29 mass units often indicates loss of an ethyl (CH_2CH_3) group.

Self test: Working the problem

Here's a sample question of the type you may see on an exam: The mass spectrum for 2-pentanone is shown in Figure 17-18. Draw the fragments responsible for the peaks in the mass spectrum at m/z 86, 71, 58, 43.

The main piece of information to remember in answering these types of questions is that each of the structures you draw *must be positively charged.* Neutral fragments are discarded and don't reach the detector.

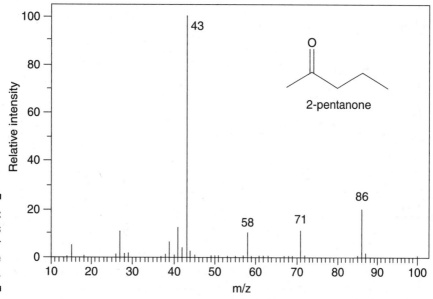

Figure 17-18: Mass spectrum for 2-pentanone ($C_5H_{10}O$).

Usually the best way to start is by looking for likely breaks in the molecule. Can you get an alpha cleavage (a break next to a heteroatom like O or N)? Can the molecule break to form a benzylic fragment or allylic fragment? Can a tertiary cation be made? If it's a carbonyl compound, does it have a gamma hydrogen that could be involved in a McLafferty rearrangement?

This example has neither a benzene ring nor a carbon-carbon double bond. It doesn't have the capacity to make a tertiary cation, either. But it does have a heteroatom (oxygen). So, alpha cleavage would definitely be a possibility. Additionally, because it's a carbonyl compound and has a hydrogen in the gamma position, you'll probably see a peak for the McLafferty rearrangement.

Start with alpha cleavage. The molecule can split on the left of the carbonyl group to make the resonance-stabilized cation that has a molecular weight of 71 amu. You may have already noticed that this peak with an *m/z* of 71 is 15 mass units below the molecular ion peak, suggesting a loss of a methyl group. This cleavage is shown in Figure 17-19.

Figure 17-19:
Alpha
cleavage.

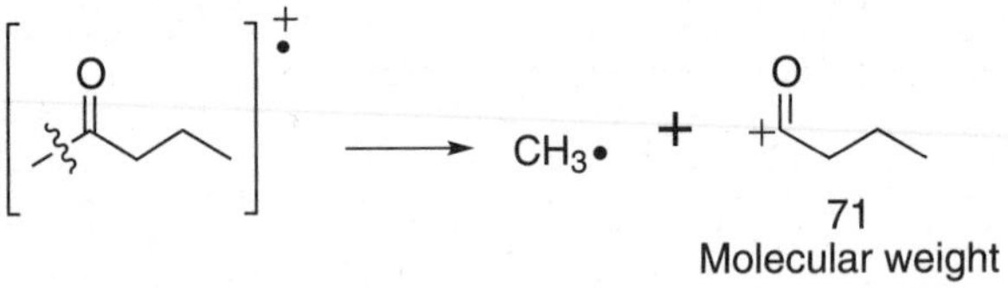

71
Molecular weight

Or the molecule could split via an alpha cleavage on the right side of the carbonyl to make a cationic fragment that weighs 43 amu, as shown in Figure 17-20.

Figure 17-20:
Another pos-
sible alpha
cleavage.

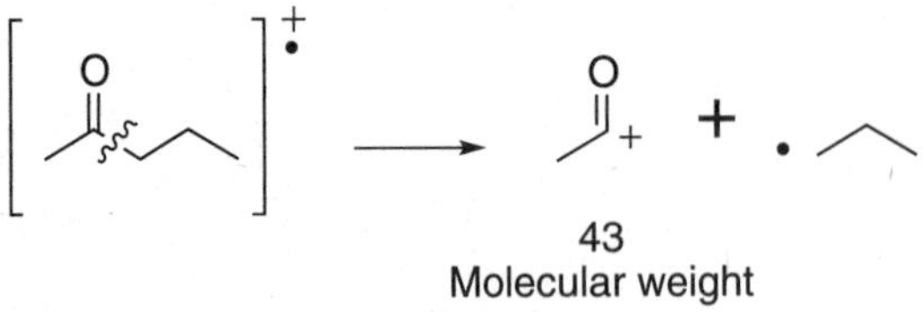

43
Molecular weight

Now try the McLafferty rearrangement. I've redrawn the molecule to make it easier to see which bonds get made and which bonds get broken. The McLafferty rearrangement gives the piece that weighs 58 amu, as shown in Figure 17-21.

Figure 17-21:
The
McLafferty
rearrange-
ment.

58
Molecular weight

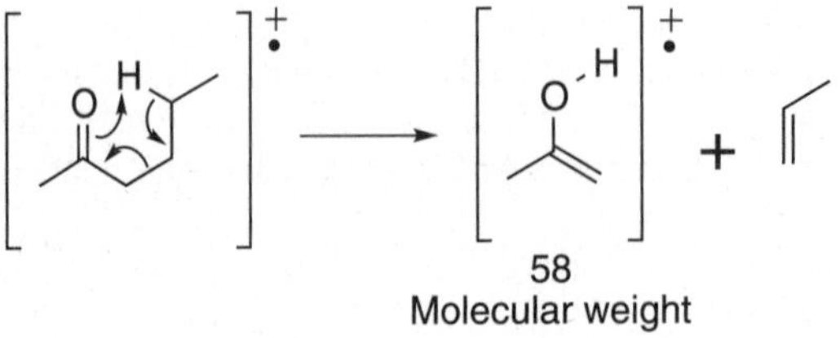

Key Ideas Checklist

For mass spectrometry, there's a good bit of detail that you need to keep track of, so it's easy to lose sight of the big picture. But here's a list of the main ideas and uses of mass spectrometry:

- Mass spectrometry is very sensitive. Only miniscule amounts of material are required to do an analysis.

- Only positively charged fragments are observed in a mass spectrum.

- The weights of fragments are plotted versus the intensity (or relative number of fragments) on the spectrum. The highest-intensity peak is arbitrarily assigned an intensity of 100 and is called the *base peak*.

- Atoms that have abundant heavy isotopes can often be spotted in a mass spectrum (particularly Cl and Br).

- Stable fragments (cations) have more intense peaks in the mass spectrum than do unstable fragments.

- The molecular formula of an unknown can often be determined using high-resolution mass spectrometers by looking at the parent ion, or M+ peak. The parent ion is a molecule that was ionized but did not break apart into smaller pieces before it hit the detector.

Chapter 18

Seeing Good Vibrations: IR Spectroscopy

- -

In This Chapter

▶ Understanding how IR spectroscopy works

▶ Seeing the IR spectrum

▶ Identifying functional groups from IR spectra

- -

*T*oday, expert infrared (IR) spectroscopists are a lot like cowboys. It's not that these chemists are cool like cowboys, or that they can hold their liquor like cowboys, or that they can beat up a buncha roughnecks in a saloon like cowboys, or even that they can attract the opposite sex like cowboys. Come to think of it, in most ways expert IR spectroscopists are the complete opposites of cowboys. But, like cowboys, expert IR spectroscopists are something of a dyin' breed (insert soulful Country/Western dirge).

Over the past 40 years, the art of gleaning structural information from an IR spectrum has been slowly going the way of record players and LPs — which is to say that this art is reminisced about over snifters of brandy by old-timers when thinking about the way life was before such terrible modern contrivances as indoor plumbing, the polio vaccine, and NMR spectroscopy. Because before NMR spectroscopy became widespread, organic chemists were usually experienced enough in the art of IR spectroscopy to squeeze detailed structural features of an unknown compound from the pulp of an IR spectrum.

Today, though, most organic chemists just extract the juicy bits from an IR spectrum, and (when no one's looking) chuck the pulpy remains in a dustbin. Nowadays, organic chemists look at an IR spectrum primarily to see what functional groups are present in a particular molecule, and then use NMR spectroscopy (see Chapters 19 and 20) to determine the structural details of the molecule. (For a review of the main functional groups, turn to Chapter 5.)

But although IR spectroscopy is somewhat less powerful than NMR spectroscopy, IR spectroscopy is still very useful as a structure-analyzing tool, and is widely used by modern organic chemists. Because the IR spectrum of a molecule tells you what functional groups are present in that molecule, and because taking an IR spectrum is a cinch (spectra can often be obtained in less than a minute), IR spectroscopy gives you a lot of bang for your buck (timewise, I mean; IR spectrometers are still somewhat expensive).

In this chapter, I discuss how IR spectroscopy works. I show you how an IR spectrum is taken, and I discuss how you can analyze an IR spectrum to determine which of the major functional groups are present in a compound.

Bond Calisthenics: Infrared Absorption

It's easy to be fooled by fixed chemical structures and tables of precise bond lengths into thinking that bonds in molecules are fixed. But bonds are dynamic things. Bonds are not of fixed lengths, but instead are constantly stretching, bending, flexing, rotating, and rocking (and if a bond is a rockin', don't come a knockin'). So when you refer to a bond length as being x angstroms long, you're talking about an average length, not a fixed distance. In fact, bonds behave kind of like springs that are in constant motion (see Figure 18-1). If you stretch a bond too far, it will apply a restoring force, snapping back in the opposite direction just like a spring.

Figure 18-1:
Bonds
behave like
springs.

Applying Hooke's law to molecules

Because bonds behave like springs, you can apply Hooke's law to determine the frequency of the vibrations of a bond. (Hooke's law is used by physicists to describe the frequency of vibrations in springs.) This frequency is described by the following equations, where v is the frequency of the vibration, c is the speed of light, and k is a constant that indicates the stiffness of the spring. The term μ is called the *reduced mass* and is a function of the masses of the two atoms involved in the bond (indicated by m_1 and m_2).

$$v = \frac{\pi c \sqrt{\dfrac{k}{\mu}}}{2} \qquad \mu = \frac{m_1 m_2}{m_1 + m_2}$$

The gist of these equations is that two main components determine the frequency of a bond stretching vibration. These components are:

- ✔ The masses of the atoms in the bond (which determine μ in the equations)
- ✔ The bond strength (or the spring constant, k)

Looking closely at this equation, you can discover that a spring attached to two light objects will vibrate with higher frequency (higher ν) than a spring attached to two heavier objects. Thus, a bond attached to a smaller atom (like hydrogen, say) will vibrate with higher frequency than a bond attached to a heavier atom (like carbon). Take an example to prove this to yourself. Imagine atom 1 (m_1) is a hydrogen atom and it's attached by a bond to another hydrogen atom (m_2). The reduced mass, μ, will be $\frac{1}{2}$ because each of the atomic weights of hydrogen is 1. If you change one hydrogen to a deuterium (deuterium is a hydrogen atom with an extra neutron, so it has a weight of 2), the reduced mass becomes $\frac{2}{3}$. Because the frequency of bond vibration relates to $\sqrt{\dfrac{1}{\mu}}$, this larger value for the reduced mass means that the frequency of vibration of the bond between H and D will be less than the frequency of the bond vibration between H and H.

Additionally, stronger springs (those with higher values of k) vibrate faster (that is, at higher frequencies) than weaker springs do, just as the stronger spring in a car suspension vibrates faster than a slinky when stretched and released. This is because the frequency is proportional to the square root of the spring strength, k, so increasing k increases the frequency of vibration. Translating this feature to molecules, triple bonds vibrate at higher frequencies than double bonds because triple bonds are stronger than double bonds. For the same reason, double bonds vibrate at higher frequencies than single bonds, because double bonds are stronger than single bonds.

Seeing bond vibration and IR light absorption

Many different motions are present in molecules — stretching motions, bending motions, twisting motions, and rocking motions. The most important motion for IR spectroscopy is stretching motion, as shown in Figure 18-2.

Figure 18-2:
Stretching
motion.

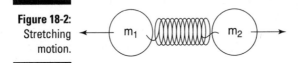

When a bond vibrates at a particular frequency (recall that this frequency is dictated by the mass of the two atoms and the strength of the bond), and light of that same frequency hits that bond, the light is absorbed by the molecule. This light absorption by the molecule can be detected. How? If you run light of a particular frequency through the sample and more light goes into the sample than comes out, that frequency of light is being absorbed by the molecules, and thus one of the bonds in the molecule must vibrate at that frequency.

In practice, this measurement is performed by running one beam of light through a blank reference cell (typically, the reference cell is an empty sample holder if the sample is a solid or liquid, or a sample container with solvent if the sample is dissolved) and one beam through the sample, and then comparing the difference in the intensity of the light beam coming out of the sample cell with that coming out of the blank, as shown in Figure 18-3. A monochrometer is a device used in an IR spectrometer that can filter out all but the desired wavelength. To generate an IR spectrum, you scan the monochrometer through all the IR frequencies (which goes on the x-axis) while simultaneously plotting the light transmittance on the y-axis. Note that even though, by convention, light transmittance (rather than light absorption) is plotted on the y-axis, peaks are usually referred to as absorptions, not transmittances!

Figure 18-3:
How an IR
spectrometer
works.

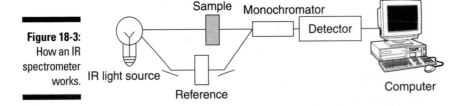

The frequencies of light required to excite the vibrations of bonds falls in — *surprise!* — the infrared region of the electromagnetic spectrum (slightly lower frequency than visible light). Hence the name "infrared spectroscopy."

Seeing absorption intensity

You now know that a bond will absorb light when it's hit with light that has the same frequency as the frequency of the bond vibration. But what about the intensity of the absorption? The intensity of the absorption of light depends on the change in the dipole moment (see Chapter 2) of a particular vibration (this absorption intensity relates to the way that electromagnetic radiation — that is, light — interacts with molecules). Bonds that have large changes in dipole moment for a particular vibration have intense light absorptions.

For example, consider the stretching vibrations for C-H, N-H, and O-H bonds. Stretching an O-H bond produces a large change in dipole moment because oxygen is much more electronegative than hydrogen (refer to Chapter 2 for an explanation of electronegativity). Nitrogen is a less electronegative atom than oxygen, and so N-H bonds have a smaller change in dipole moment when stretched than do O-H bonds. A C-H stretch has an even smaller change in dipole moment because carbon is not very electronegative at all. Thus, O-H stretches have intense light absorptions, N-H bonds have weaker absorptions, and C-H stretches are weaker still, as summarized in Figure 18-4.

Figure 18-4:
Absorption
intensity of
bonds.

C-H	N-H	O-H
Smallest dipole moment weak absorption		Largest dipole moment intense absorption

IR forbidden stretches

If a bond motion produces no change in the bond's dipole moment, the vibration will not absorb light at all. The vibrations that produce no change in dipole moment are called *IR inactive vibrations,* because they will not show up in an IR spectrum. Typically, these IR inactive vibrations occur in symmetrical molecules, because stretching these molecules does not change the dipole moment of the bond. For example, stretching the Cl-Cl bond in chlorine, or stretching the carbon-carbon triple bond in dimethyl acetylene (shown in Figure 18-5), does not change the dipole moment. These stretch modes are thus IR inactive, and do not show up in the IR spectrum.

Figure 18-5:
IR inactive bonds.

Cl—Cl $H_3C—C≡C—CH_3$

Dissecting an IR Spectrum

The IR spectrum is set up in a somewhat weird way as a result of historical conventions. Instead of plotting absorbance, by convention the IR spectrum plots light transmittance. A zero value for transmittance indicates that none of the light passes through the sample at that wavelength (and thus all of the light is absorbed), and 100 percent transmittance indicates that all of the light passes through the sample at that wavelength. To generate the IR spectrum, different frequencies of infrared light are passed through the sample, and the transmittance of light at each frequency is measured. The transmittance is then plotted versus the frequency of the light (which is presented in the somewhat unusual units of cm^{-1}).

What makes IR spectroscopy useful is that similar kinds of bonds show up in the same places in the IR spectrum, as Figure 18-6 shows. For example, the stretches associated with an alcohol (a molecule containing an OH functional group) show up generally in the same region of the spectrum more or less independently of what the rest of the molecule looks like.

Note how Figure 18-6 makes sense in terms of the two factors that contribute to the vibration frequency — the size of the atom and the bond strength. Absorptions at high frequency (like those produced by OH, NH, and CH) appear at higher frequency than the absorptions produced by C-C bonds (including single, double, and triple carbon-carbon bonds) because hydrogen is a smaller atom than carbon. Also, triple bonds absorb at higher frequency than double bonds because triple bonds are stronger than double bonds.

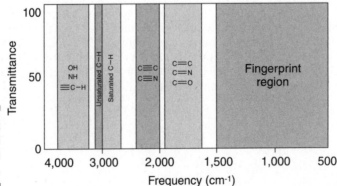

Figure 18-6:
Typical bond absorption locations.

The region of the spectrum between 500 cm^{-1} and 1500 cm^{-1} is called the fingerprint region (refer to Figure 18-6). Every molecule has a unique set of absorptions in the fingerprint region of the spectrum. Unfortunately, this region of the spectrum is usually a complicated mess and is difficult to interpret. For this reason, beginners in the art of interpreting IR spectra typically ignore this region of the spectrum. However, the fingerprint region can still be useful for identifying unknown compounds, because two molecules that have identical fingerprint regions in the IR spectrum are almost certainly identical molecules. Thus, finding that the fingerprint region of your unknown compound's IR spectrum and that of a known reference are identical is good evidence that the structure of the unknown is the same as the reference.

Identifying the Functional Groups

Table 18-1 lists the locations and intensities of absorptions produced by typical functional groups. Recognizing where the absorptions generated by the common functional groups occur will help you to interpret IR spectra.

Table 18-1	IR Absorptions of Common Functional Groups	
Functional Group	*Absorption Location (cm^{-1})*	*Absorption Intensity*
Alkane (C–H)	2,850–2,975	Medium to strong
Alcohol (O–H)	3,400–3,700	Strong, broad
Alkene (C=C) (C=C–H)	1,640–1,680 3,020–3,100	Weak to medium Medium
Alkyne (C≡C) (C≡C–H)	2,100–2,250 3,300	Medium Strong
Nitrile (C≡N)	2,200–2,250	Medium
Aromatics	1,650–2,000	Weak
Amines (N–H)	3,300–3,350	Medium
Carbonyls (C=O) Aldehyde (CHO) Ketone (RCOR) Ester (RCOOR) Acid (RCOOH)	1,720–1,740 1,715 1,735–1,750 1,700–1,725	Strong

Sizing up the IR spectrum

The IR spectrum of hexane (C_6H_{14}) is shown in Figure 18-7. Because hexane has only C-H and C-C bonds (and no functional groups), this spectrum can help orient you to the important regions in an IR spectrum.

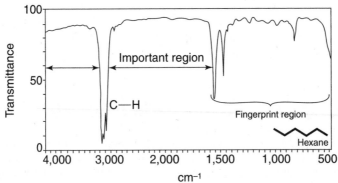

Figure 18-7:
The IR spectrum of hexane.

Note the two features of the spectrum — the C-H absorptions between 2,800 cm^{-1} to 3,000 cm^{-1}, and the fingerprint region below 1,500 cm^{-1}. You should think of this spectrum as a blank; you mentally subtract out the two regions shown on this spectrum to look for the important absorptions in unknown compounds. Using these C-H stretches as a frame of reference is also convenient, because almost every organic compound contains C-H bonds. (Note that while individual C-H stretches are weak because of small changes in the dipole moment, a typical molecule often has a lot of these bonds, which makes these absorptions appear fairly intense.) Any absorptions to the left of these C-H stretches are typically N-H,O-H, or alkynyl (triple bond) C-H stretches. The other common functional groups have bands between the fingerprint region and the C-H stretching absorptions.

Don't get too distracted by the mess in the fingerprint region. Instead, look primarily in the important places (between 1,500 and 2,800 cm^{-1}, and above 3,000 cm^{-1}).

Recognizing functional groups

Learning where the functional groups appear in the spectrum is usually not enough. You need to be able to recognize and get a feel for what the absorptions of the common functional groups look like. Are they fat and round like a sumo wrestler's belly (as in an O-H stretch), or are they small and thin like a supermodel's (as in a carbon-carbon double bond stretch)? In Figure 18-8, I show you what the common functional groups look like.

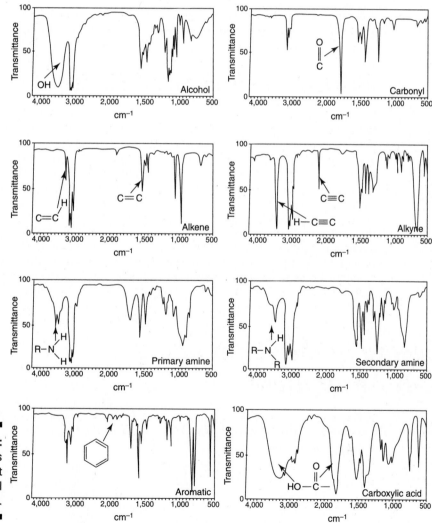

Figure 18-8: Absorptions of different functional groups.

Seeing to the Left of the C-H Absorptions

Look left and then right. In this section, I cover the features of the functional groups that appear to the left of the C-H absorptions in the IR spectrum. Refer to Figure 18-8 for examples of these functional groups.

Big and fat: The alcohols

Alcohols are very broad, fat absorptions that come to the left of the C-H absorptions. Because of their broadness, they're very easy to spot.

Milking the spectrum: Amines

Primary amines (amines substituted by only one R group and abbreviated RNH_2) are very easy to spot. They appear to the left of the C-H absorptions in the spectrum in about the same region as alcohol absorptions. Primary amines consist of two small peaks, and the overall effect is to make the absorptions of a primary amine look like a cow udder (refer to Figure 18-8). *Secondary amines* (amines substituted by two R groups and abbreviated R_2NH) consist of only a single absorption in that region. A common mistake is to confuse secondary amine absorption with alcohol absorptions, because they both occur in the same region of the spectrum. Usually, secondary amine absorptions are somewhat thinner and sharper than the broad and rounded absorptions produced by alcohols. Only practice with problems dealing with IR spectra will make you confident in telling the difference between amine and alcohol absorptions.

Seeing to the Right of the C-H Absorptions

In this section, I cover the features of the functional groups that appear to the right of the C-H absorptions in the IR spectrum but to the left of the fingerprint region.

Big and tall: Carbonyl groups

Carbonyl groups are very easy to spot in an IR spectrum. They consist of an intense, thin absorption at about 1,700 cm^{-1} that looks like a finger (not a middle finger, I hope). *Conjugated carbonyls* (carbonyl groups adjacent to double bonds) have a somewhat lower frequency of vibration than ketones (these conjugated carbonyls often have absorptions below 1,700 cm^{-1}). Ester carbonyls have a somewhat higher frequency of vibration than ketones. A carboxylic acid has a very fat carbonyl stretch and an alcohol absorption that's even fatter than a typical alcohol stretch; this alcohol stretch from the carboxylic acid often expands into the region of the C-H stretches.

Hydrocarbon stretches: Alkenes, alkynes, and aromatics

Alkene (C=C) stretches appear around 2,250 cm^{-1}, and are of either weak or medium intensity. If you're unsure whether a small blip in that region of the spectrum represents an alkene, you can look for the unsaturated hydrogen stretches (C=C-H) above 3,000 cm^{-1}. Typically, these are of medium intensity. Drawing a line down the spectrum at 3,000 cm^{-1} is often a good idea. Any stretches slightly higher than that frequency are a good indication of having an alkene (or an aromatic ring). Alkynes (carbon-carbon triple bonds) have absorptions between 2,100 and 2,250 cm^{-1}, and are of medium intensity. A terminal alkyne (one at the end of a chain) is easy to spot because of the high-intensity alkynyl C-H stretch that comes at around 3,300 cm^{-1}. Aromatic ring absorptions, on the other hand, can be tough to spot sometimes. They consist of a small series of bumps between 1,650 cm^{-1} and 2,000 cm^{-1}. The number of bumps (absorptions) changes depending on how the benzene ring is substituted.

Chapter 19

NMR Spectroscopy: Hold onto Your Hats, You're Going Nuclear!

*I*magining modern organic chemistry without the aid of nuclear magnetic resonance (NMR) spectroscopy is a lot like imagining modern life without the aid of cellphones or toilet paper: virtually unthinkable. That's because if one of the chief purposes of organic chemistry is to figure out ways to make carbon-containing stuff, then that process becomes a lot easier if you can determine what it is that you've made. And that's what NMR is for — it can help you to determine the precise molecular structure of whatever organic stuff you're interested in looking at. This stuff may include the products of reactions, an interesting metabolite that you've just isolated from rabbit muscle, a deadly neurotoxin that you've extracted from a marine shellfish, or just about anything else, really.

In this chapter, I explain the value of NMR, discuss how it works, and explain how NMR experiments are run. I also dissect the different pieces of an NMR spectrum, show you how all the different parts can give clues about a compound's structure, and discuss what each part means. In Chapter 20, I show you how to determine the structure of an unknown molecule using NMR spectroscopy.

Why NMR?

Back in the olden days when NMR spectroscopy was still in its infancy (and people had to walk 5 miles to get to school, uphill both ways), chemists had to labor sometimes for weeks to figure out the structures of even simple organic compounds, using methods that were tedious, time-consuming, and about as reliable as a fox guarding a hencoop. For example, a chemist might have compared the physical properties of a compound — its boiling point, melting point, refractive index, and so on — to compounds of similar structure previously reported in the literature to see if any of the known compounds were close matches. But because of the millions of possible structures, often the published literature contained no compounds of similar structure, so the molecule would need to be converted into a known derivative, and then the properties of this derivative would be tested.

Suffice it to say this method of structure determination was wearisome, unreliable, and about as fun as a root canal. But much to the delight of chemists today (and perhaps to the chagrin of the struggling organic student), NMR spectroscopy has now swaggered onto the scene. Hydrogen NMR can often be used to determine the structure of simple unknown compounds within minutes, eliminating most of the tiresome hassle figuring out the structures of molecules. You just pop your sample in the NMR spectrometer, run the experiment on the quick, and print out your data. Just like that — one, two, three — and you're off to the analysis, as quickly as a politician to promise a tax cut.

Couple NMR's speed with its ability to determine structure, and you've got a tool that's amazingly powerful, a tool that can nail down the structure of almost any organic molecule, from the relatively small and simple molecules like the ones that you encounter in your organic course, to the large and staggeringly complex ones like the massive enzymes that catalyze chemical reactions in the cell. Because of NMR's power and versatility, when organic chemists want to figure out exactly what that foul-smelling product from their reaction is, they turn to NMR to give them the definitive answer.

So, while mass spectrometry (see Chapter 17) can give you information about the molecular weight of a compound (and if you're good, perhaps some of its structure), and IR spectroscopy (see Chapter 18) can give you an idea of what kinds of functional groups a molecule may have, only NMR can show you reliably how a molecule is pieced together. Don't take this the wrong way: Mass spectrometry and IR spectroscopy are often integral pieces of structure determination, and they both provide first-class clues. But NMR is the Bad, Bad Leroy Brown of molecule identification. If figuring out a structure can be thought of as a chess game, then NMR spectroscopy has the moves of the queen: It's the most powerful piece on the board.

How NMR Works

NMR spectroscopy works by allowing you to "see" the nuclei in a molecule, and to see the chemical neighborhood of those nuclei. Knowing the chemical environment and the number of different kinds of nuclei present allows you to piece together the structure of a compound in most circumstances. Some nuclei, however, are *NMR active,* and some are *NMR inactive;* NMR experiments see only NMR-active nuclei and are blind to nuclei that are NMR inactive. Fortunately, both carbon and hydrogen nuclei are NMR active. For reasons I talk about in this chapter, hydrogen NMR (often called proton NMR or ^{1}H NMR) is often the most useful NMR type for organic chemists; therefore, all discussions will pertain to hydrogen NMR. After I cover hydrogen NMR, I cover the main points of ^{13}C NMR (which is pretty much the same thing as hydrogen NMR, only simpler).

Giant magnets and molecules: NMR theory

The theory behind NMR spectroscopy can be complicated — and the details are generally outside the scope of a general organic class — so the following discussion is a simplified version of how NMR works. Even simplified, it's complicated, so buckle up and fasten your thinking caps.

All nuclei with odd atomic and mass numbers have the property of spin just as electrons do, and any nucleus that has spin is observable by NMR. *Spin* is a bit of an abstract concept that doesn't really have any counterpart outside of the subatomic world, but it might be helpful to think of these spins as coming from the protons spinning rapidly inside the nucleus, or the equivalent of a positive charge break dance. This spinning involves the movement of a positive charge, and, according to the laws of physics, the movement of charged particles creates a magnetic field with a *magnetic moment.* Nuclei that have such spins include ^{1}H, ^{13}C, ^{15}N, and ^{19}F, because all these elements have odd atomic numbers. Nuclei with even atomic numbers, like ^{16}O, ^{12}C, and ^{14}N, are NMR inactive and will not show up in an NMR experiment.

The magnetic moments that come from the nuclear spins have direction. If you were to simply take a beaker full of nuclei and place it on a tabletop in your kitchen, the directions of the magnetic moments from the nuclei would be random, pointing in every possible direction. But when the nuclei are placed between the poles of a magnet — which creates what is called the *external magnetic field* — quantum mechanics dictates that the direction of the magnetic moments of these nuclei must either line up with the external magnetic field (called an α spin) or line up against it (called a β spin). You can imagine the magnet shouting a sort of politician's rallying cry to the nuclei, "Everyone take sides; you're either with me or you're against me." Thus, nuclei in the same beaker can line up with or against the external magnetic field (see Figure 19-1).

Figure 19-1:
The effect of an external magnetic field on the direction of the magnetic moments of individual nuclei.

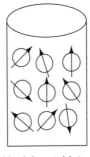 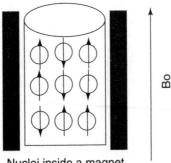

Nuclei on tabletop   Nuclei inside a magnet

Those nuclei that align with the magnetic field are slightly lower in energy than those that align against the magnetic field. How much lower in energy these nuclei are depends on the size of the magnet that you put near the nuclei. (This magnet size — or external magnetic field strength — is often labeled B_o.) As the size of the magnet increases, the difference in energy (ΔE) between the spins aligning with the magnetic field (α) and those going against the magnetic field (β) also increases (see Figure 19-2). This energy gap between the two kinds of spins — α and β — is what an NMR experiment measures. In addition to the strength of the external magnetic field, the size of the energy gap is also affected by the chemical neighborhood of particular nuclei.

Figure 19-2:
The effect of increasing external magnetic field strength (B_o) on the energy gap between α and β spin states.

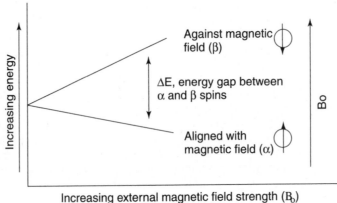

But before I can talk about how the size of the energy gap between α and β spins gives you information about the chemical environment of the nuclei, I need to tell you how the energy gap is measured by an NMR spectrometer. Like most types of spectroscopy, NMR spectroscopy uses light to effect a transition between energy levels. (For IR spectroscopy, infrared light is used to transition between vibrational and rotational energy levels. See Chapter 18 for more on IR spectroscopy.) In an NMR experiment, a radio wave (which indicates light of radio frequencies) effects a transition between the α and the β spin energy levels. When the frequency of the radio wave that hits the sample has the exact energy of the energy gap between the α and β spins (ΔE), the light is absorbed by the nuclei.

The energy from the light absorption is used as a kind of nuclear spatula to flip the α spins into β spins, a process called *spin flipping*. When a nucleus absorbs light and flips its spin, the nucleus is said to be *in resonance* (which is where the term *nuclear magnetic resonance* comes from; this resonance is not to be confused with resonance structures). A detector can measure the frequency of this absorption and plot it on a spectrum (which is the plot of an NMR experiment, which shows the intensity of the light absorption versus the frequency of the light absorbed). See Figure 19-3 for a diagram of how this works. A high-frequency light absorption indicates a large energy gap between the α and β states; a low-frequency light absorption indicates a small energy gap.

Figure 19-3:
A diagram of the steps involved in measuring the ΔE in an NMR experiment.

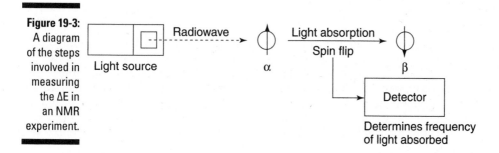

Light source — Radiowave → α — Light absorption / Spin flip → β — Detector — Determines frequency of light absorbed

To sum up, the ΔE in an NMR experiment is determined in the following way:

- ✔ The sample is placed inside a huge magnet (called the external magnetic field, or B_o).

- ✔ The direction of the magnetic moments of the nuclei line up either with the external magnetic field (an α spin) or against it (a β spin). The energy difference between an α spin and a β spin is the ΔE.

✔ Upon absorption of light that has energy equal to the energy gap (ΔE), the nuclei come into resonance, and α spins are flipped into β spins.

✔ A detector measures the frequency of the absorbed light (which is proportional to the energy gap between the α and β spins) and then plots the frequency of the light absorbed on the NMR spectrum versus the intensity of light absorption.

So, that's how an NMR spectrometer determines the energy gap. But if all the hydrogen nuclei would have the same energy gap between α and β spins, the NMR spectrum would show only a single peak (or single light absorption), representing all the hydrogens in the molecule. And that wouldn't be particularly helpful. But my discussion so far has omitted one key ingredient — electrons! So far, this discussion has been limited to bare nuclei that have no electrons surrounding them.

Grab your jackets: Electron shielding

Because they're moving charges, electrons have their own magnetic fields that come into play. The magnetic field of an electron opposes the external magnetic field (made by the big magnet that you placed the sample into) and shields the nucleus from feeling the full brunt of the external magnetic field. In other words, a naked nucleus feels more of the external magnetic field than does a nucleus that has electrons surrounding it, just as a nudist will feel more of the cold than a person wearing an insulating jacket.

Consequently, a nucleus that is jacketed with electron density shielding it from the external magnetic field will require a lower frequency (lower energy) radio wave to flip its spin than a nucleus that has less electron density around it. This is because the actual magnetic field that an electron-surrounded nucleus feels is smaller than the magnetic field felt by a naked nucleus, so the gap between the α and β spins is smaller (refer to Figure 19-2). Because the gap is smaller, less energy (lower-frequency light) is required to flip the spin.

In this way, the chemical environment of the hydrogen nuclei is revealed. A hydrogen that's next to a highly electronegative element (see Chapter 2 for a discussion of electronegativity) will have fewer electrons around it and will therefore come into resonance — or will flip its spin — at a higher frequency of light. A hydrogen that's next to an electron-donating element will have lots of electrons around it and will come into resonance at a lower frequency.

The NMR Spectrum

An NMR spectrum, then, is the plot of the resonance frequency of the different nuclei versus intensity of light absorption. One way to take an NMR spectrum is to hold the external magnetic field constant (the magnetic field created by the big magnet) and vary the frequency of the radio waves irradiated onto the sample. At whatever frequencies light is absorbed, peaks are plotted on the spectrum. This is, in fact, how old NMR spectrometers plotted an NMR spectrum, and it's a conceptually useful way of thinking about how an NMR spectrum is generated. The workings of a modern NMR spectrometer are a bit more complicated, but the general principle remains the same.

Standardizing chemical shifts

One potential problem for NMR is that if the nuclei are placed into a large magnet, their resonance frequency will be larger than when they're placed into a little magnet, because the energy gap increases as the external magnetic field increases (refer to Figure 19-2). This means that a proton will show up at a different frequency for the poor person who can only afford a small magnet than it will for a rich person who can afford a big magnet. To make the NMR spectrum the same for a given molecule regardless of how large the magnet is, the resonance frequency is plotted relative to a reference compound, usually tetramethylsilane (TMS), shown in Figure 19-4. This way, regardless of how big the external magnetic field is, the peaks in the NMR spectrum will always occur at the same location relative to TMS.

Figure 19-4:
Tetramethylsilane (TMS).

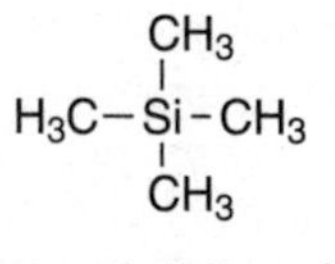

$$H_3C-\underset{\underset{CH_3}{|}}{\overset{\overset{CH_3}{|}}{Si}}-CH_3$$

Tetramethylsilane (TMS)

TMS is used as a reference because the hydrogens on the molecule are highly shielded because of the electron-donating silicon atom, and these hydrogens show up at a lower frequency than the vast majority of hydrogens in organic molecules. This TMS absorption frequency is arbitrarily assigned a value of zero on the spectrum. The frequency of a nucleus's absorption relative to TMS is given in units of parts per million (ppm), so that the frequency will be given in small, manageable numbers (between 0 and 12) rather than large numbers (0 to 12 million). For a hydrogen NMR spectrum, peaks generally fall between 0 and 15 ppm, with a peak close to 0 ppm indicating a highly shielded hydrogen (electron rich), and a peak close to 15 ppm indicating a highly *deshielded* hydrogen (electron poor). Figure 19-5 shows a hypothetical NMR spectrum.

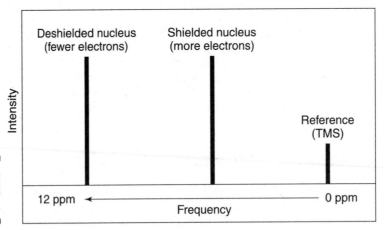

Figure 19-5:
The NMR
spectrum.

Seeing symmetry and chemical equivalency

Every hydrogen in a molecule that's in a unique chemical neighborhood will show up as a peak on an NMR spectrum. Two (or more) hydrogens that have equivalent chemical neighborhoods, though, will be represented by just a single peak. Such hydrogens that are in identical chemical environments are said to be *chemically equivalent*. For example, methanol (see Figure 19-6) has four hydrogens in it, so you might expect it to show four peaks in the ^{1}H NMR spectrum, one for each hydrogen. But it only shows two peaks. This is because there are only two different kinds of hydrogens in methanol. All the hydrogens in the methyl group (CH_3) are in identical chemical environments — all three are attached to a carbon that's bonded to two other hydrogens and an alcohol group — so each of these hydrogens sees the same chemical neighborhood. Therefore, all three hydrogens have the same resonance frequency and show up as a single peak. The second peak comes from the H on the alcohol (OH) group, which is in a different chemical environment.

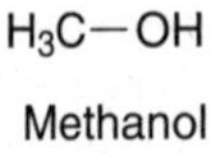

Figure 19-6: $H_3C—OH$
Methanol.
Methanol

Butane (see Figure 19-7) is another example of a molecule that has such chemical equivalency. Ten hydrogens are on the molecule, but, remembering the chemical equivalency of methanol, you might expect to see four peaks, one for each of the hydrogen sets attached to the carbons. But the NMR spectrum of butane has only two peaks. This is accounted for by a further element of equivalency that comes from the symmetry of the molecule. Because the right-hand side of the molecule is exactly equal to the left-hand side, the two CH_3 groups on the ends are in identical chemical environments, as are the two CH_2 groups in the middle. Because both CH_3 groups are in identical chemical environments, and both CH_2 groups are in identical chemical environments, you see only two peaks in the NMR spectrum.

Figure 19-7: $CH_3-CH_2-CH_2-CH_3$
Butane.
Butane

Being able to recognize which hydrogens are chemically equivalent is extremely important. Generally, hydrogens attached to the same carbon are chemically equivalent. In addition, symmetry in the molecule can contribute further to chemical equivalency. You should practice looking at organic molecules for chemical equivalency so that you can determine how many peaks its ^1H NMR will have (see the "Seeing the chemical shift" section, later in this chapter, for details).

Often, drawing out all the hydrogens on the molecule can help you see the different chemical environments each hydrogen experiences. Try this for the molecules shown in Figure 19-8.

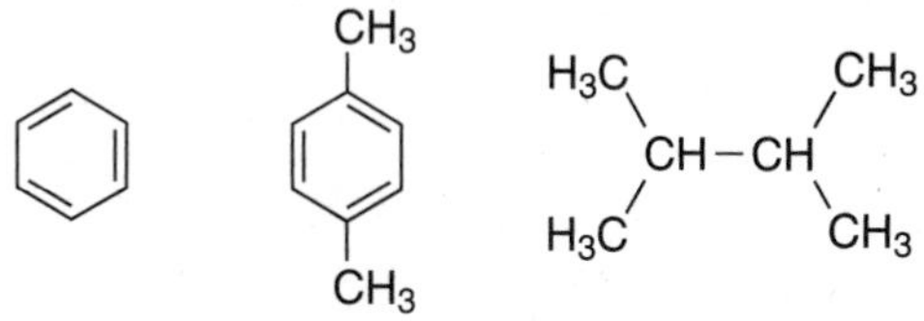

Figure 19-8:
Seeing
symmetry.

1 peak   2 peaks   2 peaks

The NMR Spectrum Manual: Dissecting the Pieces

This section dissects each of the parts of an ^1H NMR spectrum and discusses how each part of a spectrum can give clues to help determine the structure of a molecule.

Seeing the chemical shift

Where a peak is located relative to the reference TMS is called its *chemical shift* (or its δ *value,* where δ indicates the change in the absorption frequency). The chemical shift is given in units of parts per million (ppm). The two most important factors that affect the chemical shift of a given nucleus are the electron density around a hydrogen and diamagnetic anisotropy (scary phrase, huh?).

A hydrogen neighboring highly electronegative atoms — like fluorine (F), chlorine (Cl), bromine (Br), nitrogen (N), and oxygen (O) — will have a higher chemical shift than a hydrogen neighboring more electropositive elements (like silicon [Si] or carbon [C]). The derivatives of methane shown in Figure 19-9 illustrate this point. As the electronegativity of the substituent increases (recall that electronegativity increases as you go from Br to Cl to F), the chemical shift also increases. This is because as the electronegativity of the substituent increases, the electron piggishness of the substituent also increases, and the substituent steals more of the electron density away from the hydrogens. Electron-deshielded nuclei like those near electronegative elements have a larger chemical shift than those near electropositive elements.

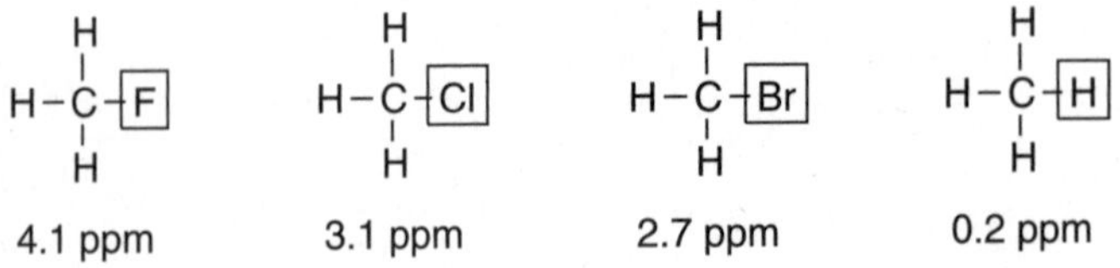

Figure 19-9:
The chemical shifts of hydrogens (protons) caused by neighboring electronegative substituents.

$$H-\underset{\overset{|}{H}}{\overset{\overset{H}{|}}{C}}-\boxed{F} \qquad H-\underset{\overset{|}{H}}{\overset{\overset{H}{|}}{C}}-\boxed{Cl} \qquad H-\underset{\overset{|}{H}}{\overset{\overset{H}{|}}{C}}-\boxed{Br} \qquad H-\underset{\overset{|}{H}}{\overset{\overset{H}{|}}{C}}-\boxed{H}$$

4.1 ppm   3.1 ppm   2.7 ppm   0.2 ppm

The second effect on the chemical shift is a local field effect caused by pi electrons; this effect is called *diamagnetic anisotropy* in organic-speak.

When placed in a magnetic field, the pi electrons in benzene (or other aromatic rings) are induced to circulate. (For more on aromatic rings see Chapter 15. For information on pi bonds see Chapter 2.) This circulation of the pi electrons creates a local magnetic field from the aromatic ring (see Figure 19-10). In the center of the aromatic ring the induced magnetic field from the benzene opposes the external magnetic field, weakening it. Outside the ring, however, the magnetic field of the benzene ring reinforces the external magnetic field, making the magnetic field stronger.

Figure 19-10:
The induced magnetic field of benzene in an external magnetic field.

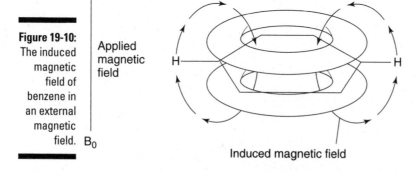

This means that any hydrogens near the center of an aromatic ring will be shielded by the aromatic ring's magnetic field and will have a lower chemical shift, while those hydrogens outside the ring will be deshielded and will have a higher chemical shift. Therefore, hydrogens attached to aromatic rings (sometimes referred to as *aromatic hydrogens,* even though the hydrogens themselves are not aromatic), have a chemical shift much higher than you might expect based on the electronegativity considerations of the aromatic ring. Figure 19-11 illustrates this effect with a complex aromatic molecule.

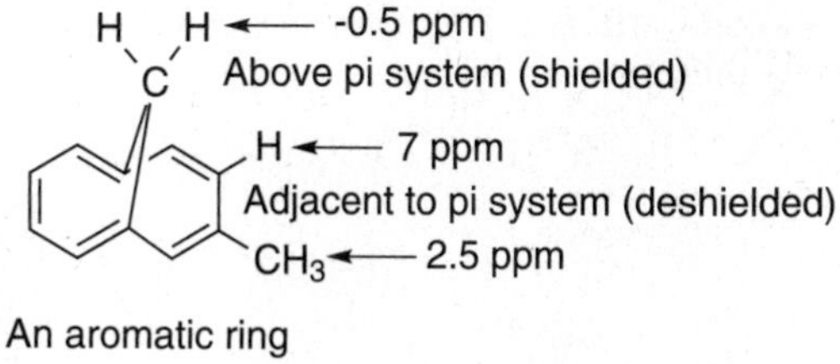

Figure 19-11:
The effect of diamagnetic anisotropy in an aromatic molecule.

An aromatic ring

The same local field effect caused by the pi electrons in benzene affects double bonds and triple bonds in a similar fashion. Hydrogens attached to double bonds or triple bonds, therefore, have higher chemical shifts than might be expected based on simple electronegativity considerations.

Figure 19-12 shows some of the approximate chemical shifts of hydrogens adjacent to the common functional groups. Keep in mind, though, that these numbers are just typical ranges and are not absolute; values may stray in an out of the ranges depending on the rest of the structure of a given molecule. Also keep in mind that the addition of electronegative elements is cumulative. A hydrogen adjacent to a carbonyl group plus a halide would have a chemical shift higher than a hydrogen neighboring just one of those groups.

Figure 19-12:
Approximate chemical shifts of hydrogens (protons) adjacent to the common functional groups.

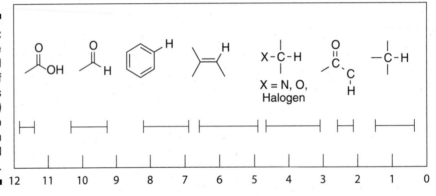

Incorporating the integration

The peak intensity — or the area underneath a peak on a spectrum — is related to the number of hydrogens that the peak represents. A computer uses the mathematical process of integration to find the areas underneath each peak. Traditionally, integration is shown on the spectrum by the addition of an integration curve (see Figure 19-13), although modern computing has made digital integration common that doesn't require you to do any measuring. The *height* of this integration curve is proportional to the area underneath a peak, so this height is proportional to the number of hydrogens the peak represents. (The width of the curve is unimportant.)

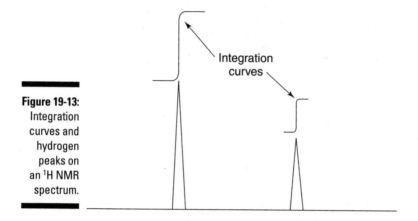

Figure 19-13:
Integration curves and hydrogen peaks on an ^{1}H NMR spectrum.

Integration curves

To measure the height of an integration, you start at the bottom of the integration curve where it's flat, and measure to where the curve goes flat again. Unfortunately, integration can't tell you how many hydrogens each peak represents — it just tells you the relative ratios of hydrogens in one type of chemical environment compared to those in another environment. This makes integration useful only for organic molecules that contain more than one kind of hydrogen (which, fortunately, is most of them).

For example, if there are two peaks, one with an integration curve that's 2 cm high and one with an integration curve that's 1 cm high, this tells you that the larger peak represents twice as many hydrogens as the smaller one (refer to Figure 19-13). This does not mean that the larger one represents two hydrogens and the smaller represents one hydrogen, necessarily — although this could be the case. It simply tells you that the ratio of hydrogens in the two chemical environments is 2:1. I show you how to measure an integration with a ruler in Figure 19-14.

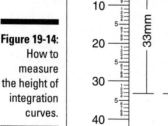

Figure 19-14:
How to measure the height of integration curves.

Catching on to coupling

Ethanol, the alcohol found in adult beverages, is a common organic compound. The NMR spectrum for ethanol is shown in Figure 19-15.

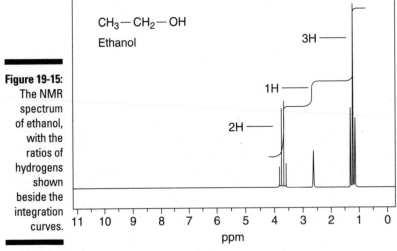

Figure 19-15:
The NMR
spectrum
of ethanol,
with the
ratios of
hydrogens
shown
beside the
integration
curves.

Ethanol has three kinds of hydrogens that are in unique chemical environments — the three hydrogens on the methyl group (CH₃), the two hydrogens on the methylene group (CH₂), and the hydrogen on the alcohol group (OH). That's why you see three peaks in the NMR spectrum and why the integration has a ratio of 2:1:3, representing the methylene group, the alcohol group, and the methyl group hydrogens, respectively. Notice, though, that some of the individual absorptions are split further into smaller peaks. This is called *coupling.* Coupling is a phenomenon that's useful in determining how each piece of the molecule is connected to the others. If integration tells you about the number of hydrogens a peak represents, coupling tells you how many hydrogens live next door to that hydrogen. Coupling comes from the interaction of the magnetic moments of the hydrogens with the magnetic moments of the neighboring hydrogens. In most cases, hydrogens only couple the hydrogens that are on adjacent carbons. Chemically equivalent hydrogens don't couple, which is why hydrogens on the same carbon don't couple most of the time and why the six hydrogens in benzene don't couple.

The "n + 1 rule" and the coupling constant

Coupling follows what is called the *n + 1 rule*. This rule says that a peak will be split $n + 1$ times, where n is the number of equivalent neighboring hydrogens. In the spectrum for ethanol (refer to Figure 19-15), for example, the methyl (CH_3) group has as its neighbor a carbon that has two equivalent hydrogens, so the methyl group signal is split into three peaks following the $n + 1$ rule (observed by the signal at δ1.2 ppm). The methylene group (CH_2) is next to three equivalent hydrogens on the adjacent methyl group, so its signal is split into four peaks following the $n + 1$ rule (see Figure 19-16). The hydrogen on the alcohol group doesn't couple, for reasons I talk about in the subsequent section on proton exchange.

Generally *will* couple

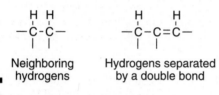

Neighboring hydrogens   Hydrogens separated by a double bond

Figure 19-16: **Generally *won't* couple**
The types of hydrogens that will and won't couple.

Chemically equivalent   Too far away to couple

A peak that does not couple and just shows a single line is called a *singlet*. The alcohol hydrogen in ethanol does not couple so it gives a singlet. A peak that's split into two is called a *doublet,* three a *triplet,* four a *quartet,* and so on. You can see the general picture for these kinds of peak splittings in Figure 19-17. The *coupling constant,* or *J value,* is proportional to the distance between the lines (or the tops of the peaks) in a signal. This value is given in frequency units (hertz, or Hz), rather than in ppm, to make the coupling constant independent of the size of the external magnetic field. The main point to remember about coupling constants is that hydrogens that couple each other have peaks that are separated by the same coupling constant. This can be a useful piece of information in determining which hydrogens are next to one another, especially when the number of peaks alone is insufficient. Returning to the ethanol example (refer to Figure 19-15), because the CH_3 and the CH_2 units couple one another, the three peaks around 1.2 ppm are the same distance apart as are the four peaks around 3.7 ppm, and both sets of peaks have the same coupling constant.

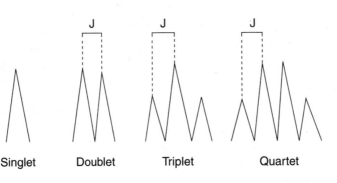

Figure 19-17:
The
coupling
constant,
or J value
for various
peaks.

Determining coupling between nonequivalent hydrogens

Many organic molecules — particularly larger molecules — contain hydrogens that are adjacent to two or more nonequivalent hydrogens. In this scenario, adding up all the hydrogen neighbors and applying the $n + 1$ rule won't work. Instead, you need to apply the $n + 1$ rule to each nonequivalent set of hydrogens separately.

Take the case shown in Figure 19-18. To determine the number of peaks that you'll see for the boxed-in hydrogen — which is surrounded by two nonequivalent sets of hydrogens — you must apply the $n + 1$ rule to each of the nonequivalent hydrogens separately. For the CH_2 unit on the right, the $n + 1$ rule tells you that this will split the signal into three peaks; for the lone hydrogen on the left, the $n + 1$ rule tells you that this hydrogen will split the signal into two peaks. To get the total number of peaks you could see, you multiply the two results together. Multiplying three by two gives a total of six peaks for the boxed-in hydrogen signal.

Figure 19-18:
The
hydrogen
(proton)
surrounded
by non-
equivalent
hydrogens
(protons).

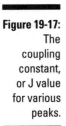

In practice, though, you often don't see all the possible peaks on a spectrum for a hydrogen that's surrounded by nonequivalent hydrogens, because the coupling constants for the nonequivalent hydrogens often have nearly the same value. When the coupling constants have nearly the same value, the signals overlap, and you observe fewer peaks than the maximum number possible.

In practice, you can often approximate the number of expected signals for a hydrogen (proton) surrounded by nonequivalent hydrogens. To do this, pretend that the nonequivalent hydrogens *are* chemically equivalent. Pretend that the CH_2 and the lone hydrogen are chemically equivalent. This gives a grand total of three hydrogens, so the $n + 1$ rule would suggest that if the coupling constant were the same (or nearly the same) the signal would be split into approximately four peaks. Although in theory you could observe up to six peaks for this hydrogen, there's a fair chance you'll observe only four.

Timber! Drawing tree-splitting diagrams

Given the coupling constants of neighboring hydrogens, you can predict the splitting pattern of a hydrogen using what is referred to as a *tree diagram*. Suppose, for example, you were asked to predict the splitting pattern for the hydrogen H_a in Figure 19-19, and you were given the coupling constant between H_a and H_b (J_{ab}) and the coupling constant between H_a and H_c (J_{ac}).

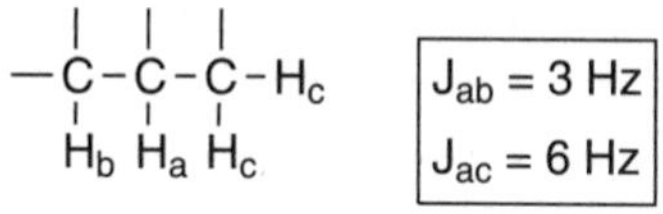

Figure 19-19:
Predicting coupling patterns.

$$-\overset{|}{\underset{H_b}{C}}-\overset{|}{\underset{H_a}{C}}-\overset{|}{\underset{H_c}{C}}-H_c \qquad \boxed{J_{ab} = 3\ Hz} \\ \boxed{J_{ac} = 6\ Hz}$$

To draw the tree diagram, start with one of the neighboring hydrogens and determine how many times that hydrogen will split the signal. You'll get the same result no matter which neighboring hydrogen you start with (in this case, whether you start with H_b or H_c), but it's often easier to start with the hydrogen that gives the larger J value. So, in this case, start with the H_c protons, because the coupling constant between H_a and H_c is larger than the coupling constant between H_a and H_b. Because there are two H_c hydrogens, these hydrogens split the H_a signal into three peaks following the $n + 1$ rule.

After you determine that the H_c hydrogens will split the signal into three peaks, begin the tree diagram by making three lines — one that goes straight down, one that goes to the right, and one that goes to the left. Take a look at Figure 19-20 to see how this works. The distance between the peaks is 6 Hz, so in making your tree diagram you may want to use a ruler to measure 6 cm, 6 mm, or 6 pinky-tip lengths to represent the separation between the lines to a proportionally correct distance.

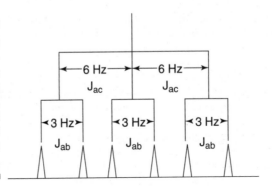

Figure 19-20:
The tree diagram for the molecule in Figure 19-19, showing the origin of the six peaks for H_a.

Instead of measuring the distance on your tree diagram, you could draw your tree on graph paper, using one or more squares of the graph paper to represent 1 Hz.

After diagramming the J value for H_a and H_c, do the same for H_a and H_b. Because there is only one H_b hydrogen, the signal will be split further into two, following the $n + 1$ rule. So, in your drawing, split each of the three lines into two additional lines, keeping a separation of 3 Hz (which would be 3 cm, 3 mm, or 3 pinky-tip lengths across) between the lines. Where each line ends, a peak would be found on the spectrum. In this case, you would expect the signal for that hydrogen to be split into six peaks in the ^1H NMR spectrum.

Swapping protons: Exchange

Hydrogens attached to either oxygen or nitrogen generally don't show any kind of coupling. These peaks are found in ^1H NMR spectra as fat singlets. This lack of coupling is caused by chemical exchange.

Small impurities of base or acid in a sample (which are often very difficult to remove) can catalyze the exchange of the hydrogens attached to either nitrogens or oxygens — in other words, the hydrogens of alcohols or amines (see Chapter 5 for more about these functional groups). In solution containing a trace of base, for example, the hydrogens attached to nitrogens and oxygens are deprotonated by the base and then reprotonated very rapidly. Because the NMR experiment takes a long time compared to the quickness of the proton exchange, coupling is not observed for these protons except in meticulously purified samples, or at very low temperatures where this chemical exchange is slower.

Because this exchange is so quick, to determine whether a peak is an alcohol or an amine, chemists often add a drop of D_2O (or heavy water) to the sample. Because *deuterons* (deuterium nuclei) are not observed in a proton NMR, when the *deuterium* (hydrogen atom with an additional neutron) replaces a hydrogen through this process of exchange, the peak disappears. This confirms that the peak was either an alcohol or an amine.

Considering Carbon NMR

^{13}C NMR is in many ways similar to ^1H NMR, only simpler (yay!). Carbon NMR, though, is less sensitive than proton NMR, because only ^{13}C nuclei are NMR active; ^{12}C nuclei are inactive. Because the ^{13}C isotope only has a natural abundance of 1.1 percent (^{12}C makes up most of the rest of the natural abundance), and because the carbon nucleus is less sensitive than a hydrogen nucleus, you generally need more samples for ^{13}C NMR than you do for ^1H NMR. For these two reasons, carbon NMR experiments can also take quite a bit longer (many minutes to hours rather than minutes).

Generally, you won't see either integration or coupling on a ^{13}C spectrum (although occasionally they do pop up). You don't see carbon-carbon coupling because the natural abundance of ^{13}C is so low — the chance that two ^{13}C nuclei will neighbor each other is so small as to be insignificant. You could see ^{13}C-^1H coupling, but most NMR experiments decouple these interactions (doing this increases the intensity of ^{13}C signals, for somewhat complicated reasons). Because of this decoupling, ^{13}C NMR peaks are often just singlets, as can be seen in the ^{13}C spectrum for butyric acid (see Figure 19-21).

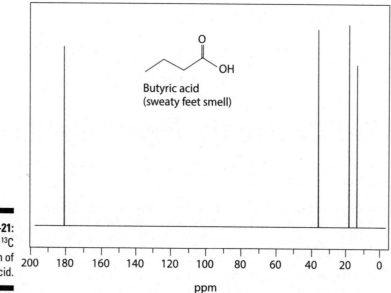

Figure 19-21: The ^{13}C spectrum of butyric acid.

Therefore, ^{13}C NMR is mostly useful for telling you the number of different kinds of carbons in a molecule from the number of peaks in the spectrum, while the chemical shift gives you an idea of the chemical environment of those carbons. The range of chemical shift values in a carbon spectrum is quite a bit broader than the shift in a proton NMR — the range is from 0 to 200 ppm, with a 200 ppm peak representing a highly deshielded nucleus. Keep in mind that hydrogen peaks don't show up on a ^{13}C NMR, just as ^{13}C peaks don't show up on a ^{1}H NMR spectrum; the ^{13}C NMR and ^{1}H NMR spectra must be taken separately (although you can typically use the same NMR spectrometer to take both spectra). Figure 19-22 gives some of the ranges of chemical shift values for particular carbon types.

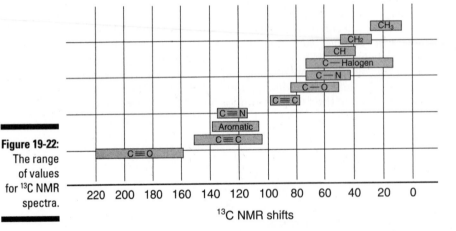

Figure 19-22: The range of values for ^{13}C NMR spectra.

Checklist: Putting the Pieces Together

Use the following definitions to help you put everything in this chapter together.

- ✔ **Chemical shift:** Refers to where a peak appears on a spectrum, relative to the reference molecule TMS (also called the δ *value*). The chemical shift is given in units of parts per million (ppm). It gives you an idea of what kinds of functional groups a hydrogen or carbon is surrounded by.

- ✔ **Integration:** Tells you the area underneath a peak, which in turn gives you information about how many hydrogens a peak represents, relative to the other peaks in the spectrum. Carbon NMR usually does not have integration on the peaks.

✔ **Coupling:** Tells you how many hydrogen neighbors a hydrogen sees, following the $n + 1$ rule. The $n + 1$ rule can be applied all at once if a hydrogen is surrounded only by chemically equivalent hydrogens. It must be done individually when it's surrounded by two or more chemically nonequivalent hydrogens. NMR tree-splitting diagrams are used to predict how many peaks will be found in the spectrum of a given hydrogen that's surrounded by two or more nonequivalent hydrogens. Hydrogens on either nitrogen or oxygen generally don't couple because of fast exchange processes — these hydrogens generally show up as fat singlets (which go away when the sample is shaken with D_2O).

✔ **Coupling constant:** Is proportional to the distance between the tops of the peaks in a multiplet. The coupling constant is referred to as the J value (given in units of Hz). This value can tell you which hydrogens are coupled to each other because coupled hydrogens have the same coupling constant.

Turning NMR to MRI

NMR has been widely used not only in chemistry, but also in medicine. To take a snapshot of your insides, doctors often perform X-rays. But X-rays have their limitations, particularly in looking at soft tissue. A more powerful technique is magnetic resonance imaging (MRI). MRI machines are huge, specially designed NMR spectrometers that can map what your insides look like in three dimensions and can detect any abnormalities such as tumor growth or bone injury. Instead of sticking a small tube into the spectrometer as you would in a typical NMR spectrometer, in an MRI machine *you* are the sample, and you're placed into a very large donut-shaped magnet where a reading can be taken.

The name was changed from NMR to MRI because patients balked when they heard the name nuclear magnetic resonance — nobody wanted to get in a machine that was nuclear! Of course, they didn't know that *nuclear* referred to the *nucleus* of the atom, not to a nuclear reaction. Still, the name was changed to *magnetic resonance imaging* to make people feel more comfortable about the procedure.

MRI machines are designed to look for the quantity of hydrogen atoms in a given section of the body. Because different kinds of tissue have different numbers of hydrogen atoms than other kinds of tissue (normal brain tissue will have a different quantity of hydrogen than tumor tissue, for example), doctors can map out the densities using a computer and then try to detect any abnormalities. MRI is so valued in modern medicine that in 2003 the Nobel Prize in medicine went to Paul Lauterbur and Peter Mansfield for the development of this technique.

Chapter 20

Following the Clues: Solving Problems in NMR

In This Chapter

▶ Using clues to determine structure

▶ Tackling NMR problems

▶ Avoiding common NMR problem-solving mistakes

I realize that spectroscopy problems are intimidating, particularly when you're first approaching them. A spectrum presents so much information, so many peaks and lines and squiggles that trying to interpret one for the first time can make you feel like a child trying to read *Moby-Dick*. The amount of information to absorb makes it hard to even know where to begin.

The trick is not to get intimidated or discouraged. You may groan when you hear this now, but after you've worked enough problems to become good at them, solving spectroscopy problems can be really fun. Think of these problems as puzzles whose rules happen to be dictated by nature.

In this chapter, I present a general guide to figuring out the structures of molecules using NMR spectroscopy (for a description of how NMR works, see Chapter 19). Unfortunately, what I present isn't a definitive guide, because chemists don't have an indisputable "right way" to solve every NMR problem. Because of this, many textbooks don't give a specific problem-solving strategy. Instead, they instruct you to "work lots of problems" and assure you that after a while and lots of hard work you'll get the hang of it. Working many problems is, indeed, excellent advice, but it leaves you in the awkward position of not knowing where to begin. As a result, students often start in the wrong places (I talk about common mistakes a little later in this chapter), and many of them never grasp how to work these types of problems. Come exam time, they're often ready to admit themselves to the local psychiatric ward.

Because of the difficulty so many students have with solving problems in NMR spectroscopy, I present the structure-solving strategy that works for me and that's worked for others. I suggest that it might work for you too, at least as a foundation until you're confident enough to alter it to fit your own learning style and adapt it to the particular problem at hand. Spectroscopy problems can be particularly intimidating because they can be solved in a number of different ways. But the way I like to determine the structure of a compound using NMR is systematically, by following the clues.

Follow the Clues

I like to think of solving spectroscopy problems as analogous to a detective in the dark with a flashlight, searching for "whodunit." Methodical clue gathering is the best way to determine the culprit — the mischievous compound in this case — instead of blindly waving the flashlight (or pencil) about, hoping for some sort of chance epiphany (which never comes during an exam, but only after you've turned it in).

You're likely to encounter many different kinds of NMR problems. Many problems will give you at least the molecular formula and the ^{1}H NMR spectrum. Some problems will also provide the IR spectrum, mass spectrum, and ^{13}C NMR spectrum as well. This clue-gathering technique I give you can be tailored to fit just about any of these problem types; simply ignore any clues that don't apply to the particular problem that you're working on. After I tell you about how to gather all the clues, I use two examples to show you how to apply this guide to real problems.

Here's the rundown of how the clue gathering works. You start by determining the degrees of unsaturation from the molecular formula. This gives a quick indication of whether the compound contains rings, double bonds, or triple bonds, or whether the molecule is saturated. A more definitive indication of what kind of molecule you're dealing with can be found by looking at the IR spectrum (if you're given one in the problem) and also by a quick peek at the ^{1}H NMR for peaks that scream out a particular functional group (specifically, carboxylic acids, aldehydes, and aromatic rings). Next, you use the integration (plus the clues from the degrees of unsaturation and the IR) to determine all the fragments of the molecule. With fragments in hand, you put the pieces together in a way that's consistent with the chemical shift and the peak splitting. Finally, you double-check your proposed structure by making sure that everything is consistent with all the clues. In short, first you determine all the pieces of the molecule, and then you try to fit them together in the correct way.

That's the big picture of how you solve a structure with this technique. Next, I cover the practical details of each of those steps. So, switch on your flashlight, detective — here's how to gather those clues!

Clue 1: Determine the degrees of unsaturation from the molecular formula

For a reminder on degrees of unsaturation (DOU), mosey on over to Chapter 9.

If you're given the molecular formula, just plug the formula into the equation:

$$\text{Degrees of Unsaturation} = \frac{\left[(\text{Number of Carbons} \times 2) + 2\right] - \text{Number of Hydrogens}}{2}$$

If you have a formula with more than just carbons and hydrogens, you also need the following conversions:

- **Halogens (F, Cl, Br, I):** Add one hydrogen to the molecular formula for each halogen present.
- **Nitrogen:** Subtract one hydrogen for each nitrogen present.
- **Oxygen or sulfur:** Ignore.

The degree of unsaturation is a valuable clue because it tells you how many double bonds, triple bonds, or rings must be present in the unknown compound (although it doesn't tell you *which* of these elements of unsaturation are present).

Double bonds and rings count for one degree of unsaturation each, and triple bonds count for two degrees each. If the molecule has zero degrees of unsaturation, you know that there are no double bonds, triple bonds, or rings in the molecule.

Clue 2: Look at the IR spectrum to determine the major functional groups present in the unknown compound

To brush up on IR spectroscopy, see Chapter 18. If possible, confirm the presence of the major functional groups in the NMR spectrum. Easy ones to confirm include aldehydes, carboxylic acids, and benzene rings.

Clue 3: Determine the peak ratios by measuring the heights of the integration curves

Determine how many hydrogens each peak represents by comparing the integration ratios to the molecular formula. Especially if you're just starting to learn how to solve ^1H NMR problems, using a ruler can be indispensable for finding out the relative integration ratios. What you need to measure with the ruler is the *height* of the integration curve over a given peak (the integration curve is the squiggle over a peak that looks like a capital *S*). These heights are related to the areas underneath a peak, which in turn are related to the number of hydrogens that the particular peak represents. Determining how many hydrogens each peak represents will help you to find the fragments of the molecule in Clue 4.

Here's an example of how to determine the integration using a ruler (see Figure 20-1; for clarity, the peaks underneath the integrations aren't shown). First, you want to establish the relative ratio between the integrations. To do this, use a ruler to measure the difference between the top and the bottom of each of the integration curves. After you measure the height of each peak on the spectrum, divide each of the heights by the smallest height. This will give you the ratios.

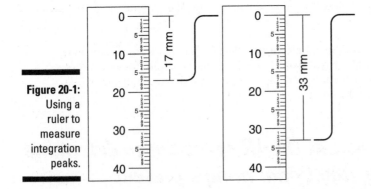

Figure 20-1:
Using a ruler to measure integration peaks.

In this example, the smallest height is 17 mm, so you divide both heights by 17 mm. In this case, this gives an integration ratio of 1:1.94. Because you can never have a fraction of a hydrogen atom, you need to convert the relative ratios into whole numbers.

As is the case in this example, you often find that the integrations don't work out to be perfect whole numbers. If the numbers are very close to whole numbers, the integrations are probably just in error by a bit, and you simply need to round them to the nearest whole number. Because 1.94 is very close to 2.0, in this case you would say that the integration is 1:2.

These imperfect integrations can be attributed to noise in the baseline, a poorly calibrated integral, impurities in the sample, and errors made while measuring the heights of the curves with the ruler.

For ratios that are not close to whole numbers, you need to multiply the ratio by the smallest number that makes both numbers in the ratios whole numbers. If the ratio were 1.0:1.5, for example, you would multiply both parts of the ratio by two to give a whole-number ratio of 2:3. If the ratio were 1.0:1.33, you would multiply both by three to give a whole-number ratio of 3:4, and so on.

So, for this example, you could say that your relative ratio of hydrogens is 1:2. Alternatively, using organic-speak, you could say that the peak on the left integrates for one and that the peak on the right integrates for two.

But does this then mean that the peak on the left represents one hydrogen, and the peak on the right represents two hydrogens? Not necessarily. Remember that this is the *relative* ratio of hydrogens, not the *absolute* ratio, which means that you know only that the larger peak has twice the number of hydrogens as the smaller peak; you don't know the actual number of hydrogens. To find the absolute number of hydrogens, you must compare your ratio to the molecular formula.

If the sum of all the numbers in the relative ratio matches the number of hydrogens in the molecular formula — in this example, if there were only three hydrogens in the molecular formula — *then* you could say that the peak on the left represents one hydrogen and the peak on the right represents two hydrogens, because then all the hydrogens in the molecule would be accounted for. But if the molecular formula had six hydrogens in it, you would have to multiply the relative ratio by two to make the number in your spectrum match the number in your molecular formula — so in that case the peak that integrates for one would represent two hydrogens and the peak that integrates for two would represent four hydrogens. If there were nine hydrogens, you would multiply the relative integrations by three, and so forth.

Remember that the sum of all the hydrogens (absolute number) that you derived from the integrations must exactly match the number of hydrogens in the molecular formula. If the numbers don't match, then you need to reconsider the values of the integrals.

Here's a summary of the steps to determine the number of hydrogens represented by each peak:

- ✔ Measure the heights of each peak using a ruler.
- ✔ Divide each of the heights by the smallest height measured.
- ✔ If one or more of the numbers are not whole numbers, convert them to whole numbers by rounding (if they are all very near a whole number) or by multiplying all the numbers in the ratio by a whole number. This gives you the relative ratio of hydrogens in the peaks.
- ✔ To determine the actual number of hydrogens each peak represents, add each of the numbers in your relative ratio and compare this number to the number of hydrogens in the molecular formula. If these numbers are not equal, determine what you need to multiply your relative ratio by in order to make the number of hydrogens in your spectrum match the number of hydrogens in the molecular formula. If they're the same, you don't need to multiply the ratio by anything.

Clue 4: Break the NMR peaks into fragments using the integration from Clue 3

After you've determined the number of hydrogen atoms that each peak represents, you can assign each peak to a fragment of the molecule. Table 20-1 shows some common fragments.

Table 20-1		Common Fragments
Number of Hs	**Likely Fragment**	**Notes**
1H	$\begin{array}{c} H \\ \mid \\ -C- \\ \mid \end{array}$	Check the IR spectrum to make sure this is not an alcohol (OH), secondary amine (NH), aldehyde (CHO), or acid proton (COOH).
2H	$\begin{array}{c} H \\ \mid \\ -C-H \\ \mid \end{array}$	Rarely: This is two symmetric CH groups.
3H	$\begin{array}{c} H \\ \mid \\ -C-H \\ \mid \\ H \end{array}$	Rarely: This is three symmetric CH groups.

Number of Hs	Likely Fragment	Notes
4H	2 H | —C–H |	Usually two symmetric methylene (CH_2) groups.
6H	2 H | —C–H | H	Rarely: This is three symmetric CH_2 groups. Often indicates an isopropyl group.
9H	3 H | —C–H | H	Often indicative of a tertiary butyl group.

As you determine the fragments in the molecule, write them all down on a sheet of scratch paper so they'll be in front of you when you're ready to solve the structure. For example, if you have three peaks — one that integrates for 1H, one that integrates for 2H, and one that integrates for 3H, you would write on your scrap paper CH, CH_2, and CH_3. After you determine all the fragments (including fragments you determined from the IR, if you're given one), add up all the atoms in your fragments to make sure that they match the molecular formula and to be certain that you're not missing any atoms.

Clue 5: Combine the fragments in a way that fits with the NMR peak splitting, the chemical shift, and the degrees of unsaturation

Often, when you're starting, the best way to go from fragments to a possible structure is to simply brainstorm all the possible molecules that could include the fragments you listed, because in many cases there will be more than one way to put all the fragments together. After you've brainstormed all the possible structures, you systematically eliminate all the incorrect structures. You can eliminate the incorrect structures based on a few different factors. Most often, the incorrect structures will not produce the peak splitting that you see in your NMR spectrum, or will have symmetry when your NMR spectrum suggests that the molecule is not symmetrical. But sometimes you may need to use the chemical shift to eliminate incorrect structures. This procedure may seem tedious, but after working a few problems, you'll develop an intuition about which structures are likely correct, and putting together the fragments to make the correct structure will take no time at all.

After working several problems, you'll notice some common patterns. A quartet that integrates for 2H and a triplet that integrates for 3H on a spectrum is often indicative of an ethyl group (–CH_2CH_3). Likewise, a multiplet that integrates for 1H and a doublet that integrates for 6H usually suggest an isopropyl group (–$CH(CH_3)_2$). A singlet integrating for 9H is usually a *t*-butyl group (–$C(CH_3)_3$).

Clue 6: Recheck your structure with the NMR and the IR spectrum to make sure it's an exact match

For this clue, look only at your proposed compound, and then predict what the spectrum should look like, going carbon atom by carbon atom. Then compare this expected spectrum of the compound you proposed to the real spectrum to see if they match. Here is where you check to see if the chemical shift for each proton (H) makes sense. ***Remember:*** Protons near electronegative atoms (like N, O, Br, F, Cl, and so on) will have a higher chemical shift than protons that that are not near electronegative atoms.

So, what if you get to Clue 5 or Clue 6 and realize, "Oh no! My structure's wrong!"?

In that case, there are two checks to spot the problem.

Check 1

The first thing to check is whether you combined the fragments in the correct way. Often, there is more than one way to put the structure together from the fragments. If you've tried all the ways and none of them checks out, then the problem is probably with the fragments themselves (see Check 2).

Check 2

A less likely possibility is that the fragments themselves are incorrect. The likely fragment in Table 20-1 will be correct in most cases — so it's certainly a good place to begin. Unfortunately, this procedure is not 100 percent fool-proof. When it does fail, it's usually with molecules that have a high degree of symmetry. If the procedure fails, you'll reach Clue 5 and be unable to figure out how to reasonably put the fragments back together again in a way that makes sense with the splitting and chemical shift. In that case, return to the table and rethink your choices of fragments, looking in the notes column to see alternative possibilities.

An example illustrating how the most likely fragment in the table would be incorrect is shown in Figure 20-2. Because of the symmetry in the molecule — that is, the left-hand side of the molecule is identical to the right-hand side of the molecule — the two CH_3 groups are identical and will show a single peak (with an integration of 6H), and the two CHs will also show a single peak (with an integration of 2H).

Figure 20-2:
Dimethyl $CH_3-CH=C=CH-CH_3$
allene.

From the 6H row of Table 20-1, the two symmetrical CH_3 fragments are predicted correctly in the "Likely Fragment" column; the 2H integration, however, is predicted to be a CH_2 fragment rather than two identical CH fragments. Here, choosing the likely fragment (and not the rare case) from the table, you would reach Clue 5 and be unable to piece together a sensible structure. Going back to the "Notes" column in Table 20-1 and looking at the integration for 2H, you see that in rare cases this integration could also represent two symmetrical CH fragments. Making this substitution, then, would lead you down the road to the correct answer.

Working Problems

Here's an example of a bare-bones problem — one where you're given just the molecular formula and the ¹H NMR and asked to solve the structure. When you're doing problems on your own, write all the clues you find on a sheet of scratch paper, just as a detective would write down all the clues on a notepad, so that they'll all be in front of you when you're ready to solve the murder, er, the structure.

Example 1: Using the molecular formula and NMR to deduce the structure of a molecule

Determine the structure of a compound whose molecular formula is $C_8H_8O_2$, using its ¹H NMR spectrum in Figure 20-3.

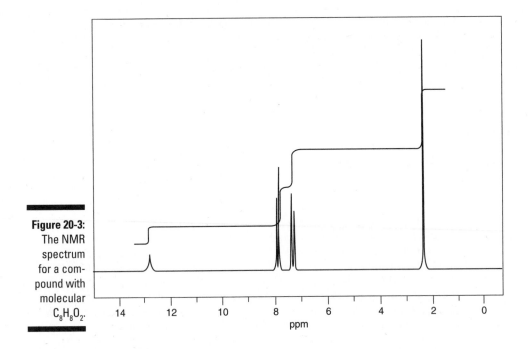

Figure 20-3:
The NMR
spectrum
for a com-
pound with
molecular
$C_8H_8O_2$.

The way to tackle a problem like this is to work through it systematically, clue by clue.

Clue 1: Determine the degrees of unsaturation

Plugging the molecular formula, $C_8H_8O_2$, into the equation gives you five degrees of unsaturation.

Whenever you see four or more degrees of unsaturation in a molecule, the molecule likely contains a benzene ring. To confirm this, look for peaks around $\delta6.5$ to $\delta8.5$ ppm in the ^1H NMR. Benzene rings account for four degrees of unsaturation (three for the three double bonds and one for the ring).

Clue 2: Determine the major functional groups

In this problem, you're not given an IR spectrum to determine the functional groups. However, you can still look at the NMR to see if any peaks indicative of functional groups are present. The three biggies to look for in the ^1H NMR are

✔ Carboxylic acid protons around $\delta12$ ppm (short, fat peaks)

✔ Aldehyde protons around $\delta10$ ppm

✔ Aromatic protons between $\delta6.5$ ppm and $\delta8.5$ ppm

Only rarely (rarely, that is, in problems seen in an undergraduate class) do any other types of protons absorb in these regions.

In the example in Figure 20-3, you have both a carboxylic acid (the short, fat peak around δ13 ppm), and a benzene ring (the two sets of peaks between δ7 ppm and δ8 ppm), which you would have suspected earlier from the number of degrees of unsaturation. (For a list of chemical shifts in NMR produced by common functional groups, see Chapter 19.) Write down those two portions of the molecule on your scratch paper as shown in Figure 20-4.

Figure 20-4:
Carboxylic
acid and
a benzene
ring.

After you've identified from the NMR spectrum the peaks that belong to a particular functional group, put a check mark next to these peaks on the spectrum so that you don't reassign them to fragments in Clue 4. In this example, you would put a check mark next to the benzene protons and the carboxylic acid proton, as they've already been assigned to fragments.

Clue 3: Determine the peak ratios

For this step, you first want to figure out the height ratios for the four peaks on the spectrum. Going from left to right across the spectrum, the ratio is 1:2:2:3. If you can't see that just by looking at the spectrum, break out a ruler and physically measure the size of the integration lines just as I did in the preceding example and prove to yourself that this ratio is correct.

Now that you have the relative ratio (1:2:2:3), you need to convert that ratio into the absolute number of hydrogens that each peak represents. In this case, the molecular formula ($C_8H_8O_2$) tells you that eight hydrogens are in the molecule. Because the relative ratio adds up to 8 (1 + 2 + 2 + 3), you don't have to multiply the ratio by anything; the relative ratio is also the absolute ratio. You now know that the first peak represents 1H, the second and third peaks represent 2H, and the fourth peak represents 3H.

Clue 4: Finding the fragments

You've already determined from Clue 2 that the peak at around δ12 ppm is from a carboxylic acid proton, and that the two peaks between δ7 ppm and δ8 ppm are from a benzene ring. So, you've accounted for all the peaks in the NMR except for the peak at δ2.3 ppm that integrates for 3H. From Table 20-1, you find that 3H represents a –CH_3 fragment. Jot that down on your scratch paper.

Now, take a closer look at the peaks for the benzene ring. From the integration, you know that the benzene has four hydrogens. This tells you that the ring must be substituted exactly twice; that is, that two non-hydrogen groups are attached to it.

If the aromatic region has five protons, then you have a benzene ring that's substituted once. If it has four protons, it's substituted twice (disubstituted). If it has three protons, then it's substituted three times (trisubstituted). If you need to, draw the structures out to prove this to yourself.

If you have a disubstituted benzene, as in this case, you must determine the orientation of the substituents relative to each other. A disubstituted benzene ring can have three different orientations. The substituents can either be *ortho, meta,* or *para* to each other (see Figure 20-5; for a review of aromatic substitution patterns see Chapter 15).

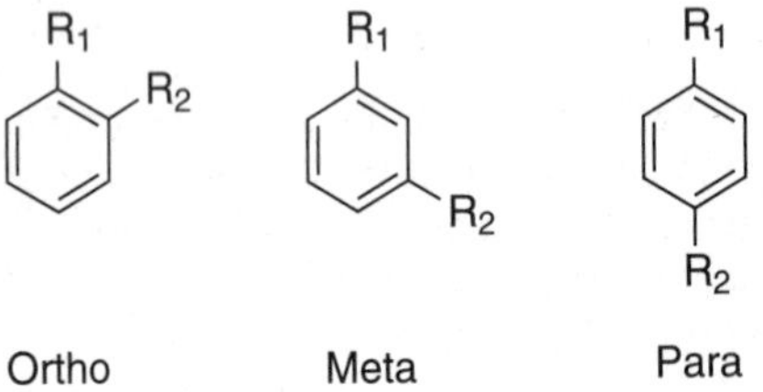

Figure 20-5: The three possible orientations for a disubstituted benzene ring.

Ortho   Meta   Para

Because the benzene ring in this compound contains four hydrogens, you might expect to see four peaks, one for each hydrogen. However, in the spectrum you can see only two peaks in the aromatic region. This suggests a plane of symmetry in the molecule. Looking back at Figure 20-4, only a para-substituted benzene has such a plane of symmetry — in this case, the left side of the ring is identical to the right side.

Because of this symmetry (see Figure 20-6), the two hydrogens neighboring R_1 (labeled H_a) are in identical chemical environments (both are one carbon away from R_1 and two carbons from R_2), and the two hydrogens neighboring R_2 (labeled H_b) are in identical chemical environments. Because of this symmetry, you would expect to see only two peaks in the ^1H NMR that integrate for 2H each. This pattern is typical for para-substituted benzene rings.

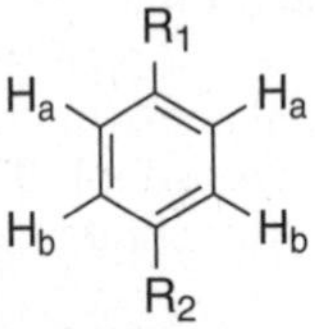

Figure 20-6: Seeing symmetry.

If you can't see the identical chemical environments from looking at the drawing, you might try building the molecule using a model so that you can turn the model over in your hands and prove to yourself that the two H_a atoms are identical, and the two H_b atoms are identical.

Now that you know that the benzene ring has two substituents that are para to each other, you can write out all the fragments (after double-checking that the number of atoms in our fragments exactly matches the number in the molecular formula). See Figure 20-7.

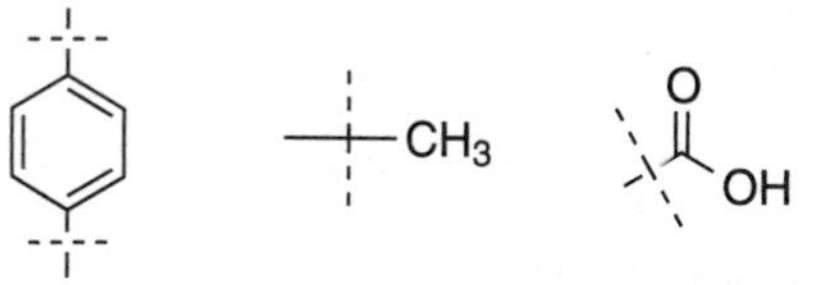

Figure 20-7:
Structure
fragments.

Clue 5: All the king's horses: Putting the pieces together again

In this case, there's only one possible way to put the fragments together. The proposed structure is *p*-toluic acid, shown in Figure 20-8.

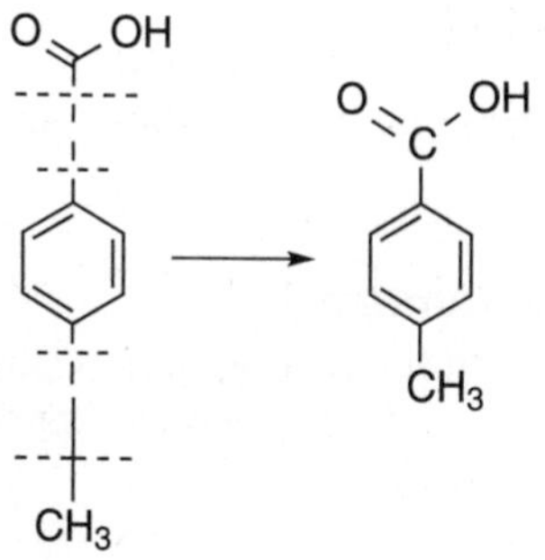

Figure 20-8:
***p*-toluic**
acid.

Clue 6: Confirm the proposed structure

Double-checking to make sure that your proposed structure is the correct one is always a good idea. You check the structure by predicting what the NMR should look like, and then comparing it to the actual spectra to make sure that it matches. So, check this structure (see Figure 20-9):

✔ The methyl group should correspond to a peak that integrates for 3H and is a singlet. It would be at a slightly higher chemical shift because it's near an aromatic ring. At δ2.3 ppm, there's a singlet that integrates for 3H. Check!

> ✔ The aromatic ring protons should show two doublets that have a combined integration of 4H between δ6.5 and δ8.5 ppm. Check!
>
> ✔ The carboxylic acid should have a peak at around δ12 ppm that integrates for 1H and is a singlet. Check!

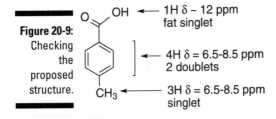

Figure 20-9: Checking the proposed structure.

1H δ ~ 12 ppm
fat singlet

4H δ = 6.5-8.5 ppm
2 doublets

3H δ = 6.5-8.5 ppm
singlet

Looks like you can lock this one up!

Example 2: Using the molecular formula, IR, and NMR to deduce the structure of a molecule

Solve the structure of compound $C_5H_{10}O$ that has the ^{1}H NMR and IR spectra shown in Figure 20-10.

Clue 1: Determine the degrees of unsaturation

Plugging the molecular formula, $C_5H_{10}O$, into the DOU equation shows that this molecule has one degree of unsaturation. This indicates that the molecule has either one double bond or one ring.

Clue 2: Determine the major functional groups

The IR spectrum shows an intense peak at 1,710 cm^{-1}, which is characteristic of a carbonyl group. (See Chapter 5 for a listing of the functional groups containing a carbonyl group.) The small peaks between 3,400 and 3,700 cm^{-1} cannot be amine stretches because, according to the molecular formula, the compound doesn't contain nitrogen, and the peak is not broad and intense enough to be caused by an alcohol (besides, the oxygen is already accounted for with the carbonyl group). This is probably just a carbonyl overtone stretch. No other peaks in the IR are very helpful. (For a list of IR peaks produced by common functional groups, see Chapter 18.) Looking in the ^{1}H NMR for distinct peaks above δ6.5 ppm turns up nothing; the NMR spectrum shows no peaks attributable to aldehyde, aromatic, or acid functional groups. (For a list of chemical shifts in NMR produced by common functional groups, see Chapter 19.)

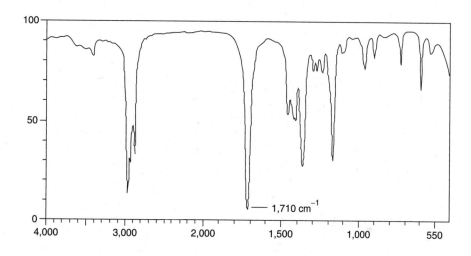

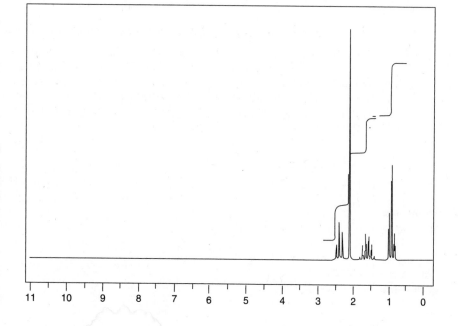

Figure 20-10:
The IR and NMR spectra for a compound with molecular formula $C_5H_{10}O$.

Clue 3: Determine the peak ratios

The NMR peaks integrate with a relative ratio of 1.0:1.5:1.0:1.5 from left to right (use a ruler if you don't believe me). Multiplying this entire ratio by two gives a whole number relative ratio of 2:3:2:3. Adding up 2 + 3 + 2 + 3 gives 10, matching the number of hydrogens in the molecular formula. So, you don't need to multiply the relative ratio by anything to get the actual number of hydrogens. Therefore, the integrations for the three peaks from left to right are 2H, 3H, 2H, and 3H.

Clue 4: Finding the fragments

From Table 20-1, the most likely fragments for 2H, 3H, 2H, and 3H are CH_2, CH_3, CH_2, and CH_3, respectively. Adding up all the atoms from these fragments to see if the number of atoms matches the molecular formula shows that one carbon and one oxygen atom are missing. Additionally, from Clue 1 you know that this molecule has one degree of unsaturation; this unsaturation could be accounted for by the double bond in the carbonyl seen in the IR spectrum (refer to Clue 2). Carbonyl groups are found in aldehydes, ketones, esters, and carboxylic acids. Here, however, esters and carboxylic acids can be eliminated because the molecular formula shows only a single oxygen. Because the NMR spectrum doesn't contain an aldehyde peak (aldehyde protons are seen around δ10 ppm) and the IR spectrum doesn't show an aldehyde C-H stretch in the at 2,700 cm^{-1} (see Chapter 15), the carbonyl group is a ketone.

Clue 5: And all the king's men: Putting the pieces together again

Because the carbonyl is a ketone, the fragments can only be pieced together in two ways, as shown in Figure 20-11. Note that one of these ways gives a symmetric molecule (possible compound A, where the carbonyl group is in the center of the molecule) while the other way gives a molecule that's not symmetric (possible compound B, where the carbonyl group is on one side of the molecule).

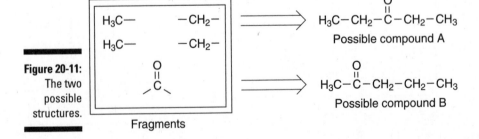

Figure 20-11: The two possible structures.

Now that you've narrowed down the choices to two, you want to check to see which one is the most consistent with the peak splitting and the chemical shift. Try possible compound A first.

You can easily rule out this compound because it has a center of symmetry — that is, the left-hand side of the molecule is identical to the right-hand side. Therefore, you would expect to see just two peaks in the ^{1}H NMR for this compound rather than the four you see, because both methyl (CH_3) groups would be in identical chemical environments, and both methylene (CH_2) groups would also be in identical chemical environments.

Clue 6: Confirm the proposed structure

Now check possible compound B (see Figure 20-12).

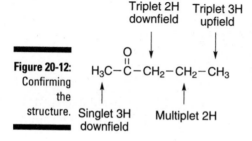

Triplet 2H
downfield

Triplet 3H
upfield

Figure 20-12:
Confirming
the
structure. Singlet 3H
downfield

Multiplet 2H

This compound has no symmetry, so you go carbon atom by carbon atom to check the structure.

✔ The far-left methyl group has no adjacent hydrogens. Therefore, it would be a singlet and integrate for 3H. Because it's next to the electron-withdrawing carbonyl group, you would expect it to be shifted downfield (higher ppm). Sure enough, the NMR shows a singlet at δ2.1 ppm that integrates for 3H. Check!

✔ The methylene (CH_2) to the immediate right of the carbonyl group would be expected to be a triplet, because it's adjacent to two hydrogens. This triplet would integrate for 2H and would be expected to be at a higher chemical shift because it's next to the electron-withdrawing carbonyl group. Sure enough, at δ2.5 ppm there's a triplet that integrates for 2H. Check!

✔ The methylene group farther to the right would be expected to be a multiplet because it's adjacent to five hydrogens (it would be approximately a sextet, although it could split into as many as 12 peaks; see Chapter 19), and would integrate for 2H. Sure enough, at δ1.5 ppm the NMR spectrum shows a multiplet (that appears to be a sextet) that integrates for 2H. Check!

✔ The methyl to the far right would be expected to be a triplet because it's adjacent to two hydrogens and would integrate for 3H. This triplet would be expected to be farther upfield (lower ppm) than the other peaks because it's the farthest away from the electron-withdrawing carbonyl group. Sure enough, at δ0.9 ppm the NMR spectrum shows a triplet that integrates for 3H. Check!

Everything matches, so you can cuff this one up.

Three Common Mistakes in NMR Problem Solving

In NMR problems, there's so much information to be considered that making a mistake is easy. But here are three common mistakes that you can try to avoid when solving NMR spectroscopy problems.

Mistake #1: Trying to determine a structure from the chemical shift

Chemical shift regions overlap, so attempting to determine all the fragments of a molecule solely on the basis of chemical shift is almost always a mistake. A few fragments can be determined in some cases — aldehydes, carboxylic acids, and aromatic ring fragments are often given away by the chemical shift — but most fragments cannot. A proton at δ2.5 ppm could be a proton adjacent to a carbonyl group; it could also be a proton on a carbon neighboring a benzene ring or even a proton next to a double bond. It would be impossible to tell the difference among these three possibilities simply by looking at the chemical shift.

Yes, there are chemical shift tables big enough to choke a small beluga whale, but your time will be much better spent understanding the general ideas behind the chemical shift than memorizing all the values for every conceivable proton that you might encounter. Such mindless memorization is a fruitless task anyway. You might know that a proton next to a chlorine will have a chemical shift of δ3 to δ4 ppm, but what about a proton next to two chlorines? Or what about a proton next to two chlorines and a carbonyl group? Too many possibilities exist to simply memorize them all.

A better approach is to learn the ranges of chemical shifts for protons next to the common functional groups (such as those found in Chapter 19; many professors are kind enough to give you a table that they want you to learn), and then understand the idea behind the chemical shift. If you understand that protons next to groups that have a local magnetic field (like double bonds, triple bonds, and aromatic rings) will have a higher chemical shift than isolated protons, and that protons next to electronegative elements will have a higher chemical shift than those that are far from electronegative elements, you'll be in pretty good shape. If you understand these concepts, you won't need a table to know that a proton next to two chlorines will have a higher chemical shift than a proton next to one chlorine. This understanding will give you greater flexibility for solving NMR problems.

Mistake #2: Starting with coupling

Unless you really know what you're doing, trying to solve a structure by looking first at the coupling will result in fistfuls of your pulled-out hair and several empty bottles of aspirin. Unlike integration, coupling doesn't refer to how many hydrogens (protons) a peak represents; instead, it indicates how many proton neighbors a certain hydrogen has. Because of that fact, determining the pieces of the molecule using the coupling is much more difficult than using the integrations.

Think of this process as analogous to the work of a census taker going house to house gathering information about the number of residents in each house. If you were taking the census, what would make your job easier: having people tell you how many people lived in the house you were visiting, or having people tell you how many people lived next door to the house you were visiting? Doing the census would be easier if the residents told you how many people lived in that particular house, right? To continue with this analogy, the integration tells you how many people (hydrogens) live in the house (the specific carbon); the coupling tells you how many people (hydrogens) live next door.

Another drawback to starting with coupling is that coupling has so many nuances (many of which are beyond the scope of undergraduate-level NMR), and peaks are not always split into the standard-fare singlets, doublets, triplets, and so on. Long-range coupling (which I don't cover in this book) can complicate things, as can the presence of chiral centers (refer to Chapter 6).

Protons that are next to nonequivalent hydrogens can give some pretty complicated splitting patterns, too. Often, you just get a haystack — a big jumble of many peaks clumped together that are hard to resolve. Sometimes the coupling will also be hard to see unless your professor was kind and magnified the peaks for you. Because of all these factors, coupling is most useful for determining the *connectivity* of the fragments and is less useful in determining the *identity* of the fragments themselves.

Mistake #3: Confusing integration with coupling

Remember that the integration tells you the relative number of protons a peak represents, but that coupling refers to how many neighbors those protons see. Therefore, the *n* + 1 rule applies only to coupling, not to integration, and has nothing to do with how many protons a peak represents. A multiplet can integrate for 1H just as a singlet can integrate for 9H. Coupling and integration are independent of one another.

Part V
The Part of Tens

In this part . . .

- ✔ Get acquainted with ten great organic chemists.
- ✔ Find out about ten cool organic discoveries and the role of serendipity in the discovery process.
- ✔ See ten interesting organic molecules.

Chapter 21

Ten (Or So) Great Organic Chemists

In This Chapter

▶ Introducing pioneers in the field of organic chemistry

▶ Finding out about discoveries that transformed organic chemistry

*T*his chapter presents the work of ten (or so) of the great organic chemists from the past. The work of these chemists reflects the diversity of organic chemistry. Some of these folks pioneered the ideas that laid the foundation of organic chemistry as a modern science, while others made more recent contributions. In all cases, the work of these great chemists made a significant, lasting impact on the field and is worthy of admiration.

August Kekulé

By some accounts, August Kekulé (1829–1866) was a bore as a lecturer and a klutz in the lab, but his ideas on the nature of organic compounds were extremely influential (and, more important, correct). He proposed, correctly, that carbon could make four bonds to other atoms, and could form complicated structures by combining with other carbon atoms to make chains and rings. Kekulé also proposed the correct structure for benzene, C_6H_6, and notoriously bragged years after the fact that his inspiration for the cyclic structure of benzene came from a dream he had of a snake biting its tail, a claim as suspect as it is entertaining.

Friedrich Wöhler

Friedrich Wöhler (1800–1882) helped lay the foundation for modern organic chemistry by unseating the established theory of *vitalism,* the theory that organic molecules contained a vital life force not present in inorganic compounds. By synthesizing the organic molecule urea from inorganic ammonium cyanate, Wöhler helped deal a blow to this notion.

Archibald Scott Couper

Couper (1831–1892) was a pioneer in proposing the structure of organic molecules and in showing how bonding worked. He suggested the tetravalency of carbon before Kekulé did, but his PhD advisor (Wurtz) held back this paper due to how forcefully the paper rejected the current theories, and Kekulé published his paper first. Couper, outraged that his advisor had sat on his paper and led to his ideas being scooped, confronted him angrily. Wurtz did not take this well and kicked him out of his lab. After this event, Couper suffered a nervous breakdown, became mentally ill, and spent the remainder of his life living in his mother's attic.

Johan Josef Loschmidt

Loschmidt (1821–1895) is one of those often-forgotten pioneers of chemistry. In 1861, he published a pamphlet, *Chemische Studien,* that provided the two-dimensional structural representations of hundreds of organic molecules. His accomplishment was remarkable because this was done at a time when the idea of a molecular structure (rather than simply a molecular formula) was still a novel idea. Looking back at his pamphlet with the knowledge of today shows how truly remarkable his ideas were, since so many of his structural guesses were accurate. Some people even give him credit for coming up with the cyclical structure of benzene before Kekulé did, because he used a circle to represent the benzene.

Louis Pasteur

Louis Pasteur (1822–1895) discovered the handedness of organic molecules — the fact that molecules can have the same atomic connectivity but different orientations of those atoms in space. Pasteur made this discovery by observing that tartaric acid formed into two distinct types of crystals. He then separated the two types of crystals with tweezers under a microscope and noticed that *plane-polarized light* (light that oscillates in a single direction) rotated in one direction when it passed through one type of crystal, and in the opposite direction when it passed through the other type of crystal. With these observations, he speculated that this difference in crystal shape and light rotation must relate to different orientations of the atoms in the two crystals. He was right, and the foundation for our current understanding of stereochemistry was laid.

Emil Fischer

Fischer (1852–1919) was one of the pioneers in organic chemistry. After being told by his father that he was too stupid to be a businessman, he attended the University of Bonn and studied science. After receiving his doctorate, he worked at the University of Munich. Fischer worked on synthesizing and determining the structures of sugars, and discovered the d/l stereochemistry of sugars. He then made significant advances in the understanding of purine nucleotides (adenine and guanine, two of the DNA bases, are purine nucleotides). Although the beginning of his career was marked by creative genius, his life ended in tragedy. He suffered from the side effects of the toxic mercury compounds he worked with in the laboratory, and when his wife died, and shortly thereafter two of his three sons were killed in World War I, he committed suicide. Though his life ended in tragedy, his accomplishments in organic chemistry are considered some of the finest by an organic chemist.

Percy Julian

Percy Julian (1899–1975) was a pioneer in the chemistry of steroids, which are complicated ring-containing molecules that make up the backbone of important biomolecules like cholesterol and sex hormones (testosterone, estrogen, and so on). He figured out ways to synthesize these complicated organic molecules on a large enough scale to make possible the treatment of hormone deficiencies, and started a company for the production of these steroids. He was one of the first African Americans to be awarded a doctorate degree and, in 1973, became the first African-American chemist to be admitted into the National Academy of Sciences.

Robert Burns Woodward

Many claim that Woodward (1917–1979) was the greatest of all organic chemists — and certainly, a case can be made for that claim. Prominent among his many accomplishments were his syntheses of astoundingly complex organic molecules such as vitamin B_{12} (in collaboration with Albert Eschenmoser), strychnine, and quinine. These syntheses are particularly magnificent considering that they were performed at a time when spectroscopy and structure determination were a shadow of what they are today. (In fact, Woodward pioneered several spectroscopic techniques for structure elucidation.) Along with Roald Hoffman, Woodward solved the puzzle of why pericyclic reactions (like the Diels–Alder reaction; see Chapter 14) occur in the manner they do; he did so by using arguments about orbital symmetry, a formulation now called the Woodward–Hoffman rules (which you may learn about if you take the second semester of organic chemistry). He won the Nobel Prize in chemistry in 1965 "for his outstanding achievements in the art of organic synthesis" and would have undoubtedly won the prize again along with Roald Hoffman for the Woodward–Hoffman rules had he not died two years earlier. (Nobel Prizes are not awarded posthumously.)

Linus Pauling

Linus Pauling (1901–1994) used quantum mechanics to describe the nature of chemical bonds and how atoms come together to form molecules. His book, *The Nature of the Chemical Bond,* is considered a classic in the field. After his theoretical work on chemical bonding, Pauling pioneered the field of chemical biology and elucidated the structures of proteins.

Pauling was an unusual person. He advocated taking massive daily doses of vitamin C, and he was so vocal against government policies that he was denied visas to attend major international science conferences. Some people have suggested that had he attended a conference in which Rosalind Franklin presented her X-ray structures of DNA, with his understanding of helices from his work on proteins, he may have elucidated the structure of DNA before Watson and Crick (who attended the conference). Pauling is the only person to win two unshared Nobel Prizes — one in chemistry, for his insights into the nature of the chemical bond, and the other in peace, for his uncompromising stand against nuclear testing and proliferation. Pauling died in 1994 at the age of 93.

Dorothy Hodgkin

Dorothy Hodgkin (1910–1994) became involved in structure determination using a technique called *X-ray crystallography*. Although she suffered from severe health problems, including crippling arthritis, she was extremely persistent and hard working, and determined the structures of many important molecules using X-ray crystallography, including the structure of penicillin and, later, the structure of vitamin B_{12}. She won the Nobel Prize in chemistry in 1964 for her work.

John Pople

John Pople (1925–2004) made practical the use of computational techniques. He took theoretical chemistry out of the ivory tower and placed it into the hands of practical wet chemists. He made theoretical calculations available to even the novice chemists, whereas before they had only been available to the hard-core theoreticians with degrees in mathematics. He was the pioneer of the computational program Gaussian, a user-friendly computer software program designed to carry out these molecular calculations. He shared the Nobel Prize in chemistry for his development of computational methods. He died in 2004, after a career that spanned almost an entire century.

Chapter 22

Ten Cool Organic Discoveries

In This Chapter

▶ Seeing ten important discoveries in organic chemistry

▶ Recognizing the role that serendipity plays in the discovery process

*I*n this chapter, I cover ten cool discoveries that relate to organic chemistry. One interesting feature of this list is that all these discoveries contain some element of the serendipitous. Although you can never discount hard work and preparation, chance and fortune have played a major role in the big discoveries in organic chemistry (although, as Pasteur would say, luck favors the prepared mind).

Explosives and Dynamite!

Alfred Nobel and his family recognized the potential of nitroglycerin as a commercial explosive — a substance with explosive properties more favorable than gunpowder. In his laboratory, however, his family discovered firsthand how dangerous the substance could be. In a massive nitroglycerin explosion, Alfred's brother Emil and several co-workers were killed.

After the accident, Alfred worked to make nitroglycerin safer (see Figure 22-1). He noticed that when he mixed the syrupy nitroglycerin with silica powder (diatomaceous earth) he obtained a stable mash that he called *dynamite*. When he commercialized his discovery — along with blaster caps to safely start the explosion — he became one of the wealthiest men of his time. When he died in 1896, he left his fortune to set up the Nobel Foundation. Each year, the Nobel Foundation grants the prestigious Nobel Prizes, of which, ironically, the Peace Prize is one of the most famous.

Figure 22-1:
Making
dynamite.

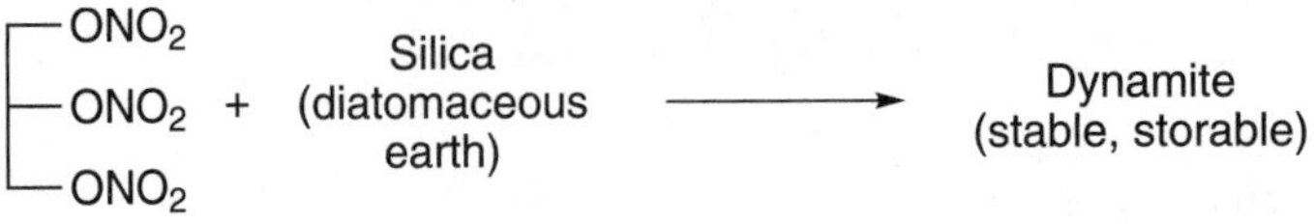

$$
\begin{bmatrix} -ONO_2 \\ -ONO_2 \\ -ONO_2 \end{bmatrix} + \begin{array}{c} \text{Silica} \\ \text{(diatomaceous} \\ \text{earth)} \end{array} \longrightarrow \begin{array}{c} \text{Dynamite} \\ \text{(stable, storable)} \end{array}
$$

Fermentation

No one knows for sure who discovered fermentation, but it was probably first observed many thousands of years ago in caskets of rotting fruit. After being bruised and left out for several weeks, fruit *ferments* (gets broken down by enzymes in yeasts) to make alcohol, as shown in Figure 22-2. Although this rotting fruit probably didn't taste very good, someone who drank the juice discovered that it had a potent and pleasurable effect on the body. This likely led to the discovery of wine making, and, some time later, beer making, which uses grains and honey as the source of sugars and carbohydrates rather than fruit.

Figure 22-2:
Making
alcohol.

$$
\underset{\text{Sugar}}{C_6H_{12}O_6} \xrightarrow{\text{yeast}} \underset{\text{Ethanol}}{CH_3CH_2OH} + CO_2
$$

The Sumerians are the first on record to begin brewing beer, some 6,000 years ago (although beer making probably was going on long before this time). In fact, some historians have suggested that many humans gave up their nomadic lifestyles and became farmers after the discovery of fermentation, just in order to grow crops to make beer. (This suggestion seems to coincide with human nature.)

The Synthesis of Urea

The synthesis of the organic compound urea from an inorganic substance by Friedrich Wöhler was one of the first major breakthroughs in organic chemistry. With this accomplishment, Wöhler showed that organic substances were not held together by a "vital life force," as had been postulated at the time.

Wöhler had attempted to synthesize ammonium cyanate as a continuation of his studies of inorganic cyanates. He isolated instead a substance that seemed to resemble urea, "a so-called animal substance," or in modern terms, an organic compound (see Figure 22-3). Comparing the properties of his synthesized compound to those of the pure urea he had extracted from urine, Wöhler found that the properties of the two were identical.

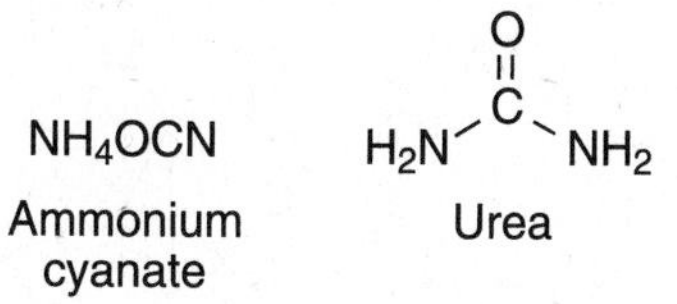

Figure 22-3:
Wöhler made urea from ammonium cyanate.

NH_4OCN

Ammonium cyanate

$$\underset{\text{Urea}}{H_2N-\overset{\overset{\displaystyle O}{\|}}{C}-NH_2}$$

The Handedness of Tartaric Acid

The handedness of tartaric acid (shown in Figure 22-4) was discovered by Louis Pasteur. Why was Pasteur studying tartaric acid? As a French chemist, he was studying wine (of course), of which tartaric acid is a component. He noticed that tartaric acid formed into two different crystal shapes. He separated the crystals with tweezers under a microscope, and noticed that crystals of one shape rotated plane-polarized light in one direction, and the other crystals rotated light in the opposite direction. He surmised (correctly) that these different crystals must relate to the handedness of the molecules themselves.

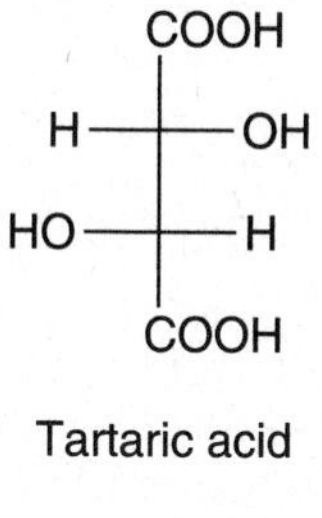

Figure 22-4:
One of the stereoisomers of tartaric acid.

```
        COOH
         |
  H —————— OH
         |
 HO —————— H
         |
        COOH
```

Tartaric acid

Diels–Alder Reaction

Two German chemists, Otto Diels and Kurt Alder, discovered the diene synthesis currently known as the Diels–Alder reaction. This reaction is just cool — it forms two carbon-carbon bonds simultaneously, and is extremely useful for the construction of strange-looking bicyclic molecules, like the pesticide Aldrin (shown in Figure 22-5) and various six-membered rings.

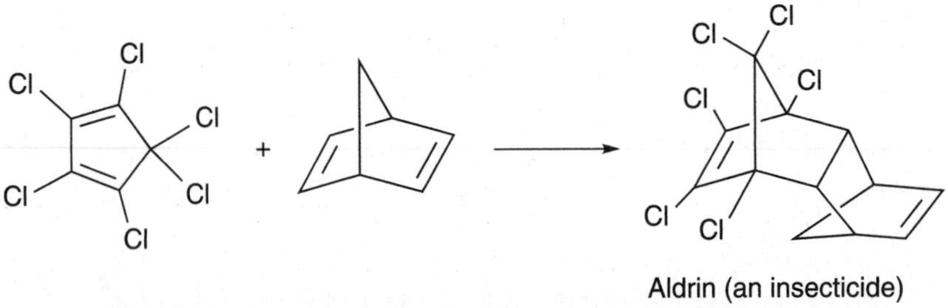

Figure 22-5:
Making
Aldrin
with the
Diels–Alder
reaction.

Aldrin (an insecticide)

Interestingly, in their classic paper describing the reaction, Diels and Alder wrote, "We explicitly reserve for ourselves the reaction developed by us to the solution of such problems [natural product synthesis]." Surprisingly, this "Back off!" warning to other chemists had the desired effect: Other chemists stayed away from this reaction until after World War II was over, and synthetic chemistry in Germany had taken a serious hit. Typically, though, such statements in the literature have the opposite effect of the intended one. Moses Gomberg found this out the hard way in 1900 when he reserved for himself the right to study the triphenyl methyl radical ($Ph_3C\cdot$) — the first radical known to chemists — and chemists trampled over each other to begin their own research on the molecule.

Buckyballs

Buckyballs are one of the most recently discovered allotropes of carbon. *Allotropes* are compounds that contain only a single element — in this case, carbon. Other carbon allotropes include graphite and diamond. Buckyballs, discovered in 1985 by Richard Smalley, Harry Kroto, and Robert Curl, look like molecular soccer balls. Buckyballs were discovered accidentally when these chemists blasted a laser at graphite powder and passed the products through a mass spectrometer. They noticed a peak in their mass spectrum that corresponded to exactly 60 carbons. Using intuition, they surmised that the structure corresponding to this peak must be a spherical molecule (see Figure 22-6). They named this molecule buckminsterfullerene, after Buckminster Fuller, who popularized the geodesic dome; because of their spherical shape, these molecules are often called *Buckyballs*. Although these molecules haven't produced any practical applications so far, they're being studied for use as superconductors, as gasoline markers (so oil spills can be traced back to a specific oil company), and as vehicles for drug delivery.

Figure 22-6: Buckyball.

Soap

Legend has it that soap was first discovered by Roman women washing their clothes along the Tiber River. They noticed that when they washed their clothes in some places of the river their clothes got cleaner than when they washed them in other places. Apparently, next to the river was a sacrificial site, where, presumably, the animal fat from the sacrifices mixed with the fire ashes (which contain lye, or potassium hydroxide), making soap (see Figure 22-7). The soap then ran down the hill to the river after it rained.

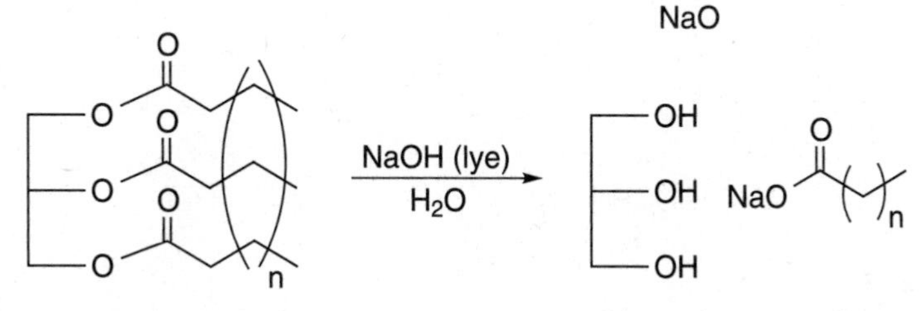

Figure 22-7: Making soap.

Animal fat (triglycerides) $\xrightarrow[\text{H}_2\text{O}]{\text{NaOH (lye)}}$ Glycerol   Soap

Aspartame

James Schlatter discovered aspartame by doing something that would make safety experts cringe — he licked his fingers after working in the lab. He found that his fingers tasted very sweet and traced back the sweetness to aspartame (shown in Figure 22-8), the compound he was working on (and he had another taste of his synthetic aspartame just to be sure he was right). And so was born one of the most popular sugar substitutes. NutraSweet and Equal all contain the sugar substitute aspartame, as do many diet soft drinks.

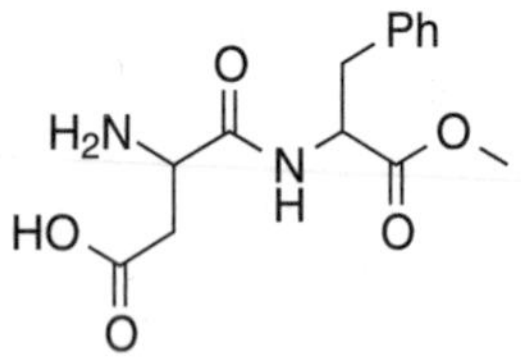

Figure 22-8: Aspartame.

Penicillin

The story of penicillin's discovery is probably the most well-known case of serendipity in the discovery process. Alexander Fleming was doing research on *Staphylococcus* bacteria, when he noticed that his culture plates had become contaminated by a *Penicillium* fungus. He discarded the plates into a sterilizing wash but didn't submerge them properly. When he returned to the lab, he noticed that on the plates where the fungus had grown, the bacteria had been killed. "Something that the fungus is producing must be killing the bacteria," he thought to himself. That something was penicillin (shown in Figure 22-9), an antibiotic that would eventually save countless lives.

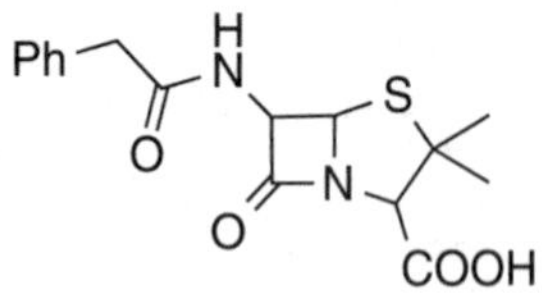

Figure 22-9: Penicillin.

Teflon

Two chemists, James Rebok and Roy Plunkett, who worked for DuPont, discovered Teflon in 1938. Teflon is the polymeric nonstick surface that keeps your eggs, bacon, and pancakes from sticking to your frying pans (see Figure 22-10 for the structure). When they discovered Teflon, Rebok and Plunkett were working with fluorinated hydrocarbon gases kept in large canisters. One day, they opened the valve on what they thought was a full canister of gas, but no gas came forth. At first, they thought that the canister had sprung a leak, until they noticed that the supposedly empty canister had the same weight as a full container. Sawing open the container, they found a sticky white substance had coated the inside of the container — Teflon!

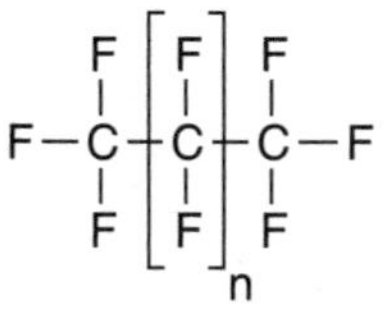

Figure 22-10:
Teflon.

$$F-\underset{\underset{F}{|}}{\overset{\overset{F}{|}}{C}} - \left[\underset{\underset{F}{|}}{\overset{\overset{F}{|}}{C}}\right]_{n} - \underset{\underset{F}{|}}{\overset{\overset{F}{|}}{C}} - F$$

Chapter 23

Ten Cool Organic Molecules

In This Chapter

- Seeing ten very cool organic molecules
- Recognizing that organic molecules can be structurally and functionally interesting

In this chapter, I cover ten organic molecules that have structurally interesting features or turn out to be functionally useful. Or they just have structures that I think look cool.

Octanitrocubane

Organic explosives are molecules that usually combine both a source of fuel and an oxidizer within the same molecule, such that upon an initiating event (such as heat, shock, or light) an uncontrollable oxidation reaction occurs that generates pressurized gases, heat, and a resulting shockwave. Often, organic structures that contain both hydrocarbons (fuel) and oxidizing groups (such as nitro groups) can serve as explosives. Trinitrotoluene, or TNT, for example, contains both hydrocarbon fuel (the benzene ring) and oxidizing nitro groups within the same structure, leading to a compound with explosive properties.

An interesting recently developed explosive is octanitrocubane, shown in Figure 23-1. The idea behind this explosive is that the structure holds a great deal of energy in its highly strained rings. As a result, explosion of this molecule leads to more heat being given off when reaction converts the structure into acyclic small gases (such as N_2, CO_2, and so on) than if it were unstrained. Consequently, this molecule has the highest detonation velocity of known explosives. A drawback to this compound is that octanitrocubane is difficult to synthesize and, as a result, is very expensive compared to more common explosives.

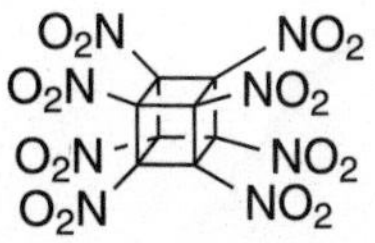

Figure 23-1: Octanitrocubane.

Fenestrane

Fenestrane (shown in Figure 23-2) is a cool-looking organic molecule that consists of four joined cyclobutane rings. This highly strained molecule is called fenestrane after the Latin word for window *(fenestra),* since the structure on paper resembles an old-fashioned windowpane.

Figure 23-2:
Fenestrane.

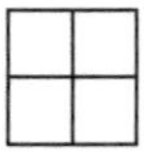

Carbon Nanotubes

Carbon nanotubes (shown in Figure 23-3) are a recently discovered allotrope of carbon, wherein planes of *graphene* (fused benzene ring sheets) are folded to form a cylinder. Carbon nanotubes have some extraordinary properties, including being the strongest materials known (many times stronger than steel) and having high thermal and electrical conductivities. New methods are still needed to make these materials on a large scale and in a cheaper way.

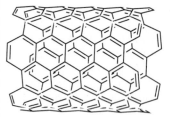

Figure 23-3:
A small
length of
a carbon
nanotube
structure.

Bullvalene

Bullvalene is a really cool organic compound because it's a molecular shape shifter. At room temperature, the molecule interconverts rapidly between numerous identical forms (the reaction mechanism is shown in Figure 23-4). A surprising consequence of these rapid rearrangements is that all ten carbons become equivalent in solution, leading this structure to have just a single peak in its ^{1}H NMR spectrum at room temperature. At lower temperatures, the shape-shifting reaction is slowed and you can see the expected multiple proton absorptions in the ^{1}H NMR spectrum. The molecule is named after William Doering (nicknamed "Bull" by his students), who did some pioneering studies on the molecule.

Figure 23-4:
The shape-shifting molecule bullvalene.

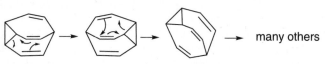

many others

The Norbornyl Cation

The determination of the structure of the norbornyl cation (shown in Figure 23-5) represents one of the biggest chemical controversies in the recent history of organic chemistry. On one side of the debate, Nobel laureate H. C. Brown thought that the norbornyl cation underwent a rapid equilibrium between two equivalent carbocation forms that interconverted by shifting atoms around. On the other side of the debate, Saul Winstein thought that the two forms of the norbornyl cation were actually resonance structures and that the norbornyl cation adopted a single "non-classical structure" that resembled a hybrid of the two cationic forms. The debate became heated at conferences and in the scientific literature, with both sides taking uncompromising stances. However, after much experimental and theoretical work, most scientists now believe that the structure of the norbornyl cation is the bizarre-looking non-classical form rather than a mix of rapidly interconverting isomers.

Figure 23-5:
The norbornyl cation, now believed to adopt a non-classical structure.

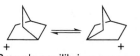

Brown's equilibrium proposal

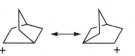

Winstein's non-classical proposal

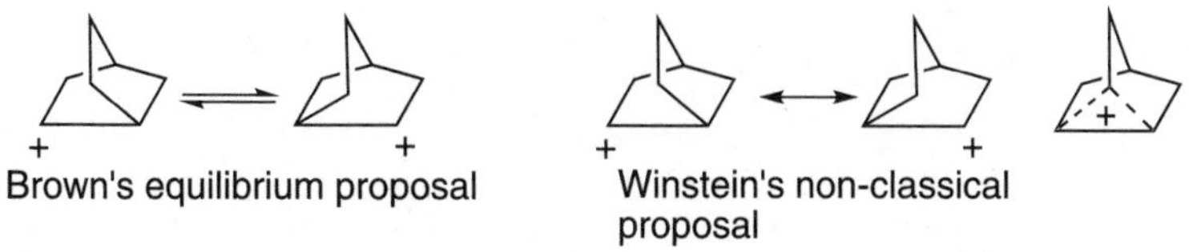

Capsaicin

Capsaicin is a cool molecule because it gives a bit of spice to life — literally. Capsaicin is the main compound that gives hot peppers their heat. The Scoville unit, the most commonly used measure of how hot a pepper tastes, scales with the amount of capsaicin found inside a pepper. Note that the structure of capsaicin (shown in Figure 23-6) contains a long, greasy, hydrophobic tail, meaning that it doesn't dissolve well in water. This gives rise to the oft-repeated advice to rinse your mouth out with milk (which contains a lot of non-polar fat) to reduce the burning sensation within your mouth rather than rinsing with water. The idea is that the hydrophobic fats within milk can solubilize the capsaicin better than water can. Capsaicin causes a burning sensation because it activates the same pain-signaling pathways as those activated by heat burns, making it feel as though your mouth were on fire.

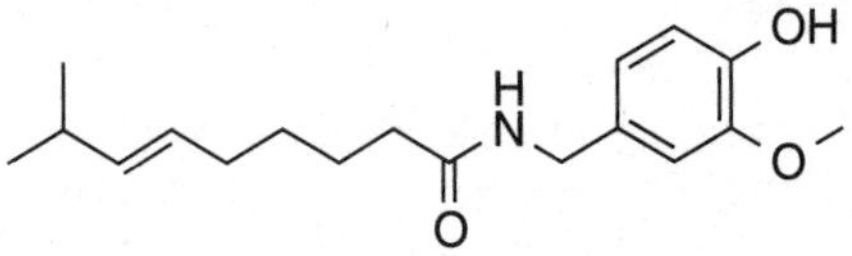

Figure 23-6:
Capsaicin.

Indigo

Indigo (see Figure 23-7) is a blue compound that was initially isolated from indigo plants. It was considered a desirable compound because blue dyes were rare at the time it was isolated. Although the dye was originally extracted from plants and used industrially, today indigo is made synthetically by the metric ton and is used mostly to dye blue jeans.

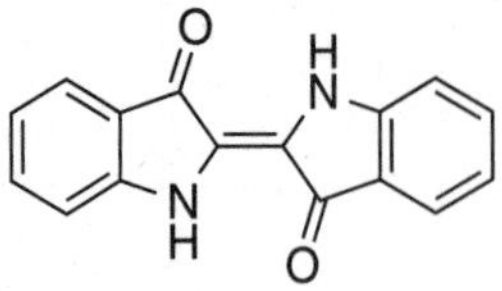

Figure 23-7:
Indigo.

Maitotoxin

Maitotoxin is a beast of an organic molecule. This monstrous organic compound, containing 32 fused ring systems and 98 *chiral centers* (carbons with four different groups attached), was isolated from a tropical fish and is the most toxic "small" organic molecule known. The LD_{50} in mice for this compound is 50 ng/kg, meaning that a gram would probably be enough to kill hundreds of thousands of people. The proposed structure of maitotoxin is shown in Figure 23-8. If you can imagine trying to solve the structure of this compound using NMR spectroscopy or proposing a multistep synthesis to prepare the molecule in the lab, it makes you grateful for the (comparatively) easy problems you encounter in your organic chemistry class. There are current efforts underway to synthesize maitotoxin, and we can only wish the best of luck for all involved — they'll probably need it.

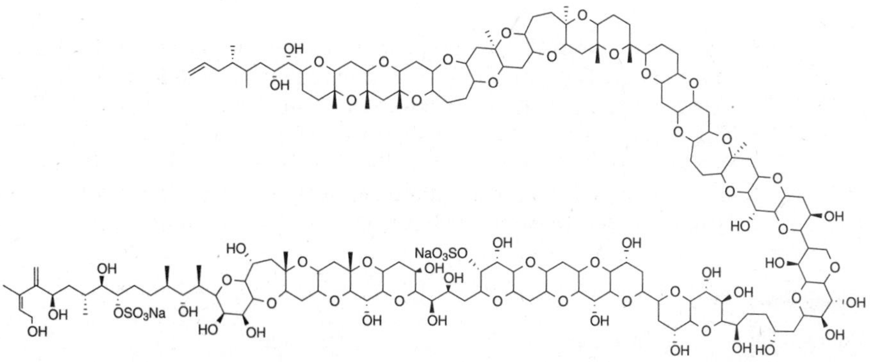

Figure 23-8:
Maitotoxin.

Molecular Cages

Molecular cages are interesting organic structures that contain cavities that can encapsulate other atoms or molecules. A few examples are shown in Figure 23-9. The crown ether cages resemble crowns that can encapsulate metal ions (like K+) within the central cavity of the structure, while calixarenes resemble shuttlecocks that can encapsulate small organic molecules within their cavities. Cyclodextrins, which consist of a ring of connected glucose molecules, are found commercially in products like Febreze, where they can encapsulate volatile compounds with foul odors within their donut hole–like cavity, preventing you from smelling them. Some scientists are testing whether related molecular cages can act as drug delivery vectors for carrying toxic drugs to the site of a disease, such as bringing a chemotherapy molecule to the site of a tumor.

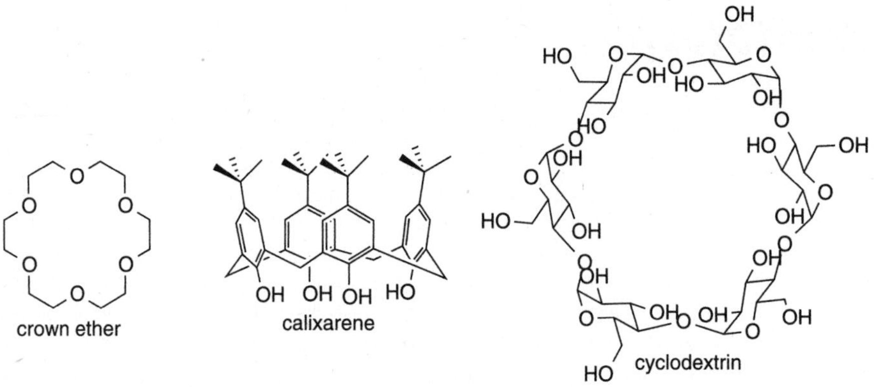

Figure 23-9:
Some
molecular
cages.

crown ether
calixarene
cyclodextrin

Fucitol

Fucose is a naturally occurring *deoxysugar* (*deoxy* means that the structure is missing an OH group). Reducing the aldehyde of aldose sugars such as fucose using sodium borohydride makes a class of reduced sugars called *alditols*. The reduced form of fucose is, thus, called *fucitol*. This compound (shown in Figure 23-10) is found naturally in seaweed, and some people speculate that a dose of fucitol can change an attitude of anxiety, stress, overwork, and fatigue related to organic chemistry to a more Zen-like outlook on life.

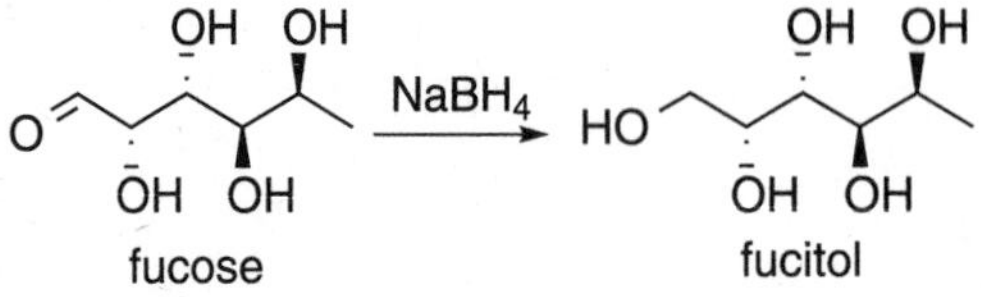

Figure 23-10:
Fucitol.
Consider
when all
else fails.

fucose

fucitol

Part VI
Appendixes

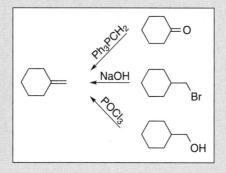

In this part . . .

✔ See how to work multistep synthesis problems.

✔ Recognizing how to use arrows to show the mechanisms of organic reactions.

✔ Get the definitions of chemical jargon and organic-speak that you should know.

Appendix A

Working Multistep Synthesis Problems

● ●

In This Appendix

▶ Understanding multistep synthesis problems

▶ Discovering tips for working multistep synthesis problems

● ●

Multistep synthesis problems are prevalent throughout organic chemistry — and they tend to show up more and more frequently as the course goes on. Many students cite these problems, along with mechanism problems, as the most difficult types in the course. This appendix details why people bother with multistep synthesis and gives several commandments for how to work these kinds of problems so that you can avoid committing the major multistep synthesis sins.

Why Multistep Synthesis?

A good illustration of the importance of organic synthesis is drug making. Throughout history, drugs have primarily been obtained from natural sources. The kings of Mesopotamia smoked poppy flowers to get high on the hallucinogenic properties of the opium alkaloids, Native Americans chewed aspirin-containing willow bark for pain relief, and Fleming discovered penicillin in a mold that grew on his *Staphylococcus* bacteria plates. Even today, natural product chemists extract compounds from living things, characterize them, and test them to see if those compounds have biological activity.

But extracting drugs from natural sources has its problems. Oftentimes, the drug comes from a source that's rare (like an exotic plant or fungus), and sometimes from a source that's endangered or near extinction.

The extraction of these compounds can also be challenging, because the active compound must be separated from the mass of inactive molecules; this can be a difficult task in some cases.

Sometimes the source doesn't produce the drug in high quantities, and starting with thousands of pounds of a source material and extracting just a few paltry milligrams of the desired compound from it is not uncommon. Nature may be a far better synthesizer of organic molecules than humans can ever dream of becoming, but that doesn't mean that nature produces the needed quantities of material on demand. Taking thousands of pounds of a species on the verge of extinction to make just a few small doses of a drug is usually impractical.

So, how can chemists get the desired compounds? By multistep synthesis. This process involves making the compounds from scratch. To do so, chemists take commercially available starting materials and build up the desired molecule using available reagents and techniques. A great many of the drugs that you can buy from pharmacies today were made by multistep syntheses, using enormous quantities of starting materials and reagents that were added with front-end loaders into gigantic reaction vessels. But all that was first planned out on paper by chemists, plotting the shortest, cheapest, and most elegant route available to the target compound (with the shortest and cheapest routes taking precedence in the pharmaceutical industry, and elegance being prized in academia).

Of course, that doesn't mean that the application of multistep synthesis is limited to drugs. Multistep synthesis is used for many different purposes, and one advantage of the organic chemist's ability to make compounds is the free reign this ability gives him or her to do pretty much anything in chemistry (which is why biochemists and physical chemists are forced to collaborate with synthetic organic chemists if they want to get anything done). In some cases, multistep synthesis is even carried out for the sake of synthesis itself. In these cases, organic chemists synthesize structurally interesting natural products that have *grant proposal activity* — that is, products that may show enough biological activity to fund a grant proposal, but likely do not have enough potential to be considered for clinical trials. In these cases, multistep synthesis is used as a framework to develop new reactions, as a way of testing known reaction methods, and more generally as a stomping ground for chemists to flaunt their synthesizing prowess.

And although synthetic chemists do their best to plan the syntheses of complicated molecules, steps invariably go wrong when trying them out in the laboratory. This failure is sometimes a good thing, however. When planned reactions in a synthesis don't work, organic chemists need to figure out new ways of doing things to get around the barriers that these molecules put up. Often, new reactions and insights are discovered in this way.

For example, the synthesis of vitamin B_{12} (discussed in more detail in Chapter 21) was carried out knowing ahead of time that any synthesis of this extremely complex molecule would be too expensive to perform on a commercial scale. Still, the synthesis of this molecule was extremely valuable to organic chemists. Many new synthetic techniques were discovered during this synthesis, in addition to one major theoretical breakthrough (understanding pericyclic reactions like the Diels–Alder reaction) that occurred while contemplating one of the steps in the reaction. So, although this synthesis of vitamin B_{12} was too impractical to carry out on a commercial scale, the total synthesis of this molecule added a great deal of knowledge to the field of organic chemistry.

The Five Commandments

Multistep syntheses are the kind of problems in which you're told what the starting material and the product are, and you're asked to figure out a synthetic route that'll convert the starting material into that product. For these types of problems, you must provide reagents to convert the starting materials into one or more intermediate compounds that you'll then further react in order to make the final product. Often, there will be several steps in the synthesis. However, *no mechanisms (arrow pushing) need to be shown for the individual reactions in a multistep synthesis problem.* This is a common mistake. For these problems, simply provide the reagents and the products of each step as you go along.

For example, suppose the question asked how to convert Compound W into Compound Z (see Figure A-1) using multistep synthesis.

Figure A-1:
Multistep   W $\xrightarrow{\quad ? \quad}$ Z
synthesis.

Here, an acceptable answer would show the individual reactions that would transform Compound W into Compound Z (an example is given in Figure A-2). Note that no mechanisms for these reactions are shown, just the reagents and the intermediates along the way. In many cases, more than one route is possible for the conversion of the starting material into the product, so these types of problems often have several acceptable answers (the shortest route is often preferable, however).

Figure A-2:
Answering
multistep
synthesis
questions.

$$W \xrightarrow{\text{reagent}} X \xrightarrow{\text{reagent}} Y \xrightarrow{\text{reagent}} Z$$

Here are five commandments to help you solve these synthesis problems.

Commandment 1: Thou shalt learn thy reactions

Learning the reactions is the most basic requirement for multistep synthesis. No matter how smart you are, you don't stand a chance of getting synthesis questions right unless you know the reactions. Memorize the reagents, use flash cards, or use whatever other techniques you find most helpful, but learn the reactions cold. This memorization is not something that you can do overnight. Learning the reactions takes time. A good technique is to add reagent learning as a portion of your daily study time.

Because organic chemistry is a cumulative course, you can't afford to forget any reactions that have been covered in previous chapters. If you use flash-cards, you should never throw away your stack of old flash cards that contain the reactions from previous chapters. Instead, keep adding to your pile. (Yes, the deck will be thick enough to choke a small whale by the end of the course!) Most textbooks have convenient end-of-chapter reaction summaries that give an overview of all the reactions that were covered in the chapter, and these overviews can be extremely helpful in making flash cards.

Commandment 2: Thou shalt compare carbon skeletons

The first thing to look for in a multistep synthesis is to compare the carbon skeleton of the starting material to the carbon skeleton of the product. Were any carbons lost or added? If so, can you identify where they were added or lost? A count of the number of carbons in the reactant and the product doesn't take long, but it can help you determine what kinds of reactions you're dealing with.

Take the following simple example shown in Figure A-3. The lighter-shaded portion of the molecule identifies where the likely carbon skeleton of the reactant is found in the product. Looking at the molecules in this way allows you to clearly see what portion needs to be added or lost during your synthesis. (This step may seem trivial in this obvious example, but in tougher problems, this process can help you organize your thoughts.)

Figure A-3:
Comparing
carbon
skeletons.

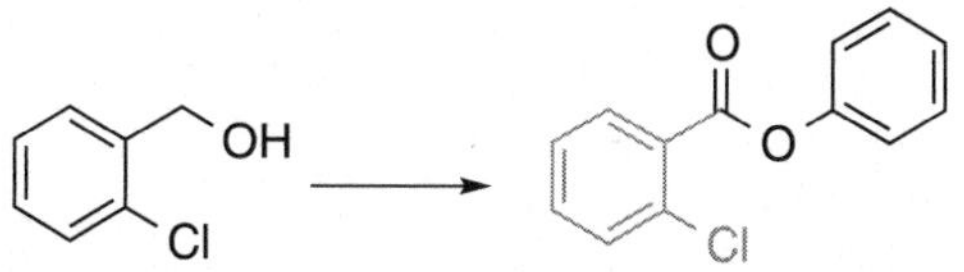

Commandment 3: Thou shalt work backward

Have you ever worked through mazes and found that you can more easily get through a maze by starting at the finish and working back to the start than you can by working from the start? The same thing applies to multistep synthesis problems. Working backward in synthesis problems is called *retrosynthesis*.

The first thing you want to look at is your product, thinking of all the reactions that you know of that could form it. For the moment, ignore your starting material. If your product is an alkene, for example, think of alkene-forming reactions like elimination reactions or the Wittig reaction. (See Chapter 10 for a summary of alkene-making reactions.) Write all of these reactions out and look at what reactant would be required for each.

After you've written out all the potential reaction candidates, return to your starting material. Which reaction has a reactant that most resembles your starting material? That one is probably the best one to select as a potential candidate.

For example, suppose you were asked to do the synthesis shown in Figure A-4.

Figure A-4:
A sample
synthesis
problem.

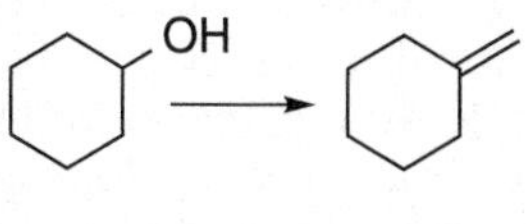

After completing the first two commandments, you would want to think of ways to make the alkene in the product. Ignore the starting material for the moment. Just brainstorm all the ways you can think of to make the alkene and write them down on your scratch paper. You should get something that looks like the following list (see Figure A-5).

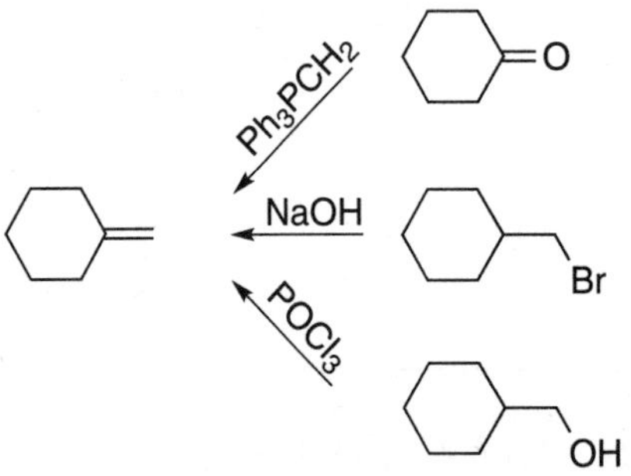

Figure A-5: Alkene-forming reactions.

Now you have three possible routes to choose from. The route to choose is the one that uses a reactant that most resembles the original starting material. If you did Step 2 (accounting for the carbon skeleton), you would know that the product has one carbon more than the starting material. Only the first reaction, the Wittig reaction, accounts for this additional carbon, so the Wittig reaction would be the reaction you would tentatively choose. If this choice turns out to be wrong, you can always go back and try another route.

Looking at the reaction scheme, you now have something that looks like Figure A-6.

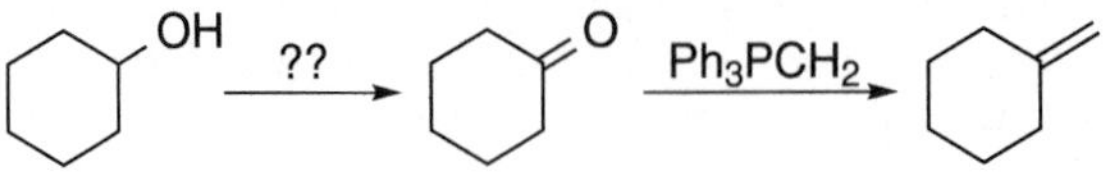

Figure A-6: Completing a retro-synthesis.

Now repeat the same procedure for cyclohexanone, thinking of all the different ways you could make this ketone. One pointer here is that the closer you get to completing a retrosynthesis — in other words, the closer you get back to your starting material — the more you should reference the starting material in your thinking. At this point, for example, you may want to modify your thinking from, "I need to think of all the ways I can make cyclohexanone," to something more like, "I need a reaction that converts an alcohol to a ketone."

If you did Step 1, you would know several ways of producing a ketone from an alcohol (for example, using Jones's reagent, PCC, and so forth). Choosing one of these methods would complete your synthesis.

If you get stuck, go back and try one of the other pathways. If the Wittig reaction in our example had lead to a dead end, you could've gone back and tried one of the elimination reactions. Choosing the correct way back is often a matter of intuition, and that only comes after working a lot of problems (see Commandment 5).

Commandment 4: Thou shalt check thyne answer

After you have a potential synthesis, go back and make sure that all your reagents are compatible with the functional groups on your molecule. Make sure, for example, if you're proposing a Grignard reaction, that there are no alcohols or other incompatible functionalities on your reagent. Undergraduate organic professors often seem to take delight in creating challenging (tricky) exam questions, so double-check every detail of your synthesis for correctness.

Commandment 5: Thou shalt work many problems

This is the greatest commandment of all. As a mere mortal, you have no way around this; you have no magic formula at your disposal to avoid this commandment. A good textbook will have plenty of multistep synthesis problems to practice on. Start with easy synthesis problems to get the feel of what's required, and then work your way to harder problems.

If you have a solutions manual to your text (something that I highly recommend you have), don't refer to it until *after* you've completed the problem. Just as it would be unreasonable to expect to be able to play the piano after only listening to someone else play and without ever practicing yourself, you can't expect to look at the study guide and be able to do synthesis problems on the exam without ever having practiced these problems by yourself. Looking at the solution manual and thinking, "Yeah, I could do this problem," is no substitute for actually doing that problem. You need experience. Get lots of it!

Appendix B

Working Reaction Mechanisms

*O*rganic chemists want to know *what* reagents convert one kind of molecule or functional group into another. But they also want to know *how* reactions happen. They want to know what exactly happened when they scooped different chemicals into a flask that converted the starting materials into the product. If organic chemists know how a reaction happens, they can more easily optimize reaction conditions (the solvent, temperature, acidity, reagent amounts, and so on) that allow the reaction to proceed most efficiently. The mechanism of a reaction shows how a reaction takes place.

By understanding the mechanism of a reaction, chemists also get a better understanding of how reactions work, and they're better informed about when a new reaction is likely to succeed or fail. A reaction mechanism is the detailed stepwise process that uses arrows to show how electrons moved during a reaction. A complete mechanism includes all the bond-making and -breaking steps that convert the starting material into product, and shows any intermediate species formed along the way.

The Two Unspoken Mechanism Types

From a very pragmatic standpoint, in organic chemistry you see two unspoken classes of reaction mechanisms — those reaction mechanisms that you can figure out on your own, and those that you simply have to learn. If you're taking a typical class in introductory organic chemistry, you'll likely see an abundance of both types. The bromination reaction of alkenes (see Chapter 10), for example, is a reaction mechanism that you simply have to learn. You'd have a very hard time figuring out on your own that the reaction mechanism involves a three-membered bromonium ion intermediate. The existence of this intermediate may even have come as a surprise to the researchers who discovered the mechanism!

Because you simply have to commit these kinds of mechanisms to memory, I can't do much to help you to learn them, except to show you the mechanisms when discussing the reactions in this book, and to give you tips on how to draw mechanisms using arrow-pushing. Where I *can* help you is in understanding the mechanisms that you can work out on your own, and in arming you with general principles for reactions whose mechanisms are not found in the text.

I think that one reason students have problems with mechanism problems is that they don't see this distinction between mechanisms that they need to learn — which are often impossible to figure out independently — and mechanisms that they need to be able to work on their own. Many look, for example, at the bromination reaction and, knowing that they have to be able to work out mechanisms for themselves, pour ashes on their heads and rend their garments and think, "I could never figure that out on my own!" They then become discouraged about the whole business, even though bromination is not a mechanism that students are expected to be able to deduce on their own.

Still, even if this distinction is understood, most students find these work-it-yourself mechanism problems quite challenging. These problems are challenging for two reasons:

- ✔ Unlike the mechanisms you need to commit to memory, memorization is useless and impossible with the kinds of mechanism problems you're supposed to be able to work out on your own.

- ✔ Professors think that asking students to draw the mechanisms for reactions they've never seen before is perfectly reasonable.

The trick for working mechanism problems is to understand the general principles of arrow pushing (see Chapter 3), and to get lots of practice (lots and lots of practice!). I detail some of the general principles here, and give you some tips on good mechanism-writing habits. I also show you some common pitfalls to avoid when writing mechanisms.

Do's and Don'ts for Working Mechanisms

Knowing what *not* to do in a mechanism problem is almost as important as knowing what *to* do. So, here's a list of do's and don'ts for writing mechanisms:

- ✔ **Don't confuse mechanism problems with multistep synthesis problems.** This is a common mistake. Mechanism problems give you all the reagents that you need to convert the starting material into the product. So, don't supply additional reagents on your own.

- ✔ **Do use all the reagents given in a mechanism problem.** Mechanism problems don't include any extraneous reagents that are not used in a mechanism, so your reaction mechanism must account for the use of all the reagents.

- ✔ **Don't confuse solvents and base scavengers with reagents.** Be able to recognize typical solvents that will generally not come into play in a mechanism — solvents such as THF, DMF, DMSO, $CHCl_3$, CH_2Cl_2, and so on. If you recognize these solvents, you won't try to incorporate the solvent into the mechanism (although it's sometimes acceptable to use the solvent for proton transfers, particularly alcohol solvents and water). Also, base scavengers (like pyridine and triethylamine) are sometimes added to neutralize any acid formed in the reaction, but often they don't have much bearing on the mechanism itself.

- ✔ **Do get in the habit of drawing out all the atoms at the reaction centers.** Drawing out full Lewis structures rather than line-bond structures for at least the portions of the molecule that change makes losing or misplacing atoms or charges less likely. Drawing out atoms becomes particularly important when charges are involved in a mechanism, because spotting which atoms have charge is much easier when all the atoms are explicitly drawn out.

- ✔ **Don't try to do two things at once in a mechanism.** Take the mechanism one step at a time. That's not to say that there won't be more than one arrow in a given step (there often will be), but don't try to do two steps at once. Don't protonate an alcohol and kick off water to make a carbocation in the same step, for example. Make these separate steps.

- ✔ **Do draw out all resonance structures for intermediates.** Although drawing the resonance structures of reactive intermediates may not be required by your professor, doing so is good practice. For example, when writing the mechanism for the electrophilic aromatic substitution reaction (see Chapter 16), draw out all the resonance structures for the cationic intermediate.

✔ **Do look where you're going.** If you want to go from Tallahassee to Tacoma, you don't just get in your car, drive in any direction, and hope you'll eventually get there. In other words, even though a step you've proposed looks like it could conceivably happen, make sure that step will get you headed in the right direction toward the product. Ask yourself what bonds must be broken and what bonds must be formed in order to take the starting material into the product. Keep the answers to these questions in the back of your mind while you work the mechanism so you can keep track of where you're going.

✔ **Don't overanalyze why you wouldn't get a different product than the one indicated.** You're given the product, so you don't need to worry too much about why the indicated product is formed rather than another. Often minor products in a reaction are shown to be the product because the professor wants you to figure out how the reaction made these minor products. Asking questions of why things happen is a sign of an excellent student, but getting too bogged down on why you form one product over another in a shown mechanism may not yield much. As a general principle, draw the mechanism first, and ask questions later.

✔ **Do ignore spectator ions.** You often see ionic reagents that include potassium (K), sodium (Na), or lithium (Li) as parts of the reagent. It's often a good idea to cross off these spectator ions so you don't get tempted to include them in the mechanism. For example, if the reagent is NaOH, cross off the sodium and make the reagent OH–. That way you won't be tempted to include the spectator sodium (Na) cation in the mechanism.

✔ **Do ask yourself if all your proposed steps and intermediates are reasonable.** Do you always show minuses attacking pluses (and never, ever, vice versa)? If you're in acidic conditions, did you make sure that you didn't form any strongly basic negatively charged intermediates? If you're under basic conditions, did you make sure that you didn't form any strongly acidic positively charged intermediates? Did you make sure to never draw a *pentavalent carbon* (a carbon with five bonds) or break any of the rules of valence? Do all the charges balance? Ask yourself these kinds of questions after you propose a mechanism.

✔ **Do be able to work the different types of mechanisms (discussed in the following section).** After working many problems, you'll spot mechanism patterns and you'll begin to notice that some mechanisms are similar to each other.

Types of Mechanisms

After working many problems, you notice that patterns in working mechanisms emerge for different kinds of reaction mechanisms. The most common different kinds of mechanisms are shown in the following list:

- **Thermal mechanisms:** These are mechanisms with no reagents. These mechanisms are seen mostly in the second semester of organic chemistry, but the Diels–Alder reaction is an example of this mechanism (see Chapter 14).

- **Nucleophile-electrophile mechanisms:** These are probably the most common types of mechanisms. If you can do nucleophile-electrophile mechanisms, you have understood a big part of organic chemistry. These mechanisms involve *nucleophiles* (nucleus lovers) attacking *electrophiles* (electron lovers). Typically, some species with a lone pair of electrons (a nucleophile) attacks an electron-deficient carbon (an electrophile).

- **Acid-base mechanisms:** These mechanisms are identifiable by seeing a strong acid (like HCl or H_2SO_4) or a strong base (like $NaNH_2$) as the reagent. With acid mechanisms, you typically don't want to form any species that have net negative charges (with the exception of an acid's conjugate base, which is usually negatively charged), and with base mechanisms, you don't want to form any species that have net positive charges. Typically, the first step in acid mechanisms is protonation, while in a base mechanism the first step is deprotonation.

- **Carbocation mechanisms:** These mechanisms take place generally under acidic conditions. (You typically won't see net positive charges — like the positive charge on a carbocation — under basic conditions.) Watch out for carbocation rearrangements in these mechanisms (like alkyl or hydrogen shifts). If an intermediate has a net negative charge and is highly basic, you probably did something wrong (you're probably missing a proton transfer step).

- **Anion mechanisms:** These mechanisms are found mostly in carbonyl reactions (ketones, aldehydes, and so on), and they usually take place under basic conditions. You don't see these reactions very often in the first semester of organic chemistry.

- **Free-radical mechanisms:** These are somewhat less common than the other types of mechanisms, but they do pop up occasionally. When you see light ($h\nu$), or peroxide (ROOR), think free-radical mechanisms.

You see radical mechanisms in the bromination of alkanes using light and bromine, for example (Chapter 8). When working free-radical mechanisms, use half-headed arrows rather than full-headed arrows to show the movement of only one electron. Drawing out all initiation, propagation, and termination steps is standard practice for many free-radical reactions.

Of course, some of these kinds of mechanisms overlap each other. For example, often in acid mechanisms you make cations, and often in base mechanisms, you generate anions. Determining the kind of mechanism can help you organize your thoughts when tackling problems.

Appendix C

Glossary

achiral: A molecule that's superimposable on its mirror image. Achiral molecules do not rotate plane-polarized light.

acid: A proton donor or an electron pair acceptor.

alcohol: A molecule containing a hydroxyl (OH) group. Also a functional group.

aldehyde: A molecule containing a terminal carbonyl (CHO) group. Also a functional group.

alkane: A molecule containing only C-H and C-C single bonds.

alkene: A molecule containing one or more carbon-carbon double bonds. Also a functional group.

alkyne: A molecule containing one or more carbon-carbon triple bonds. Also a functional group.

allylic carbon: An sp^3 carbon adjacent to a double bond.

amide: A molecule containing a carbonyl group attached to a nitrogen ($-CONR_2$). Also a functional group.

amine: A molecule containing an isolated nitrogen (NR_3). Also a functional group.

anion: A negatively charged atom or molecule.

anti addition: A reaction in which the two groups of a reagent X-Y add on opposite faces of a carbon-carbon bond.

anti conformation: A type of staggered conformation in which the two big groups are opposite of each other in a Newman projection (see Figure C-1).

Figure C-1:
A Newman projection of an anti conformation.

anti-aromatic: A highly unstable planar ring system with $4n$ π electrons.

anti-periplanar (also known as anticoplanar): The conformation in which a hydrogen and a leaving group are in the same plane and on opposite sides of a carbon-carbon single bond. The conformation required for E2 elimination.

aprotic solvents: Solvents that do not contain O-H or N-H bonds.

aromatic: A planar ring system that contains uninterrupted p orbitals around the ring and a total of $4n + 2$ π electrons. Aromatic compounds are unusually stable compounds.

aryl: An aromatic group as a substituent.

axial bond: A bond perpendicular to the equator of the ring (up or down) in a chair cyclohexane (see Figure C-2).

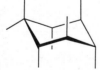

Figure C-2:
Axial bonds.

base: A proton acceptor or an electron pair donor.

benzyl group: A benzene ring plus a methylene (CH_2) unit (C_6H_5—CH_2).

benzylic position: The carbon attached to a benzene ring.

benzyne: A highly reactive intermediate. A benzene ring with a triple bond (see Figure C-3).

Figure C-3:
Benzyne.

bicyclic: A molecule with two rings that share at least two carbons.

Brønsted acid: A proton donor.

Brønsted base: A proton acceptor.

carbanion: A negatively charged carbon atom.

carbene: A reactive intermediate, characterized by a neutral, electron-deficient carbon center with two substituents ($R_2C:$).

carbocation: A positively charged carbon.

carbonyl group: A carbon double bonded to oxygen (C=O).

carboxylic acid: A molecule containing a carboxyl (COOH) group. Also a functional group.

cation: A positively charged molecule or atom.

chair conformation: Typically, the most stable cyclohexane conformation. Looks like a chair (see Figure C-4).

Figure C-4:
Chair
conformation.

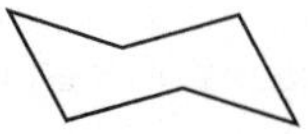

chemical shift: The location of an NMR peak relative to the standard tetramethylsilane (TMS), given in units of parts per million (ppm).

chiral center: A carbon or other atom with four nonidentical substituents.

chiral molecule: A molecule that's not superimposable on its mirror image. Chiral molecules rotate plane-polarized light.

cis: Two substituents on the same side of a double bond or ring.

configuration: The three-dimensional orientation of atoms around a chiral center. It's given the designation R or S.

conformation: The instantaneous spatial arrangements of atoms. Conformations can change by rotation around single bonds.

conjugate acid: The acid that results from protonation of a base.

conjugate base: The base that results from the deprotonation of an acid.

conjugated double bonds: Double bonds separated by one carbon-carbon single bond. Alternating double bonds.

constitutional isomers: Molecules with the same molecular formula but with atoms attached in different ways.

coupling constant: The distance between two neighboring lines in an NMR peak (given in units of Hz).

coupling protons: Protons that interact with each other and split an NMR peak into a certain number of lines following the $n + 1$ rule.

covalent bond: Bond in which the two electrons are shared between the two atoms.

dehydrohalogenation: Loss of a hydrohalic acid (like HBr, HCl, and so on) to form a double bond.

delta value (also known as δ value): The chemical shift. The location of an NMR peak relative to the standard tetramethylsilane (TMS), given in units of parts per million (ppm).

diastereomers: Stereoisomers that are not mirror images of each other.

Diels–Alder reaction: A reaction that brings together a diene and a dienophile to form cyclohexene rings.

diene: A molecule that contains two alternating double bonds. A reactant in the Diels–Alder reaction.

dienophile: A reactant in the Diels–Alder reaction that contains a double bond. Dienophiles are often substituted with electron-withdrawing groups.

dipole moment: A measure of the separation of charge in a bond or molecule.

doublet: An NMR signal split into two lines.

E isomer: Stereoisomer in which the two highest-priority groups are on opposite sides of a double bond.

E1 elimination reaction: A reaction that eliminates an acid (like HCl, HBr, and so on) to form an alkene. A first-order reaction that goes through a carbocation mechanism. It is the preferred mechanism for dehydration of secondary and tertiary alcohols to form alkenes.

E2 elimination reaction: A reaction that eliminates an acid (like HCl, HBr, and so on) to form an alkene. A second-order reaction that occurs in single step, in which the double bond is formed as the hydrohalic acid is eliminated.

eclipsed conformation: Conformation about a carbon-carbon single bond in which all the bonds of two adjacent carbons are aligned with each other (0 degrees apart when viewed in a Newman projection). (See Figure C-5.)

Figure C-5:
A Newman
projection
of eclipsed
conformation.

electronegativity: The electron piggishness of an atom. More technically, a measure of the tendency of an atom to attract the electrons in a covalent bond to itself.

electrophile: An electron lover. A molecule that can accept a lone pair of electrons (a Lewis acid).

enantiomers: Molecules that are nonsuperimposable mirror images of each other.

equatorial: The bonds in a chair cyclohexane that are oriented along the equator of the ring (see Figure C-6).

Figure C-6:
Equatorial.

ester: A molecule containing a carbonyl group adjacent to an oxygen bound to a carbon (RCOOR). Also a functional group.

ether: A molecule containing oxygen singly bonded to two carbon atoms (R-O-R). Also a functional group. Often refers to diethyl ether.

fingerprint region: Region of an IR spectrum below 1,500 cm^{-1}. The fingerprint region of the IR spectrum is often complex and difficult to interpret.

functional group: A reactivity center.

gauche conformation: A type of staggered conformation in which two big groups are next to each other (shown in a Newman projection in Figure C-7).

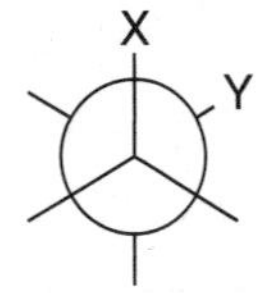

Figure C-7:
A Newman projection of a gauche conformation.

halide: A member of the VIIA column of the Periodic Table (like F, Cl, Br, I, and so on) or a molecule that contains one of these atoms. Also a functional group.

Hückel's rule: A rule that states that completely conjugated planar rings with $4n + 2\pi$ electrons are aromatic.

hybrid orbitals: Orbitals formed from mixing together atomic orbitals, like the sp^x orbitals, which result from mixing s and p orbitals.

hyperconjugation: Weak interaction (electron donation) between sigma bonds with p orbitals. Hyperconjugation helps explain why alkyl substituents stabilize carbocations.

inductive effects: Electron donation or withdrawal by electropositive or electronegative atoms through the sigma bond framework. Inductive effects help explain why alkyl substituents stabilize carbocations.

intermediate: Any species formed in a reaction on the way to making the product. Typically, intermediates are unstable.

ionic bond: A bond in which the electrons are unshared between two atoms.

IR spectroscopy: An instrumental technique that measures IR light absorption by molecules. Can be used to determine functional groups in an unknown molecule.

isolated double bonds: Double bonds separated by more than one carbon-carbon single bond.

J value: The coupling constant between two peaks in an NMR signal. Given in units of Hz.

ketone: A compound that contains a carbonyl group attached to two carbons (RCOR). Also a functional group.

kinetic product: The product that forms the fastest. (This product has the lowest energy of activation.)

kinetics: The study of reaction rates.

Lewis acid: An electron-pair acceptor.

Lewis base: An electron-pair donor.

Markovnikov's rule: A rule that states that electrophiles add to the less highly substituted carbon of a carbon-carbon double bond (or the carbon with the most hydrogen atoms).

mass spectrometry: An instrumental technique involving the ionization of molecules into fragments. Can be used to determine the molecular weights of unknown molecules.

meso: Molecules that have chiral centers but are achiral as a result of one or more planes of symmetry in the molecule.

meta: The positions of two substituents on a benzene ring that are separated by one carbon (see Figure C-8).

Figure C-8: Meta.

meta-directing substituent: Any substituent on an aromatic ring that directs incoming electrophiles to the meta position.

molecular ion: The fragment in a mass spectrum that corresponds to the cation radical (M^+) of the molecule. The molecular ion gives the molecular mass of the molecule.

molecular orbital theory: A model for depicting the location of electrons that allows electrons to delocalize across the entire molecule. A more accurate but less user-friendly theory than the valence-bond model.

multistep synthesis: Synthesis of a compound that takes several steps to achieve.

n + 1 rule: A rule for predicting the coupling for a proton in ^{1}H NMR spectroscopy. An NMR signal will split into $n + 1$ peaks, where n is the number of equivalent adjacent protons.

natural product: A compound produced by a living organism.

nitrile: A compound containing a cyano group, a carbon triply bonded to a nitrogen (CN). Also a functional group.

NMR: Nuclear magnetic resonance sprectroscopy. A technique that measures radio frequency light absorption by molecules. A powerful structure-determining method.

node: A region in an orbital with zero electron density.

nucleophile: A nucleus lover. A molecule with the ability to donate a lone pair of electrons (a Lewis base).

nucleophilicity: A measure of the reactivity of a nucleophile in a nucleophilic substitution reaction.

optically active: Rotates plane-polarized light.

orbital: The region of space in which an electron is confined (the electron "apartment").

organic compound: A carbon-containing compound.

ortho: The positions of two substituents on a benzene ring that are on adjacent carbons (see Figure C-9).

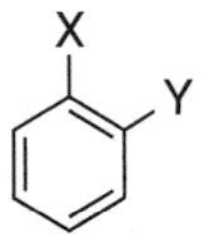

Figure C-9:
Ortho.

ortho-para director: An aromatic substituent that directs incoming electrophiles to the ortho or para positions.

oxidation: A reaction that involves the loss of electrons by an atom or molecule.

para: Describes the positions of two substituents on a benzene ring that are separated by two carbons (see Figure C-10).

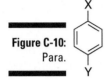

Figure C-10:
Para.

phenyl ring: A benzene ring as a substituent, sometimes abbreviated Ph.

pi bond (also known as π bond): A bond with electron density above and below the two atoms, but not directly between the two atoms. Found in double and triple bonds.

pKa: The scale for defining a molecule's acidity (pKa = –log Ka).

plane of symmetry: A plane cutting through a molecule in which both halves are mirror images of each other.

plane-polarized light: Light that oscillates in a single plane.

protic solvent: A solvent that contains O-H or N-H bonds.

proton: An H⁺ ion. Also a positively charged nuclear particle.

R group: Abbreviation given to an unimportant part of a molecule. Indicates the rest of the molecule.

Racemic mixture: A 50/50 mixture of two enantiomers. Racemic mixtures are optically inactive (they don't rotate plane-polarized light).

radical: An atom or molecule with an unpaired electron.

reduction: A reaction involving the gain of electrons by an atom or molecule.

resonance structures: Structures used to better depict the location of pi and nonbonding electrons on a molecule. A molecule looks like a hybrid of all resonance structures.

s-cis conformation: A conformation in which the two double bonds of a conjugated diene are on the same side of the carbon-carbon single bond that connects them (see Figure C-11). The required conformation for the Diels–Alder reaction.

Figure C-11:
s-cis
conformation.

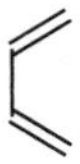

s-trans conformation: The conformation in which the two double bonds of a conjugated diene are on opposite sides of the carbon-carbon single bond that connects them (see Figure C-12).

Figure C-12:
s-trans
conformation.

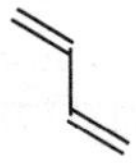

sigma bond (also known as σ bond): A bond in which electrons are located between the nuclei of the bonding atoms. Single bonds are sigma bonds.

singlet: Describes an NMR signal consisting of only one line.

S$_N$1 reaction: A first-order substitution reaction that goes through a carbocation intermediate.

S$_N$2 reaction: A second-order substitution reaction that takes place in one step and has no intermediates.

***sp*:** A hybrid orbital made by mixing one *s* orbital and one *p* orbital. The angle between sp orbitals is usually about 180 degrees.

***sp*^{2}:** A hybrid orbital made by mixing one *s* orbital and two *p* orbitals. The angle between sp^2 orbitals is usually about 120 degrees.

***sp*^{3}:** A hybrid orbital made by mixing one *s* orbital and three *p* orbitals. The angle between sp^3 orbitals is usually about 109.5 degrees.

staggered conformation: Conformation about a carbon-carbon single bond in which bonds of one carbon are at a maximum distance apart from bonds coming off of an adjacent carbon (60 degrees apart when viewed in a Newman projection). (See Figure C-13.)

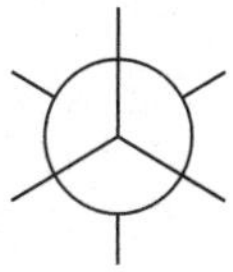

Figure C-13:
A Newman projection of a staggered conformation.

stereochemistry: The study of molecules in three dimensions.

stereoisomers: Molecules that have the same atom connectivity, but different orientations of those atoms in three-dimensional space.

steric hindrance: The way that atoms can shield a site by getting in the way of approach of a reactant.

substituent: A piece that sticks off the main carbon chain or ring.

syn addition: A reaction in which two groups of a reagent X-Y add on the same face of a carbon-carbon double bond.

tautomers: Molecules that differ in the placement of a hydrogen and double bonds and are easily interconvertible. Keto and enol forms are tautomers.

thermodynamic product: The reaction product with the lowest energy.

thermodynamics: The study of the energies of molecules.

thiol: A molecule containing an SH group. Also, a functional group.

transition state: The structure that corresponds to the highest point on the energy hill that takes one species into another.

triplet: An NMR signal split into three lines.

Z isomer: An isomer in which the two highest-priority substituents are on the same side of a double bond.

Index

carbonyl groups, 265
IR absorptions of common, 261
IR spectrum, 262
recognizing, 263
IR spectrum, 260–261
overview, 255–256
IR spectrum
deducing structure from molecular
formula, NMR, and, 302–306
determining functional group from, 291
general discussion, 260–261, 262
rechecking structure with NMR and,
296–297
isolated double bonds, 352
isomers, 107
isopropanol, 76
isopropylcyclohexane, 122
isotopes, mass spectrometry, 244–246

• J •

J value, 281–282, 353
Jones reagent, 192
Julian, Percy, 313

• K •

Kekulé, August, 311
Kekulé structure, 39
ketones
defined, 353
general discussion, 79
Grignard reaction, 190–191
McLafferty rearrangement, 249–250
oxidizing alcohols into, 192
ozonolysis, 154
permanganate oxidation, 155
reducing to make alcohols, 189
tautomerization reaction, 165
Wittig reaction, 142
kinetic product, 196, 353
kinetics
conjugate addition, 196–197
defined, 353
overview, 13

• L •

leaving groups, S_N2 reaction, 176–177
Lewis acids, 62–63, 353
Lewis bases, 62–63, 353
Lewis structures
combining with line-bond
structures, 44
condensed structures, 40–41
converting to line-bond structures,
41–43
drawing, 39
formal charges, assigning, 37–39
line-bond structures, 41, 43–44
lone pairs, determining, 44–45
overview, 37
as primary model in organic
chemistry, 36
limonene, 85
Lindlar's catalyst, 164
linear geometry, 28
line-bond structures
determining number of hydrogens on,
43–44
general discussion, 41
Lewis structures, combining with, 44
Lewis structures, converting to, 41–43
lithium aluminum hydride, 188–189
lone pairs
determining, 44–45
overview, 28
in resonance structures, 48, 50–51
Loschmidt, Johan Josef, 312

• M •

magnesium, 190
magnetic moments, 269–270
magnetic resonance imaging (MRI), 287
magnets, 22
maitotoxin, 329
Markovnikov alcohol, 188
Markovnikov enol, 165
Markovnikov product, 146, 150–151

Notes

Notes

Notes

About the Author

Arthur Winter is a graduate of Frostburg State University (located in the mountains of western Maryland), where he received his BS in chemistry. He received a PhD in organic chemistry from the University of Maryland in College Park, where his research involved studying extremely short-lived reactive intermediates (with lifetimes less than 0.000001 second) using laser spectroscopy. He did postdoctoral research at Ohio State University before becoming a professor of chemistry at Iowa State University in 2009.

In summers, Arthur is an avid fisherman and huntsman of dangerous or delicious animals; in winters, he hibernates. Year-round, he likes watching violent movies, drinking sugar-saturated tea solutions, and reading trashy fiction. He enjoys competitive weight lifting, Ironman competitions, lumberjacking, and jousting — from the comfort of a couch in front of his TV. He often boasts of his ability to make minute rice in 30 seconds (admittedly, somewhat al dente) and is the proud owner of a collection of inexpensive watches. He's a good whistler, a bad loser, and a tasteless joke enthusiast.

Dedication

To Don Weser, an inspiring teacher and a giant in the field of chemical education (if not also in real life).

Author's Acknowledgments

I thank the team at Wiley who helped put this book together. Particularly, I thank my acquisitions editor, Lindsay Lefevere, who got the ball rolling on this project, and my patient project editor and copy editor, Elizabeth Kuball, who oversaw the book's creation. I also thank the technical reviewer, Dr. Joe Burnell, who helped eliminate factual errors. I also thank Patti Boone, who I am convinced is a miracle worker and the world's greatest human being.

Finally, I thank Jonathan and Julian Winter for their support, and Brian Price, Mary Mumper, Robert Larivee, Fred Senese, Glynn Baugher, Dan Falvey, and Christopher Hadad for providing a first-class education.

Publisher's Acknowledgments

Executive Editor: Lindsay Sandman Lefevere

Project Editor: Elizabeth Kuball

Copy Editor: Elizabeth Kuball

Technical Editor: Joe Burnell, PhD

Project Coordinator: Erin Zeltner

Cover Image: ©iStockphoto.com/cb34inc

Apple & Mac

iPad For Dummies,
6th Edition
978-1-118-72306-7

iPhone For Dummies,
7th Edition
978-1-118-69083-3

Macs All-in-One
For Dummies, 4th Edition
978-1-118-82210-4

OS X Mavericks
For Dummies
978-1-118-69188-5

Blogging & Social Media

Facebook For Dummies,
5th Edition
978-1-118-63312-0

Social Media Engagement
For Dummies
978-1-118-53019-1

WordPress For Dummies,
6th Edition
978-1-118-79161-5

Business

Stock Investing
For Dummies, 4th Edition
978-1-118-37678-2

Investing For Dummies,
6th Edition
978-0-470-90545-6

Personal Finance
For Dummies, 7th Edition
978-1-118-11785-9

QuickBooks 2014
For Dummies
978-1-118-72005-9

Small Business Marketing
Kit For Dummies,
3rd Edition
978-1-118-31183-7

Careers

Job Interviews
For Dummies, 4th Edition
978-1-118-11290-8

Job Searching with Social
Media For Dummies,
2nd Edition
978-1-118-67856-5

Personal Branding
For Dummies
978-1-118-11792-7

Resumes For Dummies,
6th Edition
978-0-470-87361-8

Starting an Etsy Business
For Dummies, 2nd Edition
978-1-118-59024-9

Diet & Nutrition

Belly Fat Diet For Dummies
978-1-118-34585-6

Mediterranean Diet
For Dummies
978-1-118-71525-3

Nutrition For Dummies,
5th Edition
978-0-470-93231-5

Digital Photography

Digital SLR Photography
All-in-One For Dummies,
2nd Edition
978-1-118-59082-9

Digital SLR Video &
Filmmaking For Dummies
978-1-118-36598-4

Photoshop Elements 12
For Dummies
978-1-118-72714-0

Gardening

Herb Gardening
For Dummies, 2nd Edition
978-0-470-61778-6

Gardening with Free-Range
Chickens For Dummies
978-1-118-54754-0

Health

Boosting Your Immunity
For Dummies
978-1-118-40200-9

Diabetes For Dummies,
4th Edition
978-1-118-29447-5

Living Paleo For Dummies
978-1-118-29405-5

Big Data

Big Data For Dummies
978-1-118-50422-2

Data Visualization
For Dummies
978-1-118-50289-1

Hadoop For Dummies
978-1-118-60755-8

Language &
Foreign Language

500 Spanish Verbs
For Dummies
978-1-118-02382-2

English Grammar
For Dummies, 2nd Edition
978-0-470-54664-2

French All-in-One
For Dummies
978-1-118-22815-9

German Essentials
For Dummies
978-1-118-18422-6

Italian For Dummies,
2nd Edition
978-1-118-00465-4

Available in print and e-book formats.

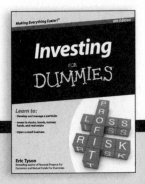
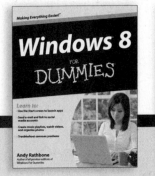

Available wherever books are sold. **For more information or to order direct visit www.dummies.com**

Math & Science

Algebra I For Dummies,
2nd Edition
978-0-470-55964-2

Anatomy and Physiology
For Dummies, 2nd Edition
978-0-470-92326-9

Astronomy For Dummies,
3rd Edition
978-1-118-37697-3

Biology For Dummies,
2nd Edition
978-0-470-59875-7

Chemistry For Dummies,
2nd Edition
978-1-118-00730-3

1001 Algebra II Practice
Problems For Dummies
978-1-118-44662-1

Microsoft Office

Excel 2013 For Dummies
978-1-118-51012-4

Office 2013 All-in-One
For Dummies
978-1-118-51636-2

PowerPoint 2013
For Dummies
978-1-118-50253-2

Word 2013 For Dummies
978-1-118-49123-2

Music

Blues Harmonica
For Dummies
978-1-118-25269-7

Guitar For Dummies,
3rd Edition
978-1-118-11554-1

iPod & iTunes
For Dummies, 10th Edition
978-1-118-50864-0

Programming

Beginning Programming
with C For Dummies
978-1-118-73763-7

Excel VBA Programming
For Dummies, 3rd Edition
978-1-118-49037-2

Java For Dummies,
6th Edition
978-1-118-40780-6

Religion & Inspiration

The Bible For Dummies
978-0-7645-5296-0

Buddhism For Dummies,
2nd Edition
978-1-118-02379-2

Catholicism For Dummies,
2nd Edition
978-1-118-07778-8

Self-Help & Relationships

Beating Sugar Addiction
For Dummies
978-1-118-54645-1

Meditation For Dummies,
3rd Edition
978-1-118-29144-3

Seniors

Laptops For Seniors
For Dummies, 3rd Edition
978-1-118-71105-7

Computers For Seniors
For Dummies, 3rd Edition
978-1-118-11553-4

iPad For Seniors
For Dummies, 6th Edition
978-1-118-72826-0

Social Security
For Dummies
978-1-118-20573-0

Smartphones & Tablets

Android Phones
For Dummies, 2nd Edition
978-1-118-72030-1

Nexus Tablets
For Dummies
978-1-118-77243-0

Samsung Galaxy S 4
For Dummies
978-1-118-64222-1

Samsung Galaxy Tabs
For Dummies
978-1-118-77294-2

Test Prep

ACT For Dummies,
5th Edition
978-1-118-01259-8

ASVAB For Dummies,
3rd Edition
978-0-470-63760-9

GRE For Dummies,
7th Edition
978-0-470-88921-3

Officer Candidate Tests
For Dummies
978-0-470-59876-4

Physician's Assistant Exam
For Dummies
978-1-118-11556-5

Series 7 Exam For Dummies
978-0-470-09932-2

Windows 8

Windows 8.1 All-in-One
For Dummies
978-1-118-82087-2

Windows 8.1 For Dummies
978-1-118-82121-3

Windows 8.1 For Dummies,
Book + DVD Bundle
978-1-118-82107-7

𝑒 Available in print and e-book formats.

Available wherever books are sold. **For more information or to order direct visit www.dummies.com**

Take Dummies with you everywhere you go!

Whether you are excited about e-books, want more from the web, must have your mobile apps, or are swept up in social media, Dummies makes everything easier.

For Dummies is the global leader in the reference category and one of the most trusted and highly regarded brands in the world. No longer just focused on books, customers now have access to the For Dummies content they need in the format they want. Let us help you develop a solution that will fit your brand and help you connect with your customers.

Advertising & Sponsorships

Connect with an engaged audience on a powerful multimedia site, and position your message alongside expert how-to content.

Targeted ads • Video • Email marketing • Microsites • Sweepstakes sponsorship

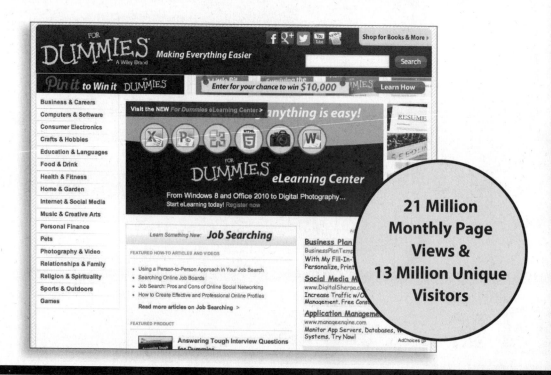

Custom Publishing

Reach a global audience in any language by creating a solution that will differentiate you from competitors, amplify your message, and encourage customers to make a buying decision.

Apps • Books • eBooks • Video • Audio • Webinars

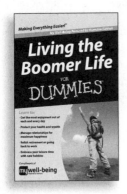

Brand Licensing & Content

Leverage the strength of the world's most popular reference brand to reach new audiences and channels of distribution.

For more information, visit www.Dummies.com/biz

Dummies products make life easier!

- DIY
- Consumer Electronics
- Crafts

- Software
- Cookware
- Hobbies

- Videos
- Music
- Games
- and More!

For more information, go to **Dummies.com** and search the store by category.

FOR DUMMIES

A Wiley Brand